ST/TCD/11

Department of Technical Co-operation for Development

Natural Resources/Water Series No. 24

GROUND WATER IN EASTERN AND NORTHERN EUROPE

UNITED NATIONS
New York, 1990

ST/TCD/11

UNITED NATIONS PUBLICATION

Sales No. E.89.II.A.11

03200

ISBN 92-1-104318-2

Explanatory notes

Symbols of United Nations documents are composed of capital letters combined with figures. Mention of such a symbol indicates a reference to a United Nations document.

The following symbols have been used in the tables throughout the report:

A dash (-) indicates that data are not available or are not separately reported.

A blank indicates that the item is not applicable.

A full stop (.) is used to indicate decimals.

A slash (/) indicates a crop year or financial year, e.g., 1976/77.

Use of a hyphen (-) between dates representing years, e.g., 1975-1978, signifies the full period involved, including the beginning and end years.

Reference to "dollars" ($) indicates United States dollars.

Details and percentages in tables do not necessarily add to totals because of rounding. Some of the data series are not homogeneous; they have been taken from various reviews and publications.

The designations employed and the presentation of the material in this publication do not imply the expression of any opinion whatsoever on the part of the Secretariat of the United Nations concerning the legal status of any country, territory, city or area or of its authorities, or concerning the delimitation of its frontiers or boundaries.

References to specific values of national currency are made in the chapters on Czechoslovakia and Finland. The exchange rates to United States dollars for their currencies, as of March 1987, were:

Country	Currency	Rate per U.S. dollar
Czechoslovakia	koruna (kcs)	9.66
Finland	markka (Fmk)	4.50

Abbreviations used

AN	Akademiya Nank (USSR)
atm	atmosphere
BITS	Swedish Commission for Technical Co-operation
BRGM	Bureau de Recherches Géologiques et Minières
dH	index of water hardness
Eh	redox potential
FDM	finite difference method
FEM	finite element method
GEOZAVOD	Institute for Geological and Geophysical Research (Yugoslavia)
GEAL	gigacalories
GIN	Geological Institute (USSR)
GUDS	Dionyz Stur Geological Institute (Czechoslovakia)
IAH	International Association of Hydrogeologists
IBRD	International Bank for Reconstruction and Development
IDA	International Development Agency
IEC	Institute of the Earth's Crust (USSR)
IGHP	Ingenyrsko-geologicky a hydrogeologicky pruskum (Czechoslovakia)
IGME	Institute of Geology and Mineral Exploration (Greece)
IWP	Institute of Water Problems (USSR)
KIH	Royal Institute of Technology (Sweden)
meq/l	milligram equivalent per litre
MOIP	Moscow Society of Nature Researchers
MS/m	microsiemens per mètre (originally micromhos per mètre)
NGU	Geological Survey of Norway
NHP	Nordic Association of Hydrology
NORAD	Norwegian Agency of International Development
NVE	Norwegian Water Resources and Electricity Board
NWA	National Water Authority (Hungary)
OFKFV	Geological Exploration and Drilling Enterprise (Hungary)
pH	acidity-alkalinity index
PVC	polyvinylchloride
SAREC	Swedish Agency for Research Co-operation with Developing Countries
SGI	Swedish Geotechnical Institute
SGU	Geological Survey of Sweden

SIDA	Swedish International Development Authority
SMHI	Swedish Meteorological and Hydrological Institute
SNV	National Environment Protection Board (Sweden)
TDS	Total dissolved solids
UNDP	United Nations Development Programme
UNESCO	United Nations Educational, Scientific and Cultural Organization
UNICEF	United Nations Children's Fund
VIKUV	Enterprise for Water Exploration (Hungary)
VITUKI	Research Institute for Water Resources Development (Hungary)
VLF	very low frequency
VSEGEI	All-Union Geological Institute (USSR)
VSEGINGEO	All-Union Institute for Hydrogeology and Engineering Geology (USSR)
WFP	World Food Programme
WHO	World Health Organization

ABSTRACT WITH KEYWORDS

This volume is the eighth in a series giving encyclopedic coverage of the hydrogeology, use and resources potential of ground water world-wide. For Albania, Bulgaria, Czechoslovakia, Finland, Greece, Hungary, Iceland, Norway, Poland, Romania, Sweden, USSR in Europe, and Yugoslavia a summary is given of the physiography and climate, the geologic setting, hydrology and stratigraphy, history of ground-water use, principal aquifers, characteristics of the ground water (including mineral and thermal spas and springs), development practices, training and research facilities, and extent of international co-operation in the field. Includes maps, tables of data and lists of supplementary reference sources.

PREFACE

The Economic and Social Council, in its resolution 675 (XXV) of 2 May 1958, requested the Secretary-General to take appropriate measures for the establishment, within the Secretariat, of a centre to promote co-ordinated efforts for the development of water resources. It also singled out ground-water problems as one of the priority subjects in the development of a programme of studies. Large-scale Ground-Water Development, published in 1960, 1/ was the first study prepared in this field by the Water Resources Development Centre (now the Water Resources Branch of the Natural Resources and Energy Division, Department of Technical Co-operation for Development of the United Nations Secretariat).

The Advisory Committee on the Application of Science and Technology to Development, in its World Plan of Action, 2/ gave priority to ground-water exploration and development. In fact, in the course of the First and Second United Nations Development Decades, more than 100 projects assisted by the United Nations Development Programme (UNDP) and other United Nations technical co-operation programmes were entirely or partly devoted to ground-water prospecting, assessment, or pilot development. (A list of ground-water projects in the subregion sponsored by UNDP is contained in the annex to the present publication.)

While such operational activities were developing, the need for a comprehensive review of the results of the projects and for the dissemination of relevant information became more evident. As a result, the Economic and Social Council, in its resolution 1761 B (LIV) of 18 May 1973, requested the Secretary-General to take the necessary measures, within budgetary limitations, to improve and strengthen the existing United Nations services for the analysis, evaluation and dissemination of world-wide data on natural resources, including water resources. The recommendations of the United Nations Water Conference (1977) 3/ placed added emphasis on these aspects.

With respect to ground water, a first comprehensive review of the African continent was published in 1972 and 1973 under the title Ground Water in Africa 4/ as a synthesis of material available in the records and files of the United Nations. The material of the second volume in this series, Ground Water in the Western Hemisphere 5/ (1976) was drawn from country papers prepared by hydrogeologists and ground-water engineers in the countries concerned; this was also the case for material presented in a third volume, Ground Water in the Eastern Mediterranean and Western Asia 6/ which includes Turkey in Europe (1982), and a

---

1/ United Nations publication, Sales No. 60.II.B.3.

2/ United Nations publication, Sales No. E.71.II.A.18.

3/ Report of the United Nations Water Conference, Mar del Plata, 14-25 March 1977 (United Nations publication, Sales No. E.77.II.A.12), chap. I.

4/ United Nations publication, Sales No. E.71.II.A.16.

5/ United Nations publication, Sales No. E.76.II.A.5.

6/ United Nations publication, Sales No. E.82.II.A.8.

fourth volume, Ground Water in the Pacific region 7/ (1983) which covers also the major Asian countries off the coast of continental Asia, including Japan, the Philippines and Indonesia, a fifth volume Ground Water in Continental Asia 8/ (1986). The first volume Ground Water in Africa being out of print and largely outdated, it was decided to publish an updated and expanded review in two volumes: Ground Water in Northern and Western Africa 9/ (1987), and Ground Water in Eastern, Central and Southern Africa 10/ (1988).

It is hoped that the two volumes on ground water in Europe which are the last in this series on ground water in the world's regions will be useful to developing countries. In addition to a description of ground-water occurrences and availability in European countries, they provide information on the conditions of ground-water development and related problems, information on technologies and methods used in the investigation, assessment and development of ground water, and on international co-operation among the countries concerned.

In particular, this volume on ground water in northern and eastern Europe deals with vast artesian ground-water basins (USSR, Hungary), ground-water exploration in crystalline rock areas (the Scandinavian Peninsula), and karstic ground water (the Balkan Peninsula); such ground-water conditions are encountered in many areas of Africa, Asia and Latin America.

It has not been possible to obtain specific information on ground water in Albania and only general information is provided herewith on its hydrology and general geology, related to aquifer potential. For the rest of the countries, abundant information has been provided by governmental organizations, and by consultants and specialists, whose valuable assistance is gratefully acknowledged by the United Nations. In particular, thanks are due to:

Messrs. L. Alföldi, E. Almassy, Dr. I. Bercia, M. V. de Geer, Professor M. Galabov, Mr. M. S. Galitsin, Mr. A. Gaut. Also Ms. G. Gioni-Stavropoulou, Messrs. B. Mijatovic, V. N. Ostrovsky, Dr. B. Paczynski, Messrs. G. Persson, G. Sigbjarnarson, F. Sigurosson, M. Taka, G. S. Vartanyan, J. Vrba, T. Wikner and L. S. Vazvin

The All-Union Institute for Hydrogeology and Engineering Geology (VSEGINGEO), Moscow, USSR, the Geological Institute of Romania, the Geological Institute of Warsaw (Poland), the Geological Survey of Finland, the Geological Survey of Norway, the Geological Survey of Sweden, the Institute of Geology and Mineral Exploration, Athens (Greece), the Institute for Geological and Geophysical Research (GEOZAVOD), Belgrade (Yugoslavia), the National Energy Authority, Reykjavik (Iceland) and the Research Centre for Water Resources Development, Budapest (Hungary)

7/ United Nations publication, Sales No. E.83.II.A.12.

8/ United Nations publication, Sales No. E.86.II.A.2.

9/ United Nations publication, Sales No. E.87.II.A.8.

10/ United Nations publication, Sales No. E.88.II.A.5.

CONTENTS

CONTENTS (continued)

CONTENTS (continued)

CONTENTS (continued)

CONTENTS (continued)

CONTENTS (continued)

CONTENTS (continued)

CONTENTS (continued)

CONTENTS (continued)

CONTENTS (continued)

Tables

Figures

CONTENTS (continued)

CONTENTS (continued)

CONTENTS (continued)

OVERVIEW

Europe is generally described as a vast peninsula, located at the western end of Eurasia, the largest land mass in the world. To the east, Europe is separated from Asia proper by the Ural Mountains (1,885 m at Mount Narodnaya) and the Ural River, and to the south by the Caucasus Mountains. From Nordkapp to the north in Norway to Cape Tainaron in the south (Greece), the distance is about 3,850 km. From Cape Roca, in Portugal to the west, to the Urals in USSR to the east, the distance is about 5,300 km. By sea, Europe is separated from Asia by the Caspian Sea, the Bosporus, the Sea of Marmara and the Aegean Sea; from Africa by the Mediterranean Sea and the Strait of Gibraltar. To the west and north, Europe ends at the Atlantic Ocean, the Norwegian Sea and the Barents Sea.

With only 7 per cent of the land mass of the world, Europe contains about one fifth of the world's population and accounts for half of the world's trade.

The influence of the sea is considerable over the continent, both from the point of view of climate and the closeness to the sea of most of the territory which contains islands and peninsula. There are few places in Europe more than 800 km from a sea.

The physiography of Europe can be summarized as follows (figure 1). A vast plain in the north crosses from west to east comprising much of France, the Low Countries, Germany, Poland and the USSR. The plain is flanked to the south by plateaus, covering central France, the Ardennes, the Black Forest, the Harz and a mass of highlands extending through Czechoslovakia, Poland and the Ukraine. A huge mountainous mass crosses Europe, mainly in the west and south, which includes the Pyrenees, the Alps, the Carpathian Mountains, the Balkan Mountains (Stara Planina) and the Caucasus Mountains. The highest point is Elbrus in the Caucasus (5,630 m). In Western Europe the highest peak, in the Alps, is Mont Blanc (4,807 m).

A great variety of climates from arctic (northern USSR) to sub-tropical (Canary Islands, Madeira) are encountered in Europe which is drained by a number of rivers, some counting among the major rivers in the world (Volga, Danube, Rhine). Geological conditions are also extremely diverse. All the geologic time periods are represented in Europe. The first geologic descriptions were made in Europe; consequently, stratigraphic terms have been chosen on the basis of the European locations where they were first described. Major orogenic, or mountain-building phases, were also named after the massifs of Europe (Caledonian, Hercynian, Alpine).

As a result of the great diversity of conditions affecting ground water in terms of morphology, climate and geology, all kinds of hydrogeologic situations are found in Europe. Also, much European expertise in managing ground water has been put at the disposal of developing countries in the form of training, transfer of technology, assignment of experts and supplying of equipment.

Such co-operative activities are described in the country papers that follow. A list of ground-water projects benefiting from participation by the United Nations Development Programme is given in the annex.

Figure 1

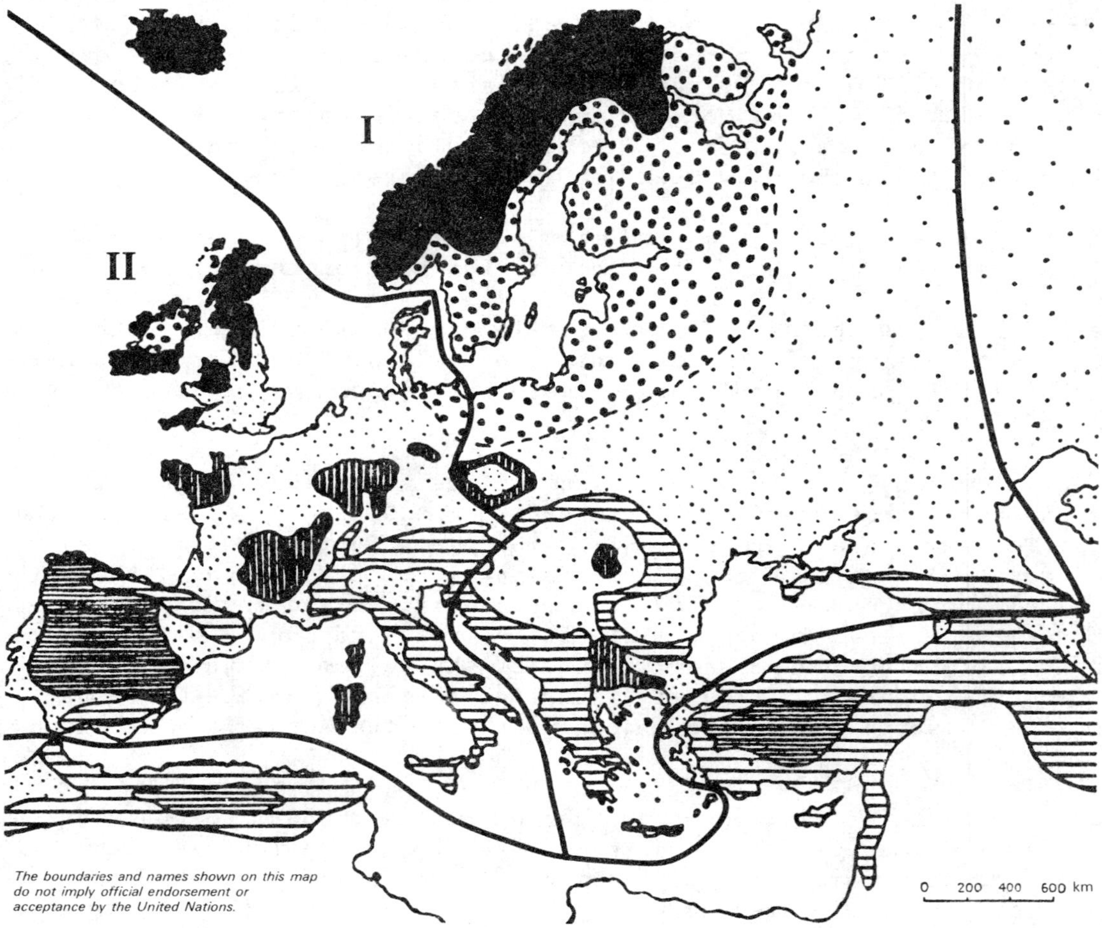

PHYSICAL DIVISIONS

- Highlands of northern Europe
- Glaciated plain
- Plain and coastal lowland
- Highlands of central Europe
- Young mountains
- Plateaus

I Area covered by ***Ground Water in Eastern and Northern Europe***

II Area covered by ***Ground Water in Western and Central Europe***

3546.1x

Source: Norman J.G. Pounds, *Europe and the Mediterranean,* New York: McGraw-Hill, 1953.

This overview of ground-water resources, knowledge and development in Europe may appear rudimentary as compared with the abundant literature available and the extensive cartographic effort carried out in recent years, such as the joint International Association of Hydrologists (IAH) and UNESCO Hydrogeological Map of Europe at the scale of 1:1,500,000 and the publication entitled Ground Water Resources of the European Community in ten volumes and nine sets of maps at the scale of 1:500,000.

The present work is intended, within the framework of a world-wide review of the subject, and primarily directed at water resources professional engineers, planners and decision makers of developing countries, to serve as a reference document, bringing nearly to completion the series initiated in 1972 with the volume Ground Water in Africa.

For reasons of format convenience, coverage of Europe is in two volumes (figure 1): I. Eastern and Northern Europe; and II. Central and Western Europe. This division is not political but rather related to geographic considerations, with the dividing line including the Adriatic Sea, the Alps, the Bohemian Forest, the Oder-Neisse River and the Skagerrak and Kattegat straits at the entrance to the Baltic Sea. Turkey in Europe is included in yet another of the series, Ground Water in the Eastern Mediterranean and Western Asia.

A. Physiography

The physiography of the area covered by this volume comprises the following three subregions:

1. To the north, Scandinavia less Denmark but including Iceland, characterized by the occurrence of hard-rock igneous formations over most of their territory, and warped by glaciers. It includes many lakes.

2. To the south, the Balkan Peninsula, an area of mostly low mountain ranges and plateaus where carbonate rocks: limestone and dolomite abound, and of plains of moderate extension, mainly in coastal areas.

3. In between, the North European plain east of the Oder River with its important lakes (Ladoga, Onega, Peipus) and the Pannonian and Romanian plains. The main relief feature in this vast area is the Carpathian Mountains (1,500 to 2,500 m) and their extensions to the west; the Tatra Mountains (up to 2,663 m), the Bohemian plateau and its mountainous fringes: the Bohemian-Moravian Highlands, the Erzgebirge (Ore Mountains, Kruoné Hory), and the Riesen Gebirge (1,000 to 1,500 m). The principal rivers, from east to west, are the Volga, Don, Dnieper and Danube.

In the seas bordering the land mass are many islands, mostly along the shores of the Scandinavian Peninsula, near the shores in the Adriatic and Ionian seas, and widely dispersed in the Aegean Sea.

The major islands are Kolguyev in the Barents Sea; Gotland, Saaremaa, Hiiumaa in the Baltic, and Crete in the Mediterranean. Peloponnesus, once a Greek peninsula, became, strictly, a major island when the Corinth Canal was built.

B. Climate

Sharp climatic contrasts are encountered in the area. However, the tropical type is not present. From north to south, six climatic zones are represented as follows:

1. Polar: North of the parallel of Arkhangelsk, climate of the tundra, temperature of the warmest month is 0° to 10° C

2. Subarctic: North of the parallel of Bergen, temperate in summer but cold in winter (coldest month: less than 0° C). During the warmest three or four months, temperatures rise above 10° C.

3. Temperate continental: North of line Prague-Belgrade, Bucharest and Kuibyshev, climate with cold winter and cool summer. Average temperature of coldest month: less than 10° C. Average temperature of hottest month: 10° to 22° C.

4. Semi-arid: South-east of the line Bucharest to Kuibyshev semi-arid steppe climate, except in Caucasus with higher altitude, temperatures are lower, rainfall is higher.

5. Temperate humid: South-west of the line Bucharest to Belgrade to Prague, no dry season; average temperature of warmest month is less than 22° C.

6. Mediterranean: Dalmatian Coast and most of Greece, temperate, mild winter with average temperature of coldest month above 0° C and less than 18° C; average temperature of warmest month, above 22° C.

C. Rainfall

Yearly rainfall averages 500 to 750 mm over most of the area (figure 2). In the extreme north of Scandinavia and to the east in the USSR, that is, north and east of a line from Tornio (north end of Gulf of Bothnia) to Perm (Ural) and Iasi (Carpathians) rainfall gradually diminishes to 250 mm, and less north of the Caspian Sea. There are also patches of less rainfall in Poland, central Sweden and Greece. Slightly higher values are encountered in the southern part of Sweden and Norway (600-800 mm) in the Balkan area (600 to 1,000 mm; up to 1,400 mm in altitude), and in the Carpathian Mountains (1,000 to 1,200 mm).

The areas of higher rainfall are mountain ridges in Norway along the Norwegian Sea (1,500 to 2,500 mm), in the Dinaric Alps along the Adriatic and the Tyrrhenian Sea (1,600 to 2,000 mm) in the Caucasus Mountains (up to 3,000 mm and more) and in Iceland (1,500 to 3,000 mm in the South).

Figure 2

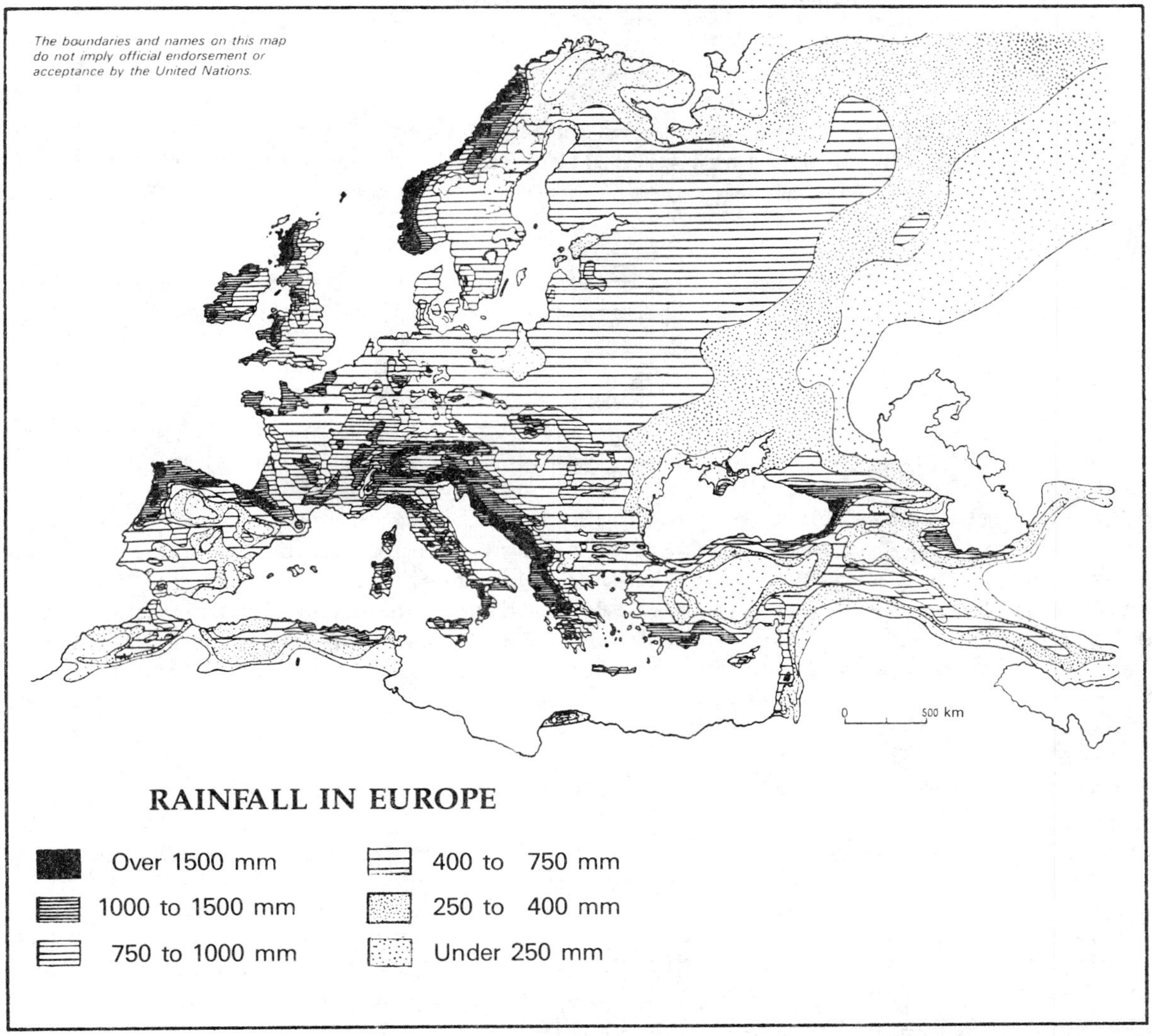

3546.2x

Source: Norman J.G. Pounds, *Europe and the Mediterranean,* McGraw Hill, 1953, p. 876.

D. Geology

Mountain-building in Europe was marked by three orogenic periods:

1. Caledonian orogeny of Devonian age,
2. Hercynian orogeny of Permian age, and
3. Alpine orogeny of late Cenozoic age.

During that last time, the enormous mass of Mesozoic and early to middle Mesozoic sediments, which had been deposited in the Tethys Sea south of Europe were buckled against the highlands of central Europe, which resulted in the Alpine system.

In the area considered here, geologic features can be summarized as follows for each of the physiographic subregions (figure 3).

1. Northwest Highlands

The Northwest Highlands (A) which include all of Scandinavia with the Norwegian Highlands (A_1) backbone of the Scandinavian Peninsula, and most of northern and western Sweden. The geologic formations are mainly of the hard-rock type of Precambrian, Cambrian, and Silurian age; (A_1); the peneplaned Swedish Highlands (A_2) with Archean schist, gneiss and granite covered in part by sandy moraine deposits and lake clays, the Smaland Highlands (A_3) similar to the preceding and the Central Swedish Lowlands (A_4) a mass of Precambrian rocks, peneplaned and covered by Paleozoic (Cambrian-Silurian) deposits; the Lapland Plateau (A_5) in north-eastern Norway and Sweden, Finland and the Kola Peninsula, a deeply dissected Precambrian crystalline mass, and the Finnish Plateau (A_6) of crystalline Precambrian rocks (granite, quartzite, schist) peneplaned and mantled by Cenozoic sediments and Pleistocene glacial deposits.

2. Great Central Plains

The Great Central Plains (B) are mostly Mesozoic and Cenozoic rocks covered by glacial drift and includes the Baltic Plain (B_1) with flat-lying Cretaceous-Tertiary belts, veneered by glacial, Recent marine and fluvial deposits. It includes two morainal zones separated by glacial outwash. The Central Russian Tableland (B_2) from the Baltic to the Urals has Paleozoic beds covered by Jurassic, Triassic and Cretaceous sediments dipping to the south, Cambrian and Ordovician rocks occur in narrow strips in southern Sweden, Gotland and Oland (islands) and also on the mainland south of the Gulf of Finland. A wide belt of Devonian and Carboniferous rocks occurs west of Moscow, and a Permian belt extends eastward to the Urals. Important coal fields (Carboniferous) occur south of Moscow. The Ural Mountains backbone is of crystalline rocks. Glacial deposits occur over most of the area. The Black Earth Region (B_3) of the Ukraine, consists of Cenozoic belts covered by loess deposits with, to the east, the Don plateau (Eocene) the Donetsk ridge (Oligocene), the Dnieper plateau (granite overlain by Oligocene strata, and the Podolian upland to the southwest (Miocene and Pliocene cut by the Dniester River (B_3). The Arctic Tundra Belt (B_4) underlain by Recent marl deposits emerged from the sea by the end of glacial times. The Caspian Depression, a lowlying steppe underlain by Cenozoic sediments with alluvium, loess and drifting sand, lies 30 metres below sea level.

Figure 3

MAP NO. 3568.1 UNITED NATIONS
NOVEMBER 1989

3. Central Massifs

The Central Massifs (C) constitute isolated blocks south of the plains and north of the Alpine System. In the area covered by this volume, only one unit is present, the Bohemian massif (C_1) with, to the south, the schist and granite of the Bohemian Forest and Bohemian-Moravian Highlands, and to the northwest the Sudetes mountains and Erzgebirge. In the center is the deformed Paleozoic Bohemian lowland with a variety of formations from Cenozoic volcanics to marine Cretaceous, Triassic and late Carboniferous rocks.

4. Alpine System

The Alpine System (D) is represented in the southern part with the Carpathian Mountains and Transylvanian Alps (D_1). Their western part is mainly flysch in the outer zone, crystalline in the central zone (Tatra Mountains), and Tertiary volcanics in the inner zone. The central part is mainly flysch and Cenozoic volcanics. The eastern part contains flysch, crystalline rocks, volcanics and Mesozoic limestone. Eight subdivisions of the Alpine System are represented:

The Transylvanian Alps (D_1) are a folded area with crystalline rocks and Mesozoic limestones.

The Balkans (D_2) constitute an extension of the Transylvanian Alps to the Rhodope. The Caucasus Mountains (D_3) have a highly granitic core with volcanic cones and includes Elbrus, the highest peak in Europe. It is flanked by Jurassic beds dipping north. The Ciscaucasus syncline (Stavropol Plateau) is filled with Cenozoic and Quaternary formations.

The Dinaric Alps (D_4) and the Pindus (D_5) are separated by the Drin River and comprise mainly Triassic and Cretaceous karstified limestone, with narrow Paleozoic belts. In the central part are folded flysch hills. The Pindus includes, in its western part, Triassic limestone and dolomite and, to the east, schist and marble.

The Rhodope (D_6) and its extension into the neighbouring Aegean islands, is a faulted block of crystalline rocks with sediment-filled basins. The highest peak is Mount Olympus.

The Hungarian Plain (D_7) is an Alpine depression. The Pannonian depression surrounded by Alpine mountains (Carpathian Mountains, Dinaric Alps). Lake Balaton is a remnant of the Cenozoic Pannonian sea and contains Cenozoic sediments covered by gravel. It has a marshy flood plain, a loess steppe and sand dunes.

The Romanian Plains (D_8) between the Balkan and Carpathian folded areas contains Miocene deposits overlain by Pliocene gravel. During the Quaternary period the Danube River moved south and dissected an alluvial upland covered by loess.

E. Ground water

The five main aquifer types are river alluvium, artesian basins, karst, fractured crystallines and permafrost:

1. The alluvial aquifers of most rivers, except in watersheds where clayey, largely sterile flysch is predominant, includes the sand and gravel of glacial and periglacial formations.

2. Complex sedimentary artesian aquifers occur in the Great Central Plains, the Hungarian Plain and the Romanian Plain.

3. The karst areas of the Balkan Peninsula, especially in Yugoslavia and Greece, contain large amounts of free-flowing ground water which, however, is difficult to locate and tap.

4. Of particular interest to many developing countries in Africa and Asia is the Scandinavian Peninsula where much progress has been made in prospecting for and developing ground water in fractured Precambrian crystalline hard rocks.

5. Spitsbergen and Greenland are not covered in this publication. Their ground water resources usually occur in the form of ice in permanently frozen layers of the ground (permafrost) but within which liquid water occurs at certain depths and at certain times.

In many parts of the Great Central Plains, in the southern plains (Hungarian, Romanian) in the coastal plains of the Adriatic, the Aegean, the Black Sea and also in the Scandinavian peninsula, ground water has always been an essential source of supply and generally tapped by wells.

I. ALBANIA

Area: 28,750 km^2 Population: 3.1 million

A. General

1. Physiography

The Socialist People's Republic of Albania is a Balkan country with about 400 km of shoreline on the Adriatic and Tyrrhenian seas (figure 4). It is located between Yugoslavia to the north and Greece to the south, thus belonging to the Dinaric zone of the north and the Pindus zone of the south. The main distinctive characteristic, when compared with neighbouring States, is the presence of a coastal plain, 5 to 40 km wide called the Western Lowland. It includes from north to south the plains of Scutari, Tirana (the capital), Shijak, Kavaje; the Myzeque plain - which contains sand dunes, swamps and lagoons - the basins east of the Guri Gomares coast hills, the Belsh plateau and the Elbasan plain. South of the Bay of Valona the coastline is mountainous.

East and south of the Western Lowlands the country includes limestone plateaus in continuation with the Montenegro plateaus (2,500 m, also called the Albanian Alps), the Mirdita Highlands, a rugged country with complex geology, the Western Border Ranges overlooking the coastal plains with limestone and flysch, the Mat basin - the floor of the main valley is covered with clay and marl - the lake region, a trough which includes the important border lakes Ohrid and Prespa (both 250 km^2) and little Prespa (50 km^2), the Drin trough and the Frontier ranges, east of it (between 2,000 and 2,500 m), and finally the central ranges between the Drin and Fan-Mat valley with serpentine and Mesozoic limestone overlain by Tertiary flysch.

The southern highlands extend, between the sea and the Greek border, over about a third of the country. They include the upper Shkumbi valley, the plain of Korcë, the Upper Vijosë valley, the Trebeshin Dhëmbel and Nemerskë highlands. These ranges consist of karstified limestone and flysch.

The coastal ranges include the Dhrino valley, karstic plateaus near the Greek border, the Eocene limestone ridges of the Delvinë district, the fertile "Albanian Riviera" of Himarë and the Dukat valley.

2. Climate

On the coast, maximum summer temperatures average from 36° to 38° C. Winters are mild; January temperatures average 10° in Valona and Durazzo, but lower inland (7° C in Tirana, the capital). Yearly average rainfall values are 1,000 to 1,100 mm in Valona, Elbasan, Tirana and Durazzo; 1,450 mm in Scutari; and 1,800 mm in Pukë. The main rivers from north to south are the Drin, the Mat, the Shkumbi, the Seman-Osum-Gjanice and the Vijosë.

Figure 4. Albania, geographic setting

MAP NO. 3417 Rev. 1 UNITED NATIONS
MAY 1989

B. Geology

It was not until the tectonic movements of the Cenozoic that the territory of Albania emerged from the sea. The stratigraphy of Albania includes upper Paleozoic through Recent formations, almost without interruption (figure 5). Emergence continued, but with less vigour during the Quaternary. This tectonic instability is reflected in the high number of earthquakes that affect the country.

The west and south-west constitute a folded autochthonous region; the north and east are occupied by mountain massifs, displaced by overthrusting. Eight principal rock types or stratigraphic divisions are distinguished (in generally ascending order):

1. Igneous rocks of various ages (Triassic, Jurassic, early Cenozoic) occupy the Serpentine zone in the north-central and eastern mountains, one of the largest serpentine complexes in Europe. They are weathered into pervious soils that are rather arid as they do not retain moisture.

2. Paleozoic strata, the oldest being Permian to Carboniferous, occur in the north and north-east, north of the Drin River. They include beds of shale, greywacke, quartzite and quartz conglomerate. East of the Drin the shale is metamorphosed into phyllite.

3. Triassic rocks are present in many locations but with wide variations in lithology. Lower and Middle Triassic include slate, eruptive rock, limestone, quartzite and conglomerate. The Upper Triassic is mainly represented by dolomite and limestone mostly in northern and central Albania.

4. Jurassic rocks are found in the north Albanian tableland which includes massive limestone and some shale. In the south the series are very thick with dolomitized limestone of lower to middle Lias succeeded by beds of marl, shale and hornstone of the upper Lias and Dogger, then by 1,000 m of marl, hornstone and bituminous limestone.

5. Cretaceous formations have about the same distribution. In the north-east they include conglomerate, sandstone, marl and limestone. The most widespread limestone strata occurring in Albania are the massive Upper Cretaceous to lower Eocene nummulite-bearing rocks which are present in most mountain ranges of the south. These rocks have been deeply karstified. Large springs occur at the contact between the limestone and the flysch upon which they have been thrust or folded.

6. Lower Tertiary flysch formations constitute about half of all sedimentary outcrops in Albania, including large areas in the central and southern parts of the country. These are beds of shale alternating with marly sandstone. The formation is impermeable and commonly lacking of springs. The lower Tertiary includes also marly conglomerate, sandstone and limestone which may hold ground-water resources of local importance.

7. Upper Tertiary formations occur mainly in the coastal belt; they include limestone, conglomerate, argillaceous sandstone, marl and clay. They also include gypsum south-west of Elbasain (lower Miocene). The upper Pliocene is represented by sand, marl and clay along the lower Vijosë River.

Figure 5

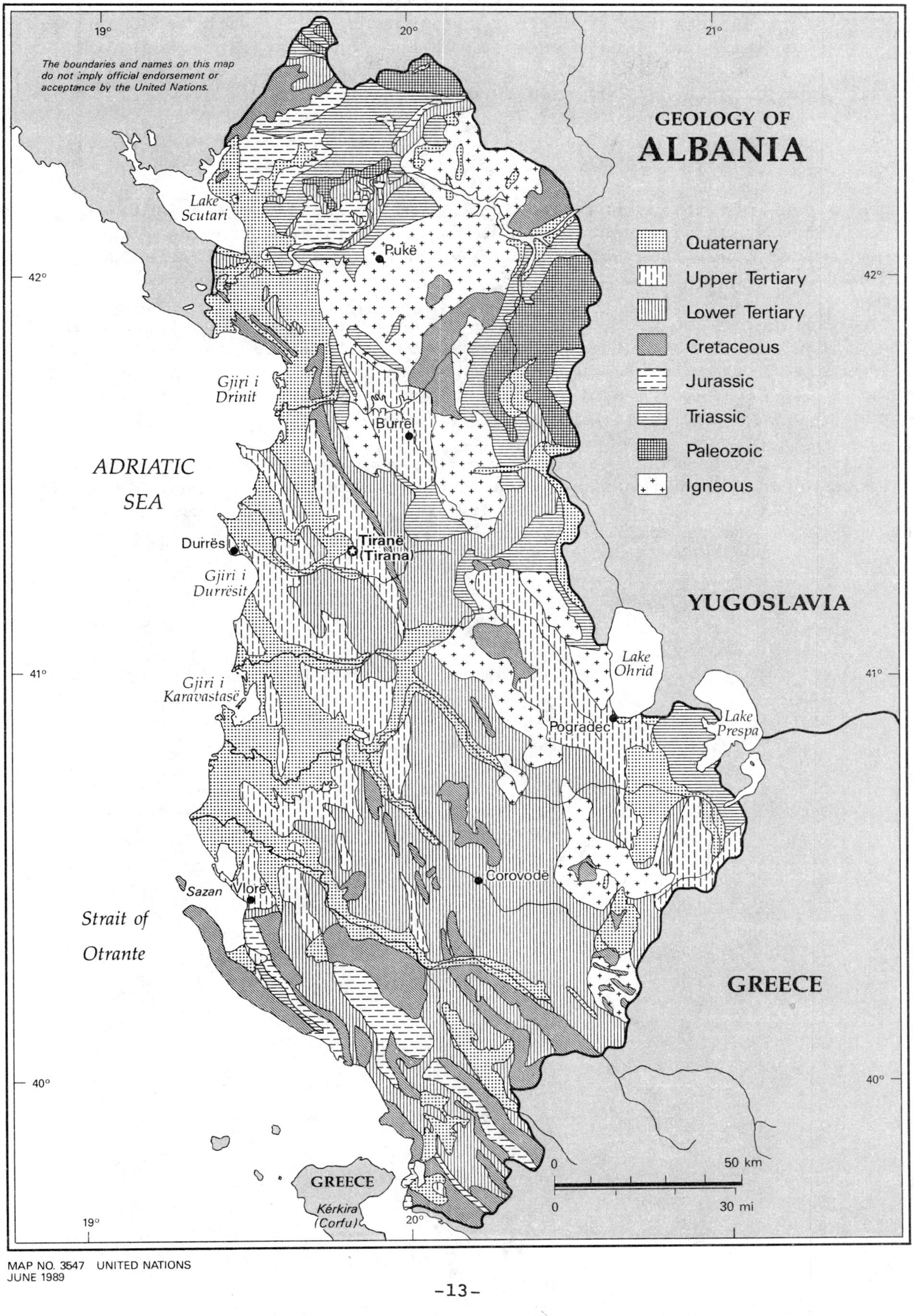

MAP NO. 3547 UNITED NATIONS
JUNE 1989

8. Quaternary formations are restricted to the coastal plains and the lowlands, with also moraine deposits in the eastern and northern highlands. The lower slopes of limestone hills are covered with rock detritus cemented as compact limestone-breccia, with important springs. Fluviatile deposits also occur in the coastal plains, including sandy clay and fine sand. The coast of lower Albania, north of Valona, is fringed by a belt of sand dunes 150 metres wide.

C. Hydrogeology and ground water

The above information has been drawn from publications available at United Nations Headquarters. Requests made for specific information regarding hydrogeological conditions and ground-water developments in Albania brought no additional information.

The explanatory booklet of the Athens sheet of the hydrogeologic maps of Europe (UNESCO-IAH), not issued as this was being prepared, should contain supplementary information. Meanwhile, the information provided in this volume on Yugoslavia and Greece may be of help in understanding the hydrogeology of contiguous portions of Albania.

Albania contains many important springs, especially in areas of karst. Before the Second World War, water supply conditions in the flysch areas, the argillaceous plains and the karstic mountains were difficult, and for many households the main source of water was rainfall collected in cisterns.

II. BULGARIA

Area: 111,000 km^2 Population: 9 million

A. General

1. Physiography and climate

The People's Republic of Bulgaria is located in the heart of the Balkan Peninsula. The Balkan Mountains (Stara Planina) cross the middle of the country from west to east. North of the range are rolling hills which grade into a loess plain.

Southern Bulgaria, rather mountainous and hilly, includes the highest mountains of the country: Rila Dagh, Pirine, Rhodope and Ossogovo. The highest peak is Musala (2,925 m). Of several plains of limited extent the largest is that of Thrace.

The climate of Bulgaria is temperate with continental influences to the north and Mediterranean to the south. Annual rainfall averages 660 mm; some dry areas get as little as 400 mm and as much as 1,000 mm falls in some of the mountains. Temperatures average 9.5° C to 11° C, depending on the area. The country is part of two large drainage basins. One, that of the Black Sea covers about 60 per cent and is drained mainly by the Danube. The other, that of the Aegean Sea covers about 40 per cent and is drained by the Rivers Maritza, Mesta and Struma.

Systematic ground-water surveys were begun after the Second World War and have continued since. Special studies were made to solve specific problems related to urban water supply, industrial development, agriculture, hydrothermalism, public works, and mining activities. In addition, synthesis reports were prepared for large areas of the country, including:

1. A map of Bulgaria's main hydrogeologic units, at scale of 1:600,000 (1962),

2. A map of thermo-mineral water, at scale 1:800,000 (1968),

3. A map showing the distribution of mineral waters, scale 1:500,000 (1970),

4. A hydrogeologic map, scale 1:1,500,000, in Atlas of Bulgaria (1973),

5. A map of ground-water natural resources, scale 1:200,000 (1980),

6. A hydrogeologic map with descriptive booklet, scale 1:500,000 (1980),

7. A map of ground-water vulnerability to pollution, scale 1:200,000 (1982),

8. Hydrochemical maps, at scale 1:1,000,000 (1982) in draft form only, and

9. Maps of fresh ground-water bodies, scale 1:25,000. Since 1968, two thirds of the country has been so mapped.

Most of the maps above include explanatory booklets. Many reports are available on ground-water resources of the country; some are listed in the References for this country paper.

2. Ground-water exploration and assessment

The number and sophistication of methods used has been increasing with time, both as regard to field surveys and laboratory work. The six most commonly used are:

1. Geologic surveys: mapping including identification of tectonic features; quantitative assessment of fracturing, geomorphologic analysis,

2. Geophysical surveys: field surveys (electric probes and profiles; seismic surveys) well logging (resistivity; gamma ray, temperature probing) airborne geophysics,

3. Hydrologic surveys: river and spring gauging, hydrograph analysis of rivers, water balance,

4. Hydrodynamic studies: pumping tests, water injections, tracer studies, long-term piezometric observations,

5. Hydrochemical methods: including gasometric measurement of natural radioactivity, analysis of stable isotopes, and

6. Modelling: analog electrical models, viscose models and mathematical models on computers.

Ground-water resources are assessed according to the following five methods:

1. Analysis of piezometric contours (according to transmissivity values and hydraulic gradient),

2. Analysis of river hydrograph (the ground-water input being considered equivalent to natural resources),

3. River gauging during low-water periods,

4. Measurement of spring yields, and

5. Determination of infiltration ratio (on the basis of piezometric data, evaporation measurements on lysimeters and water balance calculations).

Exploitable ground-water resources on regional scale are assessed by the two means of:

1. Hydrodynamic calculations (analytical or numeral), including mathematical or electrical models, and

2. Water balance calculations being an assessment of dead storage and dynamic storage; additional potential resources for recharge.

On ground-water resources maps, resource values are commonly expressed in runoff per area unit (litres per second per square kilometre).

3. Legal aspects

Ground-water development in Bulgaria is ruled by the General Law on Water (1969) which deals with all main questions related to surface and ground-water use. The law declares that as a priority, ground water is to be developed for community water supply agro-industries and balneology as they are less prone to pollution than surface water. Ground-water development from other uses such as industrial use and irrigation is decided upon according to specific cases of water availability and economic conditions.

The exploitation of new sources (wells) of water is subject to obtaining a permit from a Water Registry Office attached to the State Environment Commission. Such permits are currently required for withdrawals greater than 10 l/s; lesser withdrawals do not yet require authorization owing to the difficulties involved in control.

Authorization for the development of thermo-mineral waters is delivered by a special Committee attached to the Ministry of Public Health.

Water resources protection, including ground water, country-wide is the responsibility of the State Environment Commission which is authorized to penalize pollution cases and is prepared to issue regulations on water protection.

Ground-water protection is dealt with on both a regional and local scale. Regional protection applies to aquifers up to their whole extent, their vulnerability to various pollutants and economic activities contributing to water pollution. Local protection applies to the areas close to the wells. It is expressed in terms of zone (perimeters) of sanitary protection intended to protect the waters developed by such wells. Three zones are identified in the documents which are to be prepared for each well (field) the A zone: total interdiction to penetrate and to develop any man-made activity; B zone: activity restricted; C zone: monitoring and prevention zone. The dimensions of the zones are related to local hydrogeological conditions and the type of pollutants.

4. Training, research and international co-operation

In Bulgaria, ground-water studies are carried out by several State organizations and also by institutes for scientific research. For example, the Geologic Commission carries out hydrogeologic studies related to mineral resources (ores, coal, oil, geothermal). Vodocanalproekt deals with ground-water development for community water supply. Energoproekt deals with hydrotechnical projects, including dams, intended for industrial use. Minproekt deals with the draining of mines. The Ministry of Health deals with thermo-mineral waters for balneotherapy. To some extent, co-ordination of activities is provided by the National Council for Water and the Environment Commission.

The training of ground-water professionals is exclusively done by the Superior School for Mines and Geology in Sofia. A professorship in Hydrogeology and Engineering Geology is attached to the Faculty for Geological Research. Specialization in hydrogeology or engineering geology occurs only in the last year of study. Every year about 20 students receive a diploma of Engineer-geologist-hydrogeologist. A number of foreign students have thus far received that diploma, i.e., from Algeria, Lebanon, Jordan, Syria, Cape Verde Islands, Iraq, Mongolia, Cyprus and Mali.

Ground-water activities in foreign countries have developed over the years especially in the developing countries including Algeria, Tunisia, Morocco, Syria, Libyan Arab Jamahiriya, Congo, Zambia, Mozambique, Iraq, Islamic Republic of Iran, India, Cuba and Viet Nam. They were carried out as ground-water exploration and assessment projects, development projects, and training under either bilateral agreement or through the United Nations system. Bulgaria is active in the UNESCO International Hydrological Programme, preparation of the hydrogeologic map of Europe and other international scientific programmes.

B. Geology

The country comprises the following major structural units (figures 6 and 7):

1. Moesian Platform,
2. Stara Planina-pre-Balkan Zone,
3. Louda Kamtchia Zone,
4. Sredna Gora Zone,
5. Rhodope Zone,
6. A small Carpathian Zone, and
7. Post-orogenic depressions (Neogene-Quaternary).

All the zones above are clearly identified by their specific geologic features and by tectonic boundaries, which in turn allow identification from the hydrogeologic point of view.

The Moesian Platform includes the Danubian Plain and the coastal area on the Black Sea. Two structures can be observed. To the east, the Loudogorié Dome and to the west, the Lom Depression. The lower part of the platform consists of formations intensely folded by the Caledonian and Hercynian orogenies. They include Precambrian and Paleozoic metamorphics including Hercynian granitoids. The sedimentary cover comprises various formations including carbonate rocks, with coastal and continental facies of Paleozoic (Devonian, Carboniferous, Permian) and Mesozoic ages. In the Lom and Varna depressions the sedimentary sequence ends at the top with Cenozoic formations. At the top of the Loudogorié Dome, the Mesozoic-Cenozoic sedimentary cover ranges from 500 to 1,000 m thick but in the Lom Depression it is more than 6,000 m. The strata are practically horizontal. Faults occur mainly east of the dome (towards the Black Sea). The Paleozoic formations are folded or faulted in horsts and grabens which constitute separate ground-water units.

The Stara Planina-pre-Balkan Zone is an orogenic structure trending east west with a number of folds and tilting which create an impervious boundary with the Moesian Platform. In the pre-Balkan area, the geologic succession is known down to the Mesozoic only. The Triassic and Jurassic sequence presents a thickness and lithology similar to that of the Moesian Platform. Differences appear mainly at the Lower Cretaceous. The Valanginian and Hauterivian layers are quite thick and clayey and, therefore, have low permeability. The Barremian and Aptian series consist mainly of water-bearing carbonate rocks.

The Louda Kamtchia Zone is in a deep trough holding folded flysch sediments of Turonian to late Paleocene ages.

Figure 6

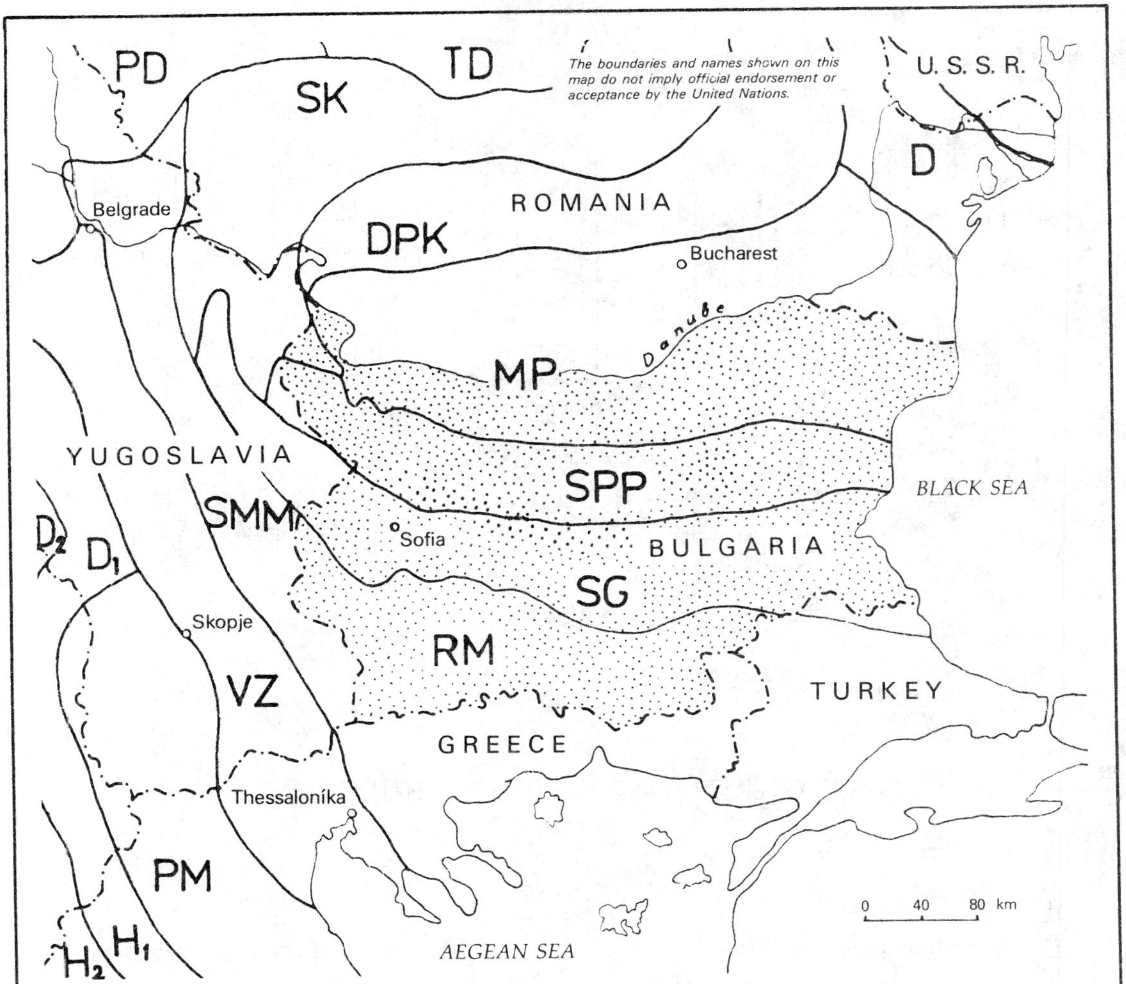

GEOTECTONIC SITUATION OF BULGARIA AND NEIGHBOURING AREAS

MP	Moesian Platform
SPP	Stara Planina pre-Balkan Zone
SG	Sredna Gora Zone
RM	Rhodopen Massif
SMM	Serbian-Macedonian Massif
VZ	Vardar Zone
PM	Pelagonian Massif
D_1 D_2 & H_1 H_2	Mountains, Dinarides and Hellenides
DPK	Pre-Carpathian Depression
SK	Southern Carpathians
PD	Pannonian Depression
TD	Transylvanian Depression
D	Dobroudja Zone

3546.3x

Figure 7

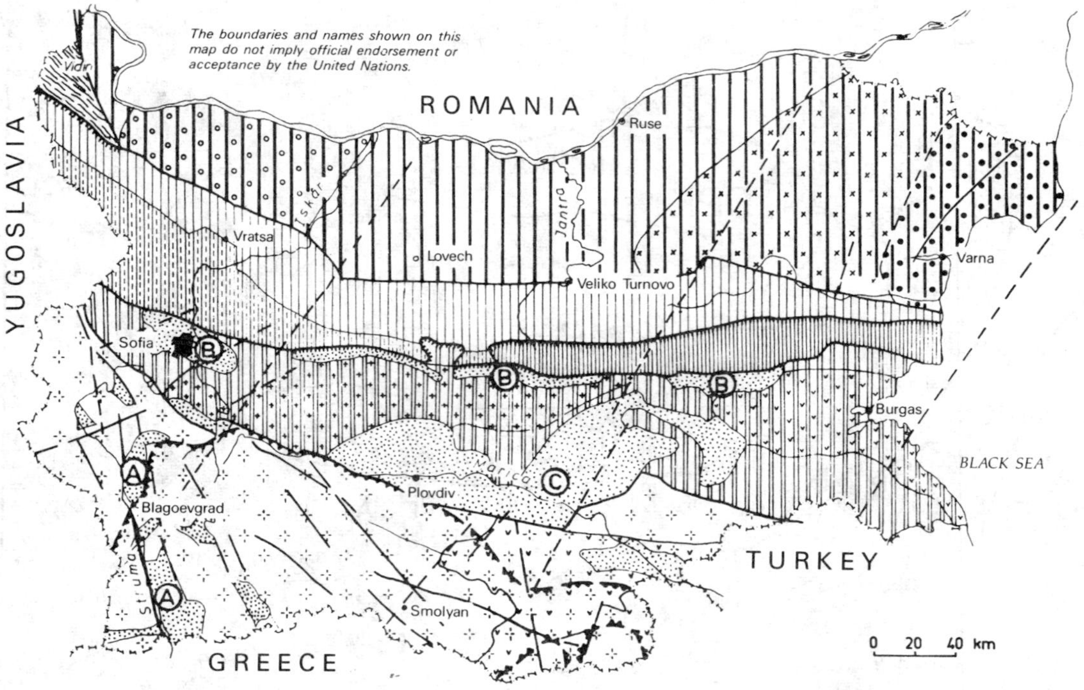

GEOSTRUCTURAL ZONES IN BULGARIA

- Moesian Platform
  - a Loudogorié Dome
  - b Lom Depression
  - c Varna Depression
- Southern Carpathians
- Pre-Balkan
- Stara Planina
- Louda Kamtchia
- Sredna Gora
  - a Granite and metamorphics
  - b Bourgas Synclinorium
- Rhodope
  - a Vulcano-sedimentary areas
  - b Front of Rhodope overthrust
- Depressions
  - (A) Strouma lineament
  - (B) Outer Balkan lineament
  - (C) High Thrace Depression
- Main Faults
- Cryptofault
- Overthrust

3546.4x

The Sredna Gora Zone is developed on a heterogeneous foundation of Pre-cambrian metamorphics, batholiths of Paleozoic granite and Triassic and Jurassic sediments. This deep trough is filled with sediments and also volcanic rock (andesite) of Late Cretaceous age. Over a major part of the zone the sediments are mainly flysch. All the complex of sedimentary and volcanic rocks is folded and dislocated. Towards its southern boundary, there are intrusions of diorite.

The Rhodope Zone includes the mountains of Rila, Pirine, Ossogovo, Ogrageden, Belassitza, Sakar, Strandja and their extensions into Yugoslavia, Greece and Turkey. The Rhodope Zone was fragmented into three parts in the early Paleozoic. The massif was raised during the Alpine orogeny. Several acidic (rhyolitic) eruptions took place during the late Paleocene and several thin depressions (volcano-sedimentary) developed, mainly to the east. The western and central parts were raised during the Neogene and Quaternary periods and a network of faults appeared which are still active. Thus a pattern of horsts and grabens developed which play a major role in the distribution and flow of ground water.

Several post-orogenic depressions contain Neogene-Quaternary sediments which range from 1,000 to 5,000 m in thickness. Some lines of depressions are oriented along faults trending 150°-160° (SE), for example, valleys of the rivers Strouma and Mesta. Another line of depressions follows the northern faulted boundary of Sredna Gora. The main depression is that of the High Thrace Plain. Predominant in its sedimentary fill are alluvium, proluvium and Pliocene lake deposits.

C. Regional hydrogeology

From the structural description above, it can be concluded that the territory of the country is a complex mosaic of platforms and raised areas, deeply tectonized with extensive lithologic variations, and multiple magmatic manifestations. Considering also the differences in climate and morphology, it is not surprising to realize the complexity and variety of conditions which rule the occurrence, distribution, flow and chemical composition of ground water. By and large, this knowledge which has been acquired on phreatic fresh water and moderately deep artesian water is fairly comprehensive. Additional studies, however, are necessary to assess the knowledge of ground water in crystalline (granite-metamorphic) areas in flysch and volcano-sedimentary troughs.

The territory of Bulgaria, from the point of view of hydrogeological conditions, is divided into six regions (figure 8).

1. Moesian

The northern boundary of this natural area is located in Romania and coincides with the faulted southern boundary of the pre-Carpathian depression in Romania. In Bulgaria, the southern boundary is the faulted area along the pre-Balkan region. Tectonic deformation and faulting have resulted in the isolation of artesian basins of both regions and the exchange of ground-water flows between the platform and the pre-Balkan is minimal or nil, although some flow occurs from one region to another through shallow (phreatic) aquifers.

Figure 8

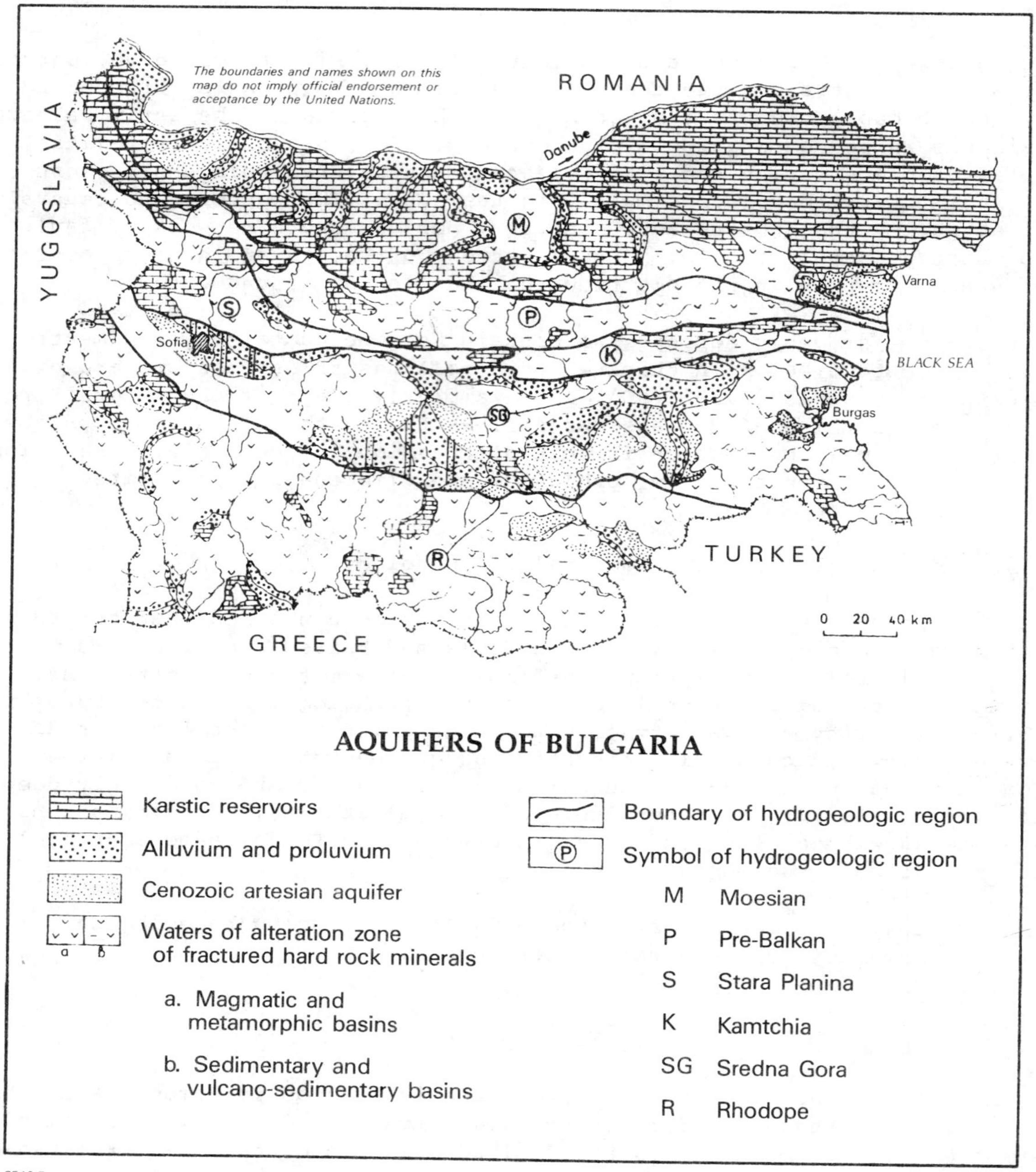

3546.5x

Aquifers include:

1. Quaternary alluvium of the Danube and its tributaries,

2. Pliocene, Tortonian, Sarmatian, Eocene Senonian and Barremian-Aptian, and

3. Upper Jurassic (Malm), Valanginian, Middle Triassic and Upper Devonian.

The distribution, stratification, hydrodynamic characteristics, hydrochemistry and geothermy of the aquifers are related to the three structural units of the platform: Loudogorié Dome, the Lom Depression and the Varna Depression. The dome divides the region into two large artesian basins; an eastern (Varna-Dobruja) and a western (Pleven-Lom). The eastern basin discharges into the Black Sea, with the estuarian valley of Lake Varna. The western basin is drained by the Danube and its adjacent plains. Loudogorié Dome can be considered the recharge zone of the artesian basins.

The most remarkable aquifer is that of Malm-Valanginian age, which covers the entire region. It includes limestone and dolomite more than 800 metres thick. Several units are highly karstified. Within the boundaries of the dome and its adjacent slopes considerable amounts of ground water are in storage. To the west the Pleven-Lom basin contains only salt water with saline concentrations of 5 to 60 g/l and temperatures reaching 40° to 60° C.

Aquifers lower than the Malm-Valanginian have not been much studied. They include carbonate layers of the Middle Triassic and Upper Devonian at 3,000 to 5,000 m depth and containing high-concentration brines with 100° to 160° C temperatures.

2. Pre-Balkan

The pre-Balkan region contains intensely and deeply folded areas which result in complex hydrogeological conditions. These are known mainly in the phreatic (shallow) aquifers which include karstified limestone of varied areal extent, non-eroded remnants of the Barremian-Aptian and Senonian massifs.

In some small synclinal areas these horizons are capped by an impervious Paleogene cover which creates artesian conditions. For most of the region underlying the Barremian-Aptian layers, a thick series (1,500 to 3,000 m) is found which includes marl and flysch of Hauterivian, Valanginian and Tithonian ages. Under these occur various fragmented aquifers of Dogger and Middle Triassic ages, which outcrop at the top of the main (Tereyen) anticline. There they contain fresh water, but at depth the waters are of the hot and concentrated thermo-mineral type. Away from the anticline all Triassic-Jurassic reservoirs are confined and contain thermal waters with high chemical content.

3. Stara Planina

The Stara Planina region includes a geostructural subzone with distinctive hydrogeological characteristics. In the heart of the Berkovitza Schipka anticlinorium rocks of low permeability are found. These include schist, phyllite, sandstone and Paleozoic conglomerate with granitoid Hercynian intrusions which contain fissure and fracture water and, in some cases, thermal water with low mineralization. On both sides of the anticlinorium one can distinguish several dislocated rock masses of limestone and dolomite of Middle to Late Triassic and Late Jurassic ages. These rocks are deeply karstified, aquiferous and give way to many springs.

4. Kamtchia

The Kamtchian region occupies the flysch trough of Louda Kamtchia. The Cenomanian-Paleocene flysch has low permeability and holds little exploitable water. It is possible that the formations include sandstone beds with better aquifer properties. Thus far, ground-water exploration and development has been concerned only with large blocks of Jurassic and Senonian limestone. It is possible that thermal saline water lies at the base.

5. Sredna Gora

The Sredna Gora region is divided into several hydrogeologic structures and basins of diverse dimensions, shape lithology and stratigraphy. Specific hydrogeologic features are related to the Upper Cretaceous volcano-sedimentary formation and the Neogene-Quaternary depressions, in particular in High Thrace. Upper Cretaceous sediments and volcanics are found in limestone structures which extend from Yugoslavia to the Black Sea, the main one being the Bourga synclinorium to the east, and another to the southern coast of the Black Sea. Most of the complex, 2,500 to 3,000 m thick, holds little ground water as it includes mainly argillite, tuffite and flysch. Several sandstone and limestone layers and bodies of andesite and basalt have some permeability. The whole is affected by Recent faults which in some cases are water-bearing. In the deeper parts of the Burgas synclinorium, small pockets of thermal saline water have been found. Neogene-Quaternary depressions of the outer Balkan overlie Mesozoic sediments. The High Thrace depression overlies metamorphics, granite and sediments of Paleogene and Late Cretaceous ages. Many depressions hold artesian aquifers which to some extent communicate hydraulically with overlying Neogene-Quaternary phreatic aquifers. That is the case of the Sofia depression.

In the middle of the region, granitoid and metamorphic rocks outcrop widely. Ground water can be found in alteration and fractured zones of these formations, which give way to thermal springs yielding slightly mineralized waters.

6. Rhodope

The main hydrogeologic features of the Rhodope region are related to:

1. The prevailing occurrence of granite-metamorphic masses holding fresh or hydrothermal waters in fractures and faults,

2. The occurrence of many outcropping marble masses, karstified and water-bearing,

3. The presence of relatively good aquifers in Paleogene rhyolite,

4. The large development of Paleogene volcano-sedimentary depressions filled with clays and tuffs of low permeability,

5. The occurrence of Neogene-Quaternary depressions; with low permeability in Neogene terrane, better in the Quaternary, and

6. A network of active faults in the central and the western part of the region with displaced blocks having an important influence on ground-water flow. Some other faults or tectonic features may act as hydraulic barriers, however, their role is not competely understood.

Up to the regional base level of erosion (300-400 m) granite metamorphic massifs also contain cold fresh water mainly related to active faults. At lower depths, hydrothermal waters are found with temperatures ranging from 25° to 110° C. Such water is discharged through outlets in river valleys or grabens. In marbles, karstification is irregular depending on mineral composition and presence of sandstone layers.

D. Qualitative hydrogeology

1. Ground-water reservoirs

Several types of aquifers have been identified; they fall into two structural classes, stratified or multilayered, and non-stratified principally fractured reservoirs. The multilayered aquifers occur exclusively in sedimentary rocks. They include sand, gravel, limestone, dolomite, and locally sandstone. They occur mainly in post-Paleozoic rocks except for Devonian aquifers in the Moesian platform, and are separated by impervious or semipervious beds of clay or marl. Such reservoirs are found in the Moesian and pre-Balkan hydrogeologic regions and the post-orogenic depressions. In most other regions, such aquifers are minor and in the Rhodope region, minimal. Non-stratified reservoirs occur in magmatic and metamorphic formations constituting a large fractured system in mountainous areas. This type of reservoir is typical in the Stara Planina, Sredna Gora and almost all the Rhodope.

On the basis of the type of voids, distinction is made between the porous, karstic, and fracture reservoirs. Porous reservoirs, mainly sand, gravel and friable sandstone, are found only in Cenozoic and Quaternary formations. The karstic reservoirs occur at many stratigraphic levels from Proterozoic marble to Sarmatian limestone. They can be seen in all hydrogeologic regions and are predominant in the Moesian and pre-Balkan regions. The fractured reservoirs, which occur widely in silicate rocks of all regions, are of great importance in the Rhodope but negligible in the Moesian region.

A distinction is made also between unconfined, water table, or free aquifers on the one hand and confined artesian aquifers on the other. Artesian systems occur at moderate depths (some tens of metres) in post-orogenic depressions, deeper in mountain areas. Pore water, fissure and fracture water, and karstic waters are encountered in Bulgaria in nearly equal proportions. Ground water is considered to be warm or hot when temperature exceeds 25° C.

Sweet or fresh ground water, which contains totally dissolved solids (TDS) under 1,000 mg/l is found throughout the country. Locally it has been found at depths of 1 to 1.5 km, in phreatic aquifers, and in the upper part of artesian aquifers. Its presence depends largely on active water flow, recharge by rain and drainage by the hydrographic network.

Thermomineral waters are found mainly in the lower part of the artesian basins and in deep faults of fractured crystalline massifs.

2. Cold fresh ground water

Cold fresh ground water occurs in karstic reservoirs, alluvium and proluvium, Cenozoic artesian sandy aquifers, and the alteration zone of fissured non-carbonate crystalline massifs. The distribution of these general types is shown in figure 8. A limited amount of fresh ground water is also present in loess.

Large karstic reservoirs in Mesozoic and Cenozoic limestones occur in the Moesian region and in the artesian basins of Varna and Pleven-Lom. From top to bottom, the following aquifers, by age, are found: Sarmatian, Upper Cretaceous (mostly Senonian) Barremian-Aptian, and Malm-Valanginian, the most important.

In the pre-Balkan region, large areas are underlain by the Senonian and Barremian-Aptian aquifers.

In other hydrogeologic regions, karstic water occurs in isolated basins of modest dimension, mainly in Middle and Upper Triassic dolomite and Upper Jurassic limestone, some in Senonian limestone. In the Rhodope several isolated karstic aquifers are found in Proterozoic marble. It is estimated that in the areas of karst, depending upon morphology, vegetation and soil conditions, 20 to 40 per cent of the rainfall infiltrates the aquifer system. Underground flow may vary from 2 or 3 $l/s/km^2$ up to 25 or 30 $l/s/km^2$. Karstic aquifers are, in fact, the most productive in the country. They discharge through many outlets including Bulgaria's largest springs, i.e., Devmia, Iskretz and Zlatna Panega, with an average yield of more than 1.5 m^3/s. Karstic water produces some 75 m^3/s, or more than 40 per cent of all fresh ground water extracted in the country.

The largest pumps are located in the Moesian and pre-Balkan regions. The chemical composition is mainly HCO_3-Ca (calcium bicarbonate), in some cases HCO_3 Ca-Mg. Hardness is less in water from the Proterozoic marbles than that from sedimentary limestone. The highest hardness values are found in Sarmatian aquifers and pH varies little (7.2 to 7.8).

Karstic waters are largely used for industry, irrigation and community needs after physical and antibacteriologic treatment. Most ground water is taken from springs, the rest from boreholes, mainly in the north-east.

3. Alluvial and proluvial

Recent alluvium of rivers and piedmont fans occur mainly in the Moesian and Sredna Gora regions. The most important units are alluvial terraces along the Danube and Maritza rivers. Their thickness ranges from 2 or 3 metres, most commonly 10-20 metres, up to 40 or 50 metres. In general, the deeper alluvial layers contain gravel and sand, whereas the upper layers contain the most silt and clay. The alluvium filling the central part of large plains (Thrace, Sofia) have a complex structure of alternating layers of gravel, sand and clay where, in some cases, semi-artesian conditions occur. The proluvium of piedmont areas contains mostly coarse material with minor clay filling.

The alluvial aquifers contain much water and are widely tapped. Transmissivity is normally in the range of 500 to 2,000 m^2/day for the Danube and Maritza alluvium, and 300 to 1,000 m^2/day for the alluvium of other rivers.

Recharge of alluvial aquifers occurs primarily through infiltration of rain but also from irrigation systems. Most rivers tend to drain the aquifers although the Danube during high-water periods feeds the aquifer along its banks.

Proluvial aquifers receive additional recharge from perched rivers which cross the piedmont alluvial fans.

The rates of ground-water flow in alluvial aquifers are generally 3 to 5 l/s/km^2, in some cases slightly higher (5 to 7 l/s/km^2).

Alluvial aquifer resources as a whole amount to 60 m^3/s or about a third of the country's fresh ground-water resources. Water quality is in general adequate, although there are significant differences. Whereas pH does not vary much (7 to 7.8), hardness ranges considerably from 2.5 to 12 meq/l, exceptionally up to 17 meq/l. Many wells, boreholes and drains tap the alluvial water for community water supply, industrial use and, in some cases, for irrigation. In recent years, water quality of alluvium has been affected by human activity. In some areas alluvial water has higher concentrations of dissolved solids (1,000 to 3,000 mg/l) as the result of intense evaporation and related salinization of the soil.

4. Cenozoic artesian aquifers

Sandy, Cenozoic artesian aquifers occur in several layers, mainly the Pliocene but also, Miocene, Eocene and Oligocene. To the north, the aquifers occur in folded structures whereas to the south (Sredna Gora and Rhodope regions) they are related to grabens containing Tertiary sediments. Most important to the north are Pontian aquifers in the Lom depression and an Eocene aquifer in the Varna depression. To the south the main artesian aquifers are Pliocene in the depressions of Thrace, Sofia and Haskovo.

These aquifers are recharged by direct rainfall infiltration on the sides of the depressions. Discharge is rendered difficult within the closed structures which have little or no communication with the outside. Most discharge occurs by seepage through semipermeable layers and boreholes. The Pontian aquifer (Lom) drains to the Danube River bed. The total resources of such aquifers do not exceed 7 m^3/s, or about 4 per cent of the country's ground-water resources. Hydrochemical composition is of the HCO_3Ca type and in some cases HCO_3 Ca-Na; pH is 6 to 5.8, hardness from 4 to 9 meq/l. These aquifers are exploited exclusively by means of tubewells.

5. Alteration zone of fractured massifs

Magmatic and metamorphic rocks with water-bearing zones occur in all regions except the Moesian. They include the granites of Sredna Gora, Rila, Pirine, the western Rhodope and other mountains; the andesite, rhyolite and metamorphics of Rila, Rhodope, Ossogovo, Sredna Gora; and various sequences of gneiss, sandstone, marl and flysch.

The alteration zone is normally several tens of metres thick. Recharge occurs through infiltration of rainfall then drainage by many low-yield springs, in many cases scattered. Aquifers are mostly of the thermo-mineral type. Granites and rhyolites at Rila and the Rhodope show more abundant yields with ground-water flow rates reaching 2 to 3 $l/s/km^2$ and above; the norm, however, is 0.1 to 1 $l/s/km^2$.

Some widespread fissured schist, marl, tuff and tuffite have little permeability and their water resources are negligible.

Total resources of the fissured or fractured massifs amounts to 37 m^3/s, or about 21 per cent of the country's fresh ground-water resources. In general, their water quality is excellent.

Hardness does not exceed 7 meq/l, the pH varies from 6 to 7 in granite and other intrusive rocks to 7.3-8 in andesite and 6.7-7.2 in gneiss. Spring water is developed for the community water supply of small towns and villages.

In summary, the ground-water potential in Bulgaria is about 178 m^3/s, or about 8 per cent of rainfall. The infiltration rate may vary from 1 per cent to 40 per cent depending upon the terrain. Average ground-water flow, country-wide, is 1.6 $l/s/km^2$. Ground water is drained by the hydrographic network and contributes 27 per cent of the total runoff. Ground-water reservoir resources are as follows: karstic, 42 per cent; porous, 37 per cent; and fissured or fractured, 21 per cent.

E. Thermo-mineral water

Thermo-mineral waters are both abundant and diverse due to sharp geological contrasts and tectonization. They can be divided into four groups:

1. Nitrogenous thermal waters with low mineralization,

2. Mineral and thermal waters with mainly carbon dioxide (or bicarbonates),

3. Mineral and thermal waters with sodium-sulphate bicarbonate composition and high mineralization (e.g., Karlovy Vary), and

4. Salty thermo-mineral waters and brines of marine origin.

The above groups differ as a result of one or several of the following elements:

1. Genesis of dissolved or dissolving elements,

2. Composition of solids or dissolved gas,

3. Temperature, pH and related characteristics, and

4. Geologic, spatial and geothermal position.

The country is divided into four hydrothermal provinces:

1. Nitrogenous hydrothermal

Nitrogenous thermal waters are of meteoric origin (figure 9). The total concentration of dissolved solids normally ranges from 100 to 1,000 mg/l. They occur in deep (400 to 3,500 m) aquifers which do not contain evaporites, marine saline waters nor endogenous carbon dioxide. The temperature ranges from 25° to 115° C. They occur at depth in all granito-metamorphic massifs, in most vulcano-sedimentary depressions in the artesian areas of the Loudogorié Dome and in many anticlinal or monoclinal structures of the pre-Balkans, Stara Planina and Sredna Gora. The total area concerned is about 50,000 km^2.

Three types of thermal waters are found in this province:

(a) In non-stratified (fractured-fissured) reservoirs of granites, metamorphic or volcanic rocks. Such waters are markedly alkaline (pH 8-10). The main components are HCO_3, SO_4 and Na with high concentrations of fluorine, silica, W, Ge, Mo and Ga and in some cases H_2S and radon. The area is about 23,000 km^2, mainly in the Rhodope and Sredna Gora regions. Total resources amount to 4 m^3/s. Average temperature is 55°-60° C. Higher temperatures are encountered in Separeva Banya (101° C), Velingrad (94° C) and Sandanski Levounovo (82° to 90° C). The rate of hydrothermal flow in the Rhodope area is in the range of 0.3 $l/s/km^2$ and geothermal energy is 60 to 70 kW/km^2. Dynamic resources of various locations are 10 to 80 l/s. In Velingrad, five different thermal locations yield a total of 170-180 l/s with temperatures of 48° to 94° C.

(b) In artesian karstic reservoirs with chemical composition HCO_3, Ca, Mg, low alkalinity and no specific components. In areas of contact with saline water of the marine group, an increase in chloride and sulphur concentration is observed. Temperatures are 25° to 80° C, the highest values being observed in the Sofia depression. Resources of hydrothermal karstic reservoirs amount to at least 8 m^3/s mostly in the artesian Varna basin (deep part of the Valanginian aquifer, near the Black Sea) and in the Triassic-Jurassic reservoirs of the pre-Balkan and western Sredna Gora regions.

(c) In artesian reservoirs of clastic and flysch sediments in depressions and volcano-sedimentary formations of the Rhodope and Sredna Gora regions, the composition is of the HCO_3-Na type with high alkalinity and temperatures ranging from 25° to 90° C. At depth, higher temperatures may be encountered. Resources average 2 to 3 m^3/s. Total hydrothermal resources of the province amount to 14-15 m^3/s of which only 3 m^3/s have been developed.

Figure 9

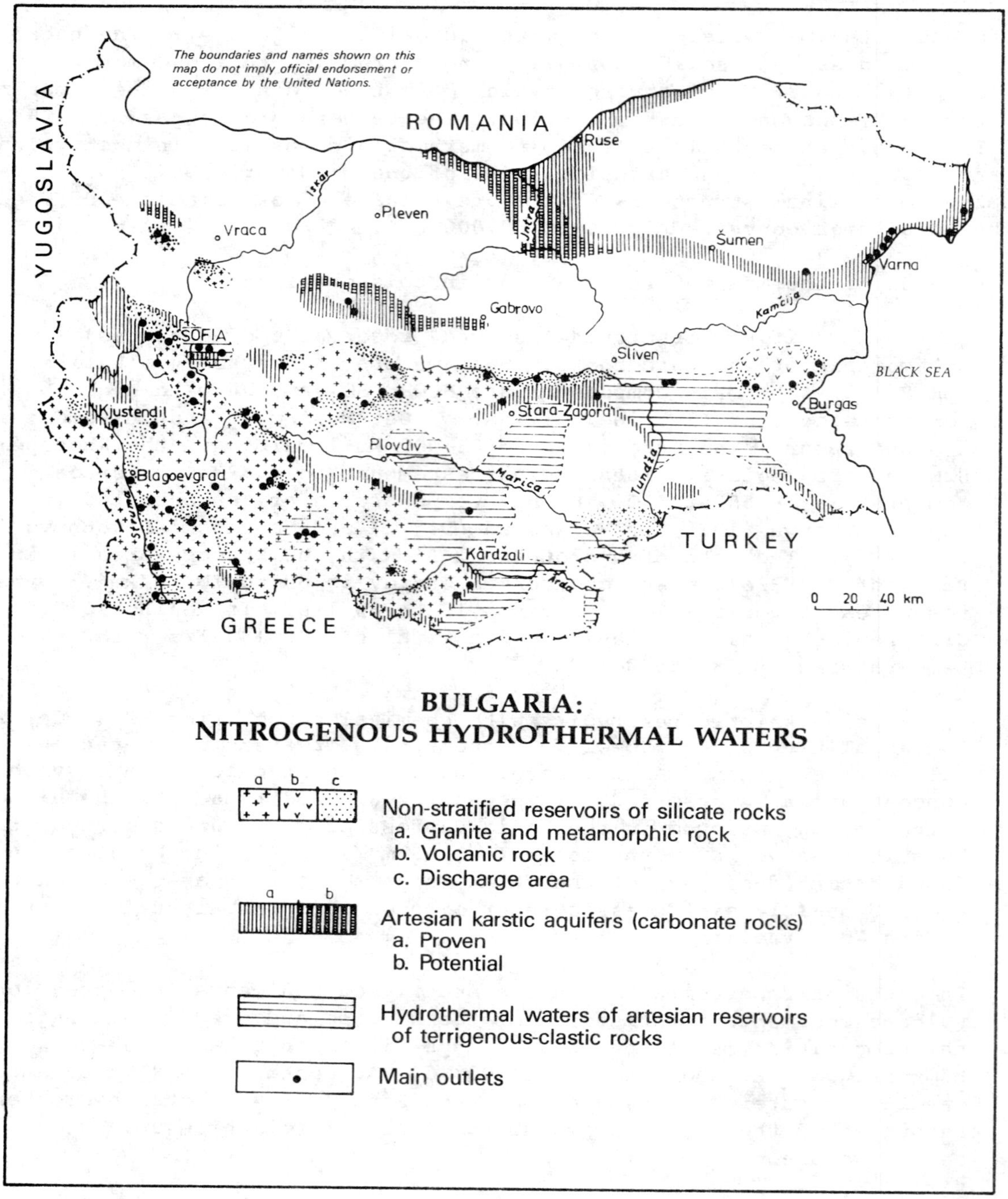

3546.6x

2. Mineral and thermal carbo-gaseous

This province consists of territories where endogenous CO^2 occurs, or where CO^2 reserves of ancient origin are stored. In its movement upwards, CO^2 may penetrate into different waters of meteoric or marine origin. In figure 10 are shown areas, both proved or potential, where waters of such meteoric origin have absorbed bicarbonates by dissolution and leaching of various rocks, silicate, carbonate or others. The TDS content ranges from 2,000 to 10,000 mg/l. The country does not contain much water of this type.

3. Sulphated sodium bicarbonate

The waters of this province have higher concentrations in sodium, sulphate and bicarbonate. This kind of water is found at the base of some Neogene depressions, which contain lake deposits. The chemical composition may be related to arid climatic conditions during the Miocene period.

4. Brackish water and brines

The province includes thermo-mineral saline waters and brines (figure 11). Its boundaries include the Moesian platform, the pre-Balkan, Louda Kamtchia and the eastern end of Sredna Gora. The waters contain mainly sodium chloride with high concentrations of iodine, bromide and boron. TDS is in the range of 5 to 10 g/l, with maxima of 100 to 150 g/l and even more in deep isolated reservoirs. The highly concentrated waters are relatively rich in calcium. Methane and nitrogen dioxide are the main gases. In bicarbonate reservoirs the occurrence of H_2S is also noted. At greater depths, higher concentrations of CO_2 resulting from the decomposition of dolomite are also found. Reservoirs are located at depths ranging from 700 to 6,000 m. Temperatures range from 30° to 180° C. Most productive are the bicarbonate reservoirs in rocks of Malm-Valanginian, Middle Triassic and Late Devonian ages.

5. Use of thermo-mineral water

In Bulgaria the utilization of thermo-mineral waters and balneotherapy dates to the Roman period and earlier. Pantalia (Kustendil), Serdika (Sofia), Augusta (Hissar), Djermanea (Séparéva Banya), Kilopolis ad Nestum (Ognyanovo) were well known balnear centres during the Roman and Byzantine periods.

Present geothermal projects are applied to the heating of some sea resorts, greenhouses and some industrial installations. The use of heat pumps has been envisaged. The thermal waters of the marine province are to be developed by means of boreholes. The first geothermal projects in Bulgaria were started in the early 1960s (greenhouses and sea resorts). More important projects are under study for the needs of Varna, Vidin and other cities.

In Bulgaria, many means of using hydrothermal resources are applied: medical treatment, recreation, geothermal energy, public baths, table water, soft drinks, extraction of chemical components and others. A number of new projects are under consideration for therapeutic centres, balneology establishments, swimming pools, heating systems and greenhouses.

Figure 10

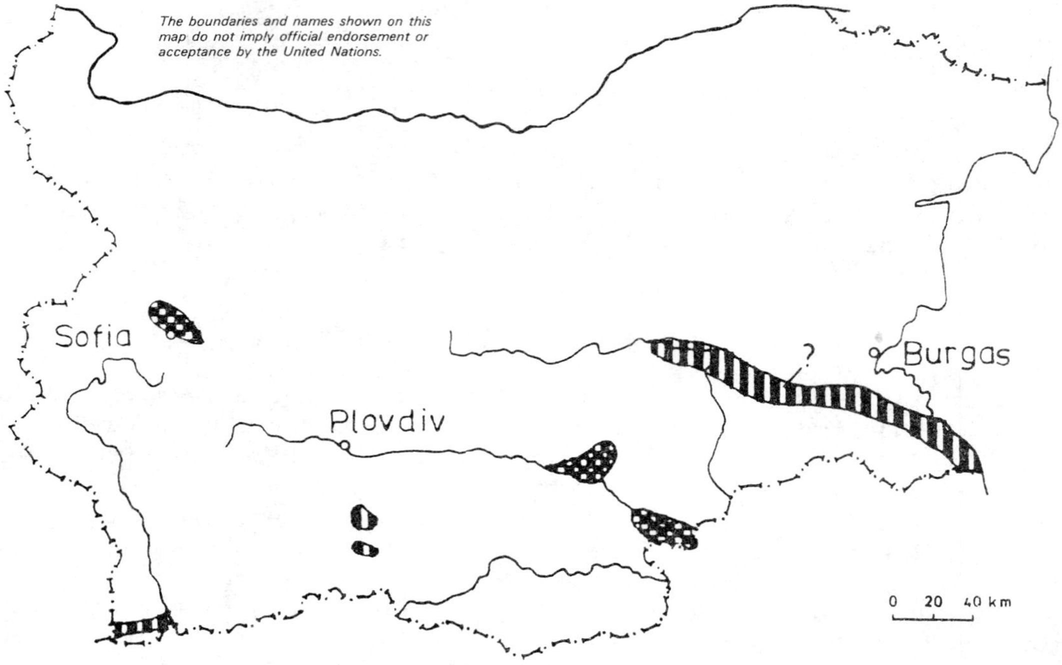

BULGARIA:
CARBONATE AND SULPHATE-BICARBONATE WATERS

Carbonate waters
Sulphate-bicarbonate waters

3546.7x

Figure 11

3546.8x

F. Problems and prospects

In Bulgaria there is a need to improve the methods of assessing ground-water resources under difficult natural conditions. There is also a need to strengthen the protection of ground-water resources through legislative and regulatory measures, and to further optimize the exploitation and management of ground water.

Improved means are needed in water well drilling in terms of equipment, well screens, study and control, monitoring, and computer modelling at all stages of development.

Specifically, as regards country-wide operations, the knowledge of ground-water conditions in several areas should be improved. Co-operative efforts should be undertaken with neighbouring countries in the study of transnational aquifers. Detailed studies of the most important aquifers of the country should be carried out in order to determine acurately the possibilities for their exploitation. A special effort is to be made regarding the study of granito-metamorphic massifs, the hydrological conditions of which are quite different from those prevailing in other European countries with similar rock types.

A more detailed study should be made of the vulcano-sedimentary Rhodope and Sredna Gora regions which are relatively poor in ground-water resources.

Additional research should be done to develop geothermal energy from open and closed reservoirs, for the utilization of heat pumps and also the dewatering of deep mines (e.g., Dobroudja and Erma Reca).

Closer co-operation should be sought when developing community water supplies for more rational use and protection of aquifers.

References

Antonov, H. Ground water investigations. Sofia, Ed. Technika, 1970.

Antonov, H. and D. Dantchev. Ground water in the People's Republic of Bulgaria. Sofia, Ed. Technika, 1981. (In Bulgarian)

Antonov, H. and others. Hydrogeological Division of Bulgaria. Sofia, Academy of Sciences, 1962. (In Bulgarian)

Bindemann, N. N. and others. Main features of ground water occurrences in Bulgaria. Moscow, Ed. Nauka, 1974. (In Russian)

Bontchev, E. Geotectonic problems in Bulgaria. Sofia, Ed. Technika, 1971. (In Bulgarian)

Galabov, M. Dynamics of ground water. Sofia, Ed. Technika, 1981. (In Bulgarian)

__________. Hydrodynamics of wells and drains. Ed. Technika, 1985. (In Bulgarian)

Galabov, M. and P. Pentchev. Use of computers in solving hydrogeological problems. Sofia, Ed. Technika, 1986. (In Bulgarian)

Hsu, K. I., I. K. Nachev and V. Vuchev. Evolution of Bulgaria in light of plate tectonics. Tectonophysics, v. 40, p. 3-4, 1977.

Ivanov, J., S. Moskovski and K. Koltchevak. Main features of the central part of Rhodopian massif. Geologica Balk, v. 9, No. 1, 1979. (In Russian)

Katz, D. M. and others. Prevision of the regimen and of the balance of ground water in irrigation areas. Sofia, Ed. Technika, 1980. (In Bulgarian)

Kehayov, T., M. Galabov and L. Berov. Basic regularities of ground water runoff distribution in Bulgaria and its mapping. Journal of Hydrol. Sciences, v. 4, No. 2, 1977.

Petrov, P. S. and others. Hydrological study of mineral waters in Bulgaria. Sofia, Ed. Technika, 1970. (In Bulgarian)

Schterev, K. Mineral waters in Bulgaria. Sofia, Ed. Naouka i Izkoustvo, 1964. (In Bulgarian)

__________. Atlas of thermomineral waters in Bulgaria. Sofia, Ed. NIIKF, Cartography Institute, 1971. (In Bulgarian)

Schterev, K., S. Stankovic and C. Garagunis. Map of mineral and thermal waters. Sheet No. 4C, Central Part of Balkanic Peninsula, v. 2, Heise Hannover, 1985.

III. CZECHOSLOVAKIA

Area: 127,900 km^2 Population: 15.6 million

A. General

1. Physiography

The Czechoslovak Socialist Republic is divided into two geomorphologic units differing in geologic genesis, hydrogeological conditions and type of relief.

The western part of the country is covered by the Bohemian Highlands formed by Variscan folding during the late Paleozoic, and platforms from the Permian to Neogene. The eastern part belongs to the Alpine Carpathian system and to the lowland subsystem of the Pannonian Basin.

The Bohemian Highlands are old, complex mountains of polygenetic relief, a result of neotectonic movements that culminated in the Neogene, and of changing climatic conditions from the Permian to Recent. In the Variscan orogeny, late Paleozoic and older metamorphic and volcanic rocks predominate. By the end of the Paleogene, the Variscan mountains had been eroded to a virtual peneplain. At the outset of the Neogene intensive volcanism produced volcanos and isolated cones in northern and western Bohemia and northern Moravia. Later, during folding of the Outer Carpathians, the border parts of the Bohemian Highlands were lifted, its central part maintaining its original low-level position. The post-Variscan platforms comprise mostly Permian-Carboniferous, Cretaceous and Neogene sediments, and the above-mentioned volcanic rocks. Quaternary deposits are morphologically indistinct. In the Quaternary Period, however, the morphology of the Bohemian Highlands was markedly affected by periglacial climate.

The Bohemian Highlands are classified into seven basic geomorphologic units: (1) the South Bohemian Highlands, (2) Ore Mountains, (3) Sudetes Mountains, (4) Oder (Odra) Lowland, (5) Berounka River Highland, (6) Brno Highland, and (7) Bohemian Cretaceous Plateau. Hills and uplands predominate rising 400 to 700 metres above sea level. Less widespread are sedimentary plateaus in the central part of the Bohemian Highlands and mountainous areas along their borders. The highest point of the Bohemian Highlands is Snezka Mountain (1,602 m), and the lowest a location near Hrensko (115 m) where the Elbe River leaves Czechoslovakia.

The Carpathian Mountain subsystem is a large morphologic structure with a specific genesis and relief. Its characteristic features are heterogeneous basic morphostructural forms, manifest in two contrasting formations: mountain ranges and depressions. The lowland subsystem of the Pannonian Basin extends into Czechoslovakia with its northern fringes: the South-Moravian basin, the Záhorská Lowland, the Danube Lowland and the East Slovakian Lowland.

2. Hydrographic network

Czechoslovakia is a land-locked country which includes part of the watersheds of the Elbe and Oder rivers (draining to the Baltic Sea) and the Morava and Danube rivers draining to the Black Sea (figure 12).

Figure 12. Czechoslovakia, geographic setting

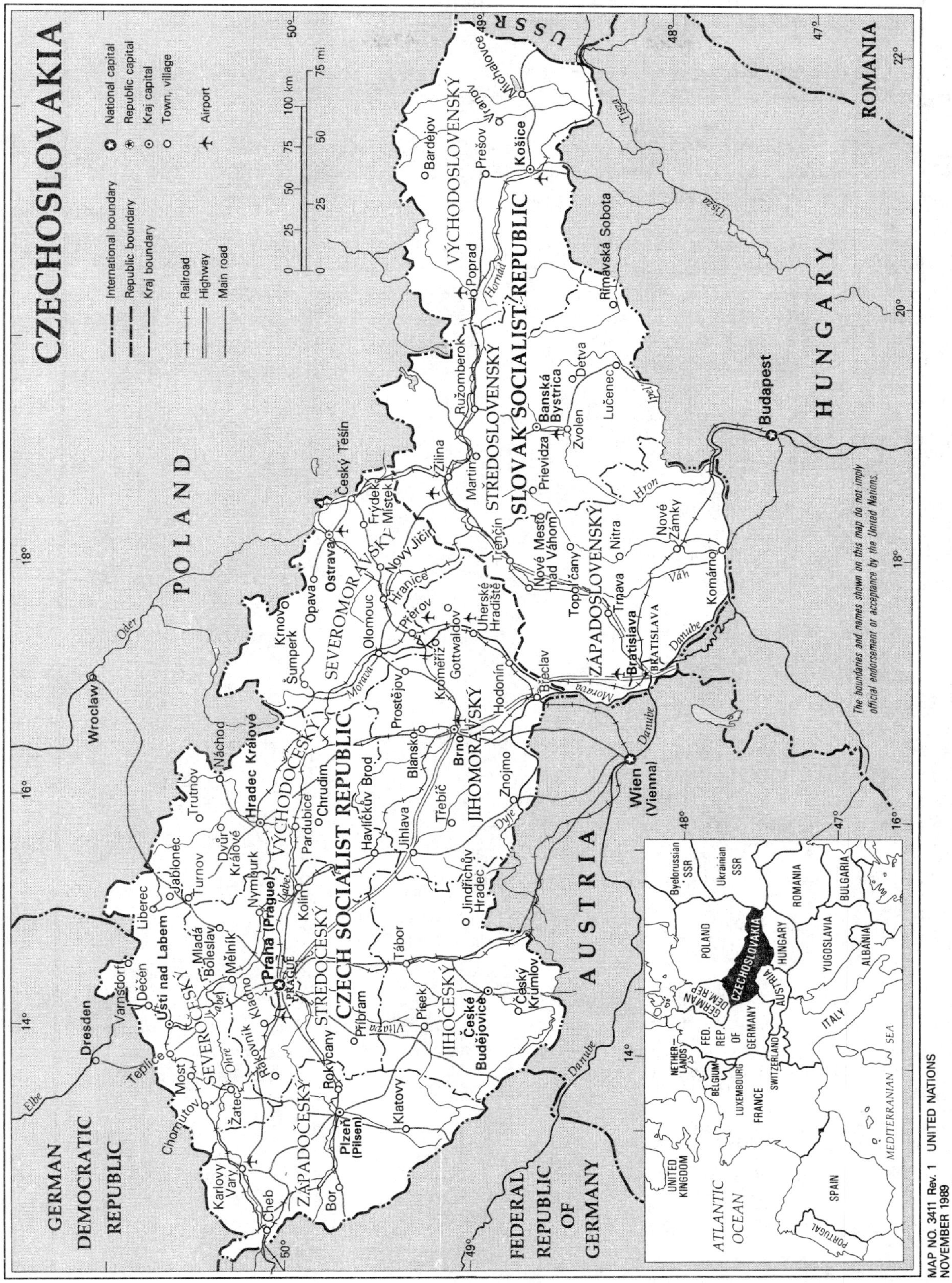

Bohemia is a hydrologically closed region. The continental divide between the Baltic and Black seas almost coincides with the south-east and south-west borders of Czechoslovakia, whereas the north-east borders are the continental divide between the Baltic and North seas. The north-west border forms the divide between the Czech Elbe basin and the Elbe River tributaries on German territory.

The river network is uniformly distributed; the two main rivers, Vltava and Elbe, have drainage basins similar in size.

Through northern Moravia pass the main European divides between the Baltic and Black seas, that is, the rivers Oder and Morava respectively. The drainage network of the Morava River is asymmetrical.

The Elbe basin at the outlet from Czech territory has a drainage area of 51,400 km^2, and the length of the river up to that point is 357.3 km. The Oder basin at the border with Poland has a drainage area of 5,835 km^2, the length of the stream being 120.1 km. The Morava basin at the confluence with the Danube has a drainage area of 26,579 km^2 and a river length of 329.9 km. Basic data on the main rivers in the Elbe, Oder and Morava basins are given in table 1.

The largest part of Slovakia's river network drains to the Black Sea. The western section of Slovakia's territory belongs to the Danube basin (with Morava, Váh, Nitra, Hron and Ipel as tributaries); the eastern section to the basin of the Tisza River (tributaries: the Bodrog system of Slaná, Bodva, Hornád and Torysa). The small basin of the Dunajec River is a tributary of the Visla River which flows to the Baltic Sea. All of the streams feature dominant high discharges in the spring, low discharges in the summer and winter, and a slightly increased discharge in the autumn.

Czechoslovakia lies within the temperate zone with oceanic and continental influences. For the greater part of the year, the temperate air mass is spread over the country, occasionally replaced by the humid tropical air in summer and by arctic air masses in winter.

In Bohemia, the temperature (mean) rises from January, reaching maximum value in July (figure 13). The average monthly temperatures are -1° to -6° C in January and 12° to 20° in July. The temperature decreases with rising altitude. The temperature gradient is 0.4° to 0.7° C per 100 m. Thermal inversions occur frequently during winter in the mountains. Extreme recorded temperatures are, minimum, -42.2° C at Litvínovice and maximum, 38.6° C at Teplice.

Weather data have been collected at the Klementinum Station in Prague since 1771.

In Slovakia the average annual temperature is 5° to 9° C, reaching a maximum of 9.9° C in the Danube Lowland. In the mountain valleys of the northern part of the West Carpathians, the average annual temperature goes to 6° C and drops by 0.56° C per 100 metres of altitude. The peaks of the High Tatra Mountains are the coldest points (-3.7° C on average).

In Bohemia, rainfall is of the continental type, with maxima in July and minima in January or February. In some mountainous regions, for instance the Jizerské Mountains, Krkonose (Giant Mountains), an irregular régime is occasionally recorded.

Table 1. Czechoslovakia, hydrology of principal rivers of Bohemia and Moravia

River	Location	Drainage area (km^2)	Precipitation (mm)	Runoff (mm)	Discharge (m^3/s)	Discharge (in m^3/s) likely to be exceeded in 30 days	90 days	180 days	270 days	355 days	364 days	Severity of floods (in m^3/s) likely in 1 year	10 years	50 years	100 years
Orlice	Mouth to Elbe	2 036.85	795	331	21.32	50.6	24.0	13.4	9.09	4.21	2.41	130	340	493	565
Jizera	Mouth to Elbe	2 193.39	794	345	23.96	50.9	28.8	17.3	11.6	6.28	5.45	188	375	570	668
Malse	Mouth to Vltara	979.10	723	230	6.92	15.9	7.89	4.62	2.58	1.05	0.68	73	215	362	442
Luznice	Mouth to Vltara	4 226.17	667	181	24.3	54.2	29.1	16.5	9.55	2.95	1.81	107	316	480	565
Orava	Mouth to Vltara	3 788.22	681	216	26.0	55.0	31.0	18.4	12.0	5.85	4.25	136	550	910	1 080
Sázava	Mouth to Vltara	4 349.19	664	183	25.23	59.6	28.5	16.0	9.29	3.54	1.90	195	493	701	810
Brounka	Mouth to Vltara	8 861.39	586	128	36.0	83.2	39.8	22.9	14.2	5.38	3.21	255	702	1 250	1 545
Vltara	Mouth to Elbe	28 090.00	638	169	149.96	330	177	106	67.2	26.4	18.4	760	2 295	3 627	4 172
Ohre	Mouth to Elbe	5 613.69	651	213	37.94	90.9	45.6	26.1	15.0	4.5	2.14	238	626	1 032	1 250
Bilina	Mouth to Elbe	1 070.88	596	162	5.51	13.0	6.64	3.98	2.42	1.14	0.67	20	42	93	128
Ploucnice	Mouth to Elbe	1 193.94	698	227	8.60	15.3	9.09	6.65	5.24	3.65	2.47	46	133	195	222
Elbe	State borders	51 393.51	654	189	308.16	673	365	221	148	63.0	44.4	1 343	3 224	4 555	5 150
Opava	Mouth to Oder	2 088.84	741	227	15.01	34.8	17.0	8.80	4.85	2.28	1.80	127	310	478	555
Oder	Above Olse	4 720.59	795	289	43.32	106	49.4	25.7	14.7	6.02	4.77	391	974	1 415	1 605
Svratka	Mouth to Dyje	7 118.65	592	120	27.14	62.6	27.3	15.9	9.89	4.52	2.72	190	500	675	725
Dyje	Above Morava	13 418.67	585	103	43.89	95.7	46.6	28.3	19.2	14.3	8.12	167	506	767	876
Morava	Mouth to Danube	26 579.69	-	142	120.00	283	143	81.5	48.0	20.7	14.3	395	920	1 210	1 350

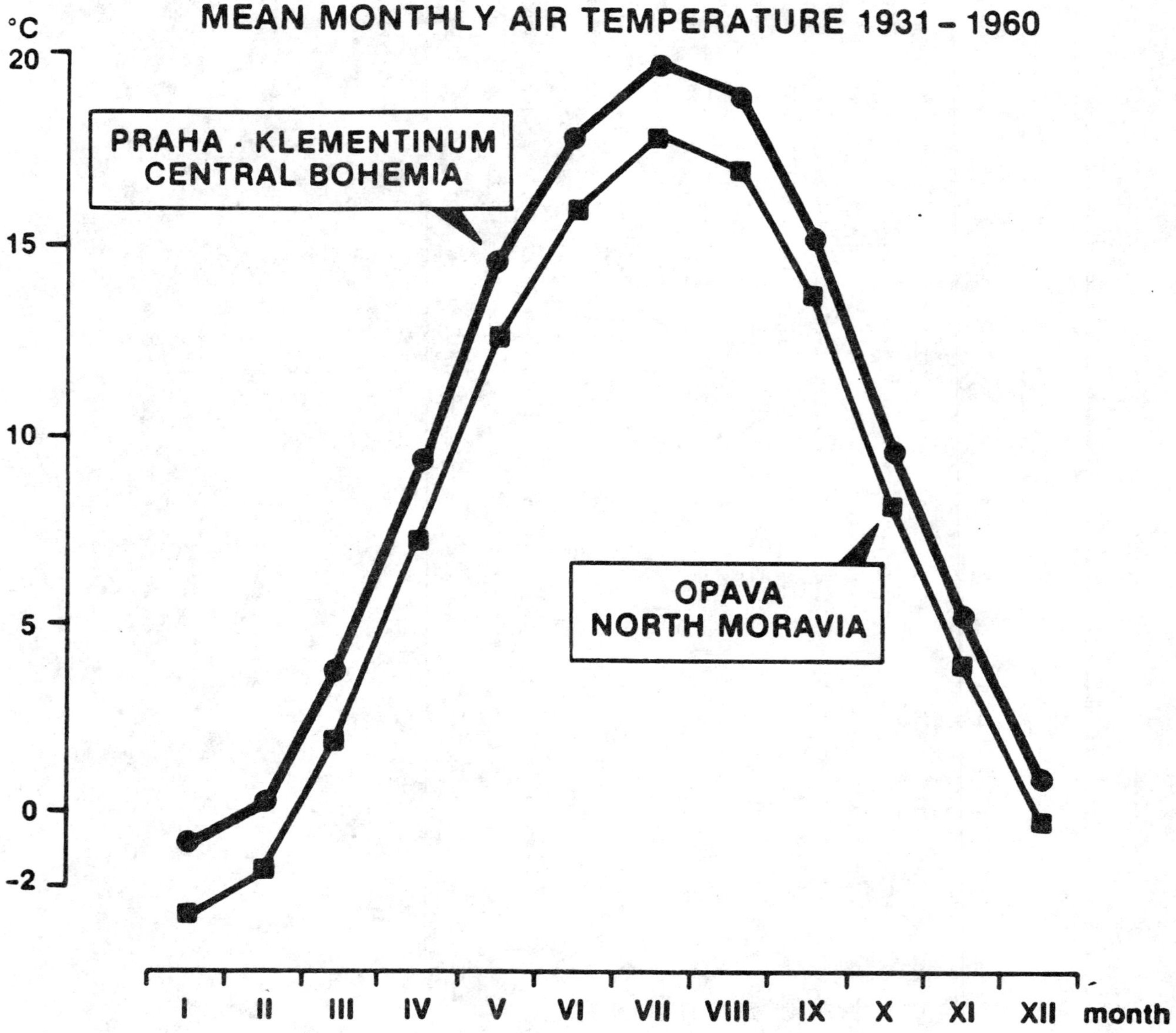

Figure 13. Czechoslovakia, mean temperature ranges

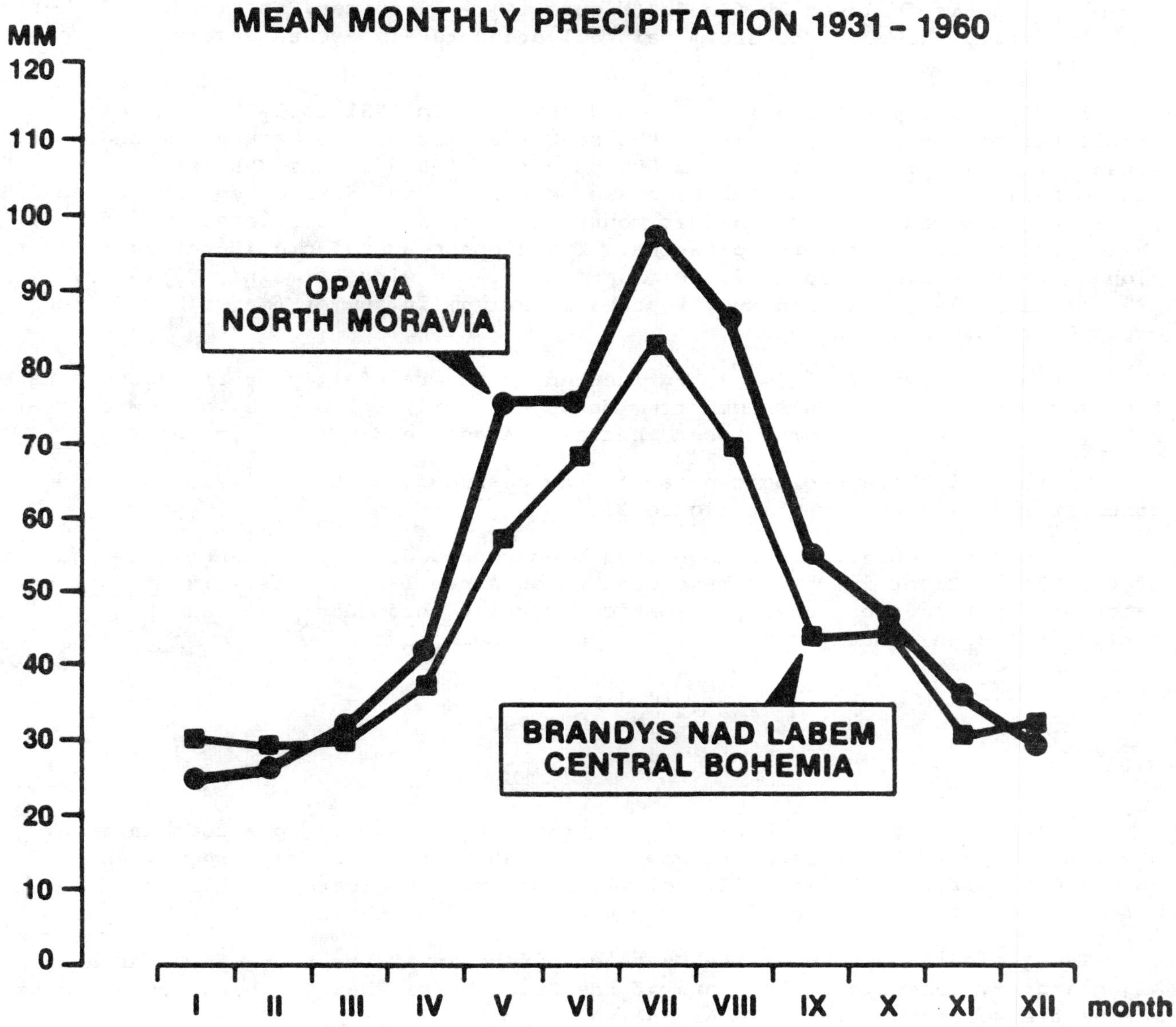

Figure 14. Czechoslovakia, mean rainfall ranges

The lowest rainfall is observed in north-western Bohemia (about 450 mm), and southern Moravia (550 mm). The highest annual rainfall in Bohemia is recorded in the Jizerské Mountains (above 1,700 mm), and in Moravia in the Beskydy Mountains (above 1,500 mm).

In the mountains, snow accumulates from December to March (50 to 150 cm and more). The snowmelt period occurs in February and March after the inflow of warm air mass, and is followed by floods (figure 14). In summer, storm rainfalls of high intensity abound. The storm rainfall areas rarely exceed several square kilometres.

In Slovakia, precipitation averaged 780 mm from 1931 to 1960. The minimum precipitation (below 550 mm) is in the southern part of the Danube Lowland. Arid zones with precipitation less than 600 mm can be found in the Záhorská Lowland, the Danube Lowland, in parts of the Rimavská Depression and Kosice Depression, and all the East Slovakian Lowland. Higher mountain peaks and some uplands record more than 1,000 mm of annual precipitation. The highest annual precipitation is in the High Tatra Mountains - up to 2,000 mm and more. In winter (mainly February), precipitation is at its minimum, reaches a maximum in summer (May to June) and grows less intensive in autumn.

The annual runoff régime is influenced by the rainfall pattern and seasons of the year. Most flood peaks come from March to April, followed by a slow discharge decline until September or October when discharges tend to rise again.

Characteristic hydrologic data for the period from 1931 to 1960 for selected Bohemian rivers are listed in figure 15.

Hydrometric observations have long been recorded. For instance, observations of the Vltava River in Prague have been taken since 1825. A network of surface-water observatories has been systematically built throughout the country since the early nineteenth century.

B. Geology

1. Czech Massif

The Czech Massif is a horst higher than the surrounding platform units of central Europe. Its development was terminated by the Variscan orogeny and folding in the Carboniferous Period. It comprises Bohemia, a greater part of Moravia, northern Austria, a part of both German States and southern Poland.

The geologic history of the Czech Massif began at the turn of the Archean and Algonkian (Proterozoic). The core of the Czech Massif has not been subjected to marine transgression since the Algonkian.

The Assyntian orogeny which followed includes a folded sedimentary envelope of older Paleozoic. At that time was formed, among others, the Barrandian synclinorium. The older Caledonian orogeny culminated by the so-called Czech Phase which took place between the Cambrian and Ordovician. The development of sedimentation in Central Bohemia then came to a turning point.

The later Caledonian orogeny, from the end of the Silurian and beginning of the Devonian, is manifest only in the west Sudeten area. In the Barrandian, sedimentation between the Silurian and Devonian was not interrupted.

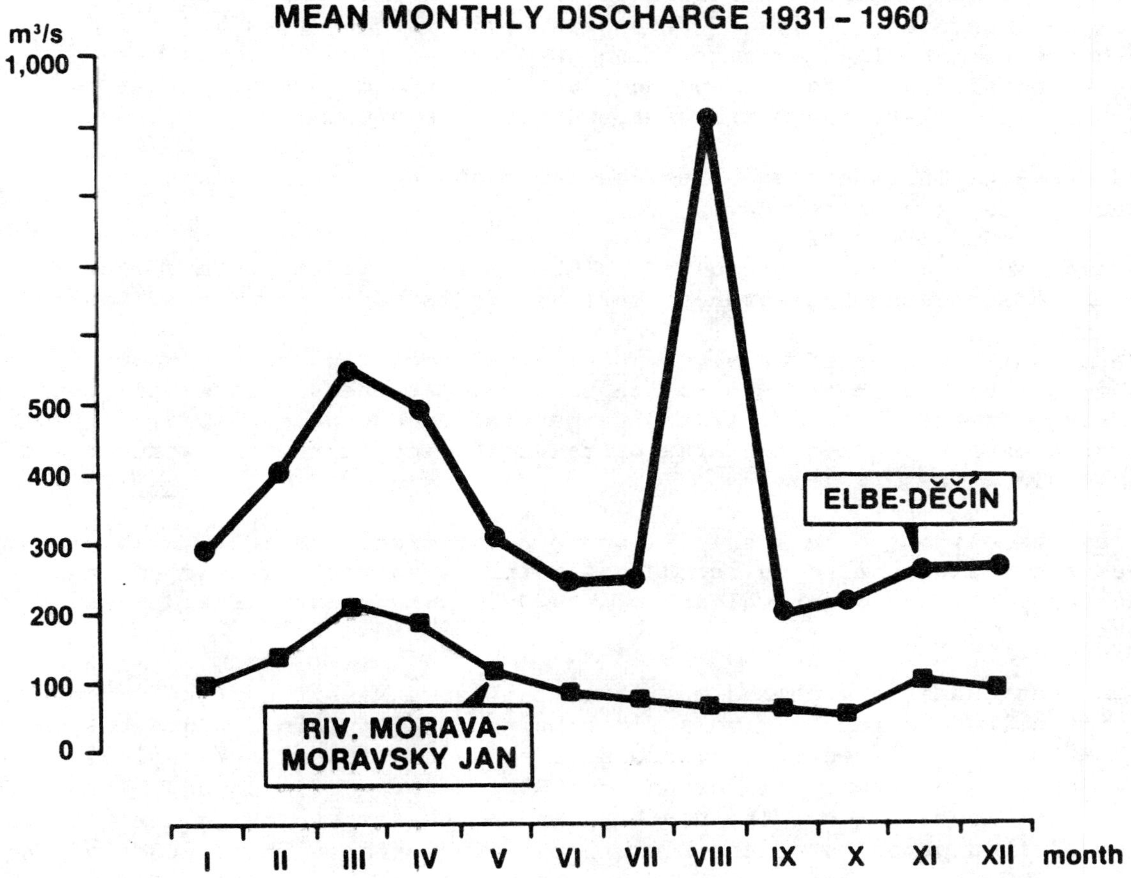

Figure 15. Czechoslovakia, mean river discharges

The Variscan orogeny affected the Czech Massif in several phases from Late Devonian until the Early Carboniferous, and consolidated it completely. During Variscan folding, numerous granitoids intruded the Czech Massif. In the last phase of the Variscan orogeny, Permian-Carboniferous basins were formed in the Czech Massif, predominantly of the continental type; the Sálská Phase terminated development of the Czech Massif in the Paleozoic.

In the Mesozoic Era began a long-lasting and extensive peneplanation of the Variscan mountains. Marine transgression in the Triassic and Jurassic periods was short-lived and regionally limited. Denuding processes predominated, connected with intensive tropical weathering to clay and laterite.

As a result of the Carpathian orogeny, an extensive marine transgression occurred in the Czech Massif, particularly its northern half, in the late Cenomanian and Senonian. A large Cretaceous basin (longer than 300 km and 100 km wide) with its longitudinal axis trending NW-SE was affected by Saxonian tectonics - radial faults following the direction of old tectonic lines (NW-SE, WSW-ENE, N-S), and flat, asymmetric furrows with flexures.

In south Bohemia, where some lake basins trend NW-SE, mid-Senonian sedimentation has been preserved.

Neogene radial movements resulted in the renewed uplift of the Czech Massif. Rift faults (graben) and automorphous horsts were formed, and alkalic lava effused.

Lake basins are located mainly in northern Bohemia in the Ore Mountains and in south Bohemia (the Trebon and Budèjovice basins). At the southern and south-eastern fringe of the Czech Massif the Carpathian foredeep developed. An extensive marine transgression reached the area of the Boskovice furrow and western and central Moravia.

During the Oligocene, Miocene, Pliocene and Pleistocene ages, Saxonian tectogenesis was accompanied by intensive, mainly surficial alkalic volcanism including the prominent stratovolcano of the Doupovské Mountains and Ceské Stredohori.

During the Quaternary period uplift of the Czech Massif continued; but in spite of intensive erosion by rivers, it remained a peneplain. A greater part of the Czech Massif is covered by loam and debris. Along rivers are typical terraces from which originate, especially around the Elbe River, drift sands. There are thick layers of loess. Along the northern and north-eastern fringe of the Czech Massif are located deposits of the Pleistocene glaciation. The surface of the Czech Massif in areas containing ore deposits has been extensively affected by human activity.

2. Carpathian Mountains

The principal feature of the West Carpathians is their nappe structure comprising Paleozoic and Cenozoic rock complexes. On the basis of their geographic distribution and the period of folding, the West Carpathians are divided into the Outer Carpathians (flysch belt) and the Inner Carpathians (central belt). There is a narrow klippen belt between them.

The <u>Inner Carpathians</u> are characterized by abundant Carboniferous crystalline schists, Variscan granitoids, later Paleozoic sediments and volcanic rocks, and the mostly carbonate Mesozoic, the pre-Senonian origin of the nappe system, Alpine metamorphism and magmatism (formation of granitoids), and abundant post-nappe sedimentary and volcanogenic formations.

The pre-Senonian system consists of nappe units of two categories. The first are built of the pre-Upper Carboniferous basement and overlying later Paleozoic and Mesozoic sediments. They are the basement nappes with dominant crystalline complexes. The second category units contain less thick, surficial, rootless nappes built of the Mesozoic and partially of the later Paleozoic. They are the nappes of the Fatricum (Krízna and Vysoká nappes), Hronicum (Choc and Sturec nappes), and the Silicicum (Mesozoic of the Slovak Karst, the Murán Plateau, the Stratenská Upland, Galmus, and the Strázov, Nedzov, Havranica, Veterník and Drienok nappes).

During the Laramide orogeny, the pre-Senonian nappe system was deformed into folds and slices. The Laramide folding was later (mainly in the Neogene) followed by another orogenic process, characterized mainly by vertical movement. The fold structure was broken into the present orographic units, namely mountain ranges, depressions or lowlands. With respect to this, the Inner Carpathians may be divided into three regions: Core Mountains, Slovak Ore Mountains, and spurs of the Pannonian Basin with a ring of neovolcanic rocks.

Marine Paleogene sediments deposited widely in the northern and southern areas of the Inner Carpathians are separated by the geanticlinal Slovak Ore Mountains, a section of the Velká Fatra Mountains, and the Low Tatra Mountains. The northern Inner Carpathian Paleogene consists at its base of transgressive conglomerates and organo-detrital limestones and, higher up, of a thick complex of flysch sediments. The southern Inner Carpathian Paleogene consists mainly of beds of clay, sand, limestone, marl, sandstone and conglomerate.

After the Savic orogeny, marine and brackish sediments formed and intensive volcanic activity began. The Lower Badenian era was followed by formation of a system of intermontane basins - analogous to the present system - completed mostly in the Pliocene and Quaternary periods. The Vienna basin, the Danube basin, the East Slovakian basin and the inner depressions received brackish-water and continental sediments. The subsequent volcanism culminated in the Badenian and extended into the Sarmatian, Pliocene and, sporadically, Quaternary. Its feature is a rhyolite-andesite-basalt rock association and their pyroclastics. During the "Romanian era", thick gravels and sand beds were deposited in basins and inner depressions. In the Quaternary Period continental sedimentation continued, mainly glacial and glaciofluvial sediments in high mountains, fluvial and proluvial sediments in the uplands, and fluvial, proluvial and eolian sediments in the lowlands.

The <u>Outer Carpathians</u> consist of the Pieniny klippen zone and a flysch belt. The Peininy klippen zone extends in a 500-km arc from the Vienna basin to the East Carpathians. It is only a few kilometres wide. A characteristic is the absence of pre-Mesozoic rocks and presence of Triassic rocks, variable amounts of Jurassic and Cretaceous rocks, the klippen tectonic style, and its position between the Outer and Inner Carpathians.

The flysch belt extends along the outer side of the Carpathians. It is characterized by an absence of pre-Mesozoic rocks, the presence of mostly terrigenous Cretaceous and Paleogene sediments, the Tertiary origin of the nappe structure, and few post-nappe formations. The flysch belt is divided into western and eastern parts. In the western part are the Silesian and the Magura nappe groups, mostly represented by Cretaceous and Paleogene beds. In the eastern part are the Magura and Dukla nappes.

C. Ground-water investigations

1. General

Regional hydrogeologic investigation of Czechoslovakia has been systematically carried out as part of the State Water Economy and Management Plan since 1945.

Of particular importance was the hydrogeologic zoning of Czechoslovakia completed early in the 1960s. The country was divided into regions delimited by geotectonic criteria (Vrba, 1966). A map of hydrogeologic regions on a scale of 1:500,000 was also drawn in 1966. The regions were studied systematically, and hundreds of wells bored, tested and monitored. Ground-water resources were calculated and basic conditions for their protection formulated. The results of more than 30 years of investigation will be issued as hydrogeologic syntheses of the country's principal aquifers. They were to be presented to the Government by the end of the 1980s.

In the second half of the 1950s, hydrogeologic investigation of relations between lignite deposits in a Tertiary basin in north-western Bohemia and mineral springs at Karlovy Vary (Carlsbad), Teplice, Frantis-kovy Lázné (Franzensbad) was started. Drainage and lowering of the water table are necessary in mining lignite where the overburden is 50 to 80 metres or more thick. The effects of the cone of depression on the hydrogeology and environment of the forested slopes of Krusné Hory (Ore Mountains), is under study. Mining at Teplice required that mineral water be drawn from wells more than 600 metres deep.

A number of hydrogeologic maps on different scales have been issued. The first, at scale of 1:500,000, includes a map of the hydrogeologic regions. A 1:1,000,000 map is part of the Atlas of Geological Maps of Czechoslovakia (1967). For the needs of the State Water Economy and Management Plan, the Hydrogeological and Ground Water Management Map, at scale of 1:200,000, was published. A hydrogeology-only map at the same scale was also issued. A 1:1,000,000 map of ground-water runoff in Czechoslovakia (1982) was the latest of the major hydrogeologic maps to be published. Maps showing the methodology of providing for ground-water protection, at scale of 1:500,000, have been prepared. The compilation of 1:50,000 hydrogeologic maps has begun.

In the mid-1950s, a network of observation wells in shallow aquifers was started, and hundreds of wells were bored in fluvial deposits throughout the country. The network was later extended to include deeper aquifers. Many springs had been measured regularly even before the Second World War. As a result, data covering a period of 30 years are available for ground-water resources evaluation.

Hydrogeologic investigation has been carried out in many areas in connection with artificial infiltration.

Important projects implemented between 1970 and 1980 include (1) hydrogeologic investigation of the basement of the Czech Cretaceous Basin east of the Elbe River with regard to coal deposits in Permian-Carboniferous sediments, and (2) hydrogeologic investigation of a ground-water system in the northern part of the Cretaceous Basin where uranium ores exist.

2. Legislation

Legislation has played an important role in the history of regional hydrogeologic investigation and in ground-water protection. The Czechoslovak Water Law has a long tradition, traceable to the early sixteenth century in Bohemia. The present Water Law, which also covers ground water, was updated in 1973 and is valid in both the Czech and Slovak Socialist Republics. However, water planning and management at the federal and state levels falls under the competence of the Ministry of Forest and Water Management. Ground-water investigation and the calculation of ground-water resources, and ground-water protection are dealt with by the separate Czech and Slovak State Departments for Geology.

From the point of view of legislation, surface, ground and special (e.g., mineral and curative) waters are recognized. According to law, ground water is preferably to be used for drinking water supplies, and is subject to obligatory calculation, protection and registration. Ground-water sources are managed within protection areas that are usually delimited in three degrees. In some mountain areas (e.g., Beskydy and Jeseníky mountains) covering several hundred square kilometres and serving as the recharge areas of important ground-water basins, human activity is restricted by law.

The State Water Inspection is a specialized supervisory and control office for water protection in Czechoslovakia.

3. Hydrogeology of Czech Massif

The history of ground-water investigation within the Czech Massif began in the 1930s (Hynie, 1963) with projects to obtain drinking water for Prague. Also at that time, investigation started for ground water and springs in the Czech Cretaceous Basin, hydrogeologically the most important basin of the Czech Massif.

Problems connected with the captation of mineral water and the mining of coal and kaolin in north-western Bohemia have been considered for several centuries. In the case of the Karlovy Vary springs, around which a spa was founded in the fourteenth century, hydrogeologic data is available from the sixteenth century. Outbursts of hot gaseous mineral water in Karlovy Vary were mentioned by the poet J. W. Goethe in 1809. A hydrogeologic description of the Karlovy Vary springs was presented by E. Suess in 1903. In 1879, cessation of thermal-spring flow in Teplice, caused by coal mining (Döllinger mine), was reported. However, investigation on a regional scale was begun in the early 1930s, and systematic study of the mineral and thermal waters in the Czech Massif dates back to the 1950s.

Ground-water resources in the Czech Massif are mainly associated with post-Variscan sedimentary formations, principally Upper Cretaceous and Neogene sediments, and Quaternary fluvial and glaciofluvial deposits (figure 16).

Figure 16

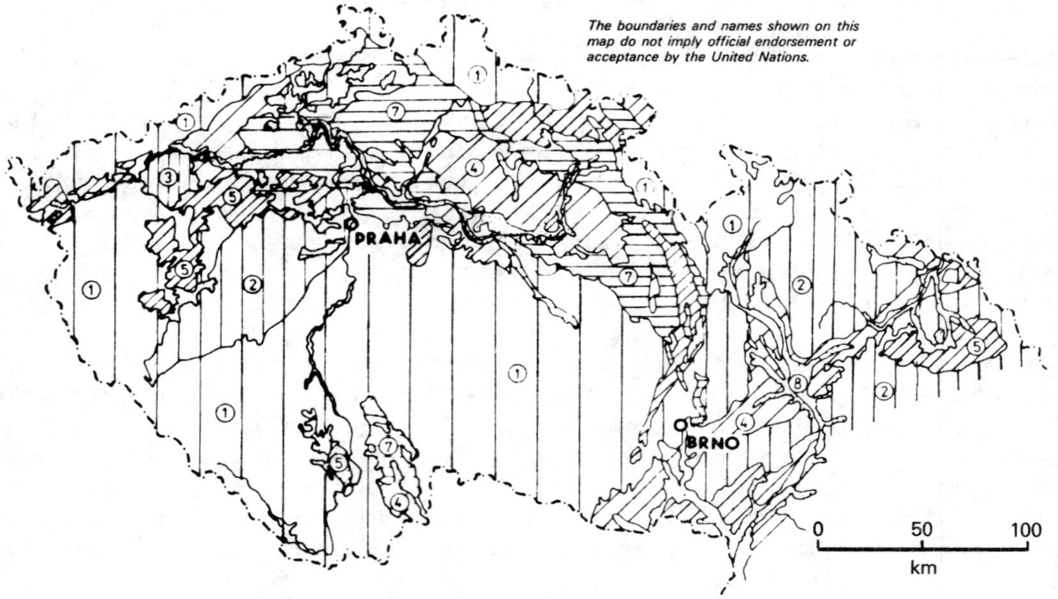

CZECHOSLOVAKIA: PRINCIPAL AQUIFERS OF THE BOHEMIAN MASSIF

Fissured aquifers with specific yield of less than 1.5 l/s/m drawdown

- Pre-Variscan metamorphic, magmatic and sedimentary rocks
- Tertiary volcanic rocks, northern Bohemia
- Flysch belt, Moravia

Fissured and porous aquifers with specific yield of less than 5 l/s/m drawdown

- Carboniferous and Permian basins and Tertiary basins in northern Bohemia
- Budejovice basin in southern Bohemia
- Central part of Bohemian Cretaceous basin, Tertiary sediments of Outer Carpathian basin and Cretaceous of Outer Carpathians

Fissured and porous aquifers with specific yield of less than 10 l/s/m drawdown

- Bohemian Cretaceous basin, central part of Trebon basin in southern Bohemia and Devonian limestones in Moravia

Porous aquifers with specific yield of more than 5 l/s/m drawdown

- Fluvial and glaciofluvial deposits of Elben and Morava rivrs and of their tributaries, Quaternary

NOTE: Hydrogeologic importance increases with progression through sequence shown above.

3546.9x

The hydrogeological importance of pre-Variscan and Variscan metamorphic and magmatic rocks of low permeability should not be neglected.

In the Czech Massif, the following aquifers can be recognized: crystalline rocks with a platform basement; Devonian and Permian and Carboniferous structures and basins; the Czech Cretaceous Basin, Neogene basins and structures; shallow aquifers in fluvial and glaciofluvial deposits; and the Carpathian depression and flysch belt in Moravia.

Crystalline basement rocks are widespread in the border regions of Bohemia and Moravia where they form morphologically distinct mountains. However, large areas of magmatic (igneous) and metamorphic rocks can also be found in the central part of the Czech Massif. Crystalline schists of Proterozoic age and of different intensities of metamorphosis, and large intrusive bodies of acidic plutonic rocks of Variscan age predominate. On the whole these rocks have low transmissivity.

Higher-yield springs occur only occasionally. The ground water in the crystalline rock has been utilized until today only for local water supplies.

The decisive part in the water-bearing of crystalline rock is played by the thickness and the lithology of the weathered zone, the intensity of the rocks tectonic exposure, geomorphology, altitude above sea level, and the quantity of precipitation. Aquifers in the true sense of the word are not found in crystalline rocks. The preferential planes of ground-water flow are associated with tectonic and fractured zones of regional importance. Favourable conditions for the recharge of a hydrogeologic system in crystalline rock exist in those areas where the tectonic zones reach the surface or are covered with permeable weathered rock, and in contact areas of different lithology. Really limited and moderately thick (5 to 10 m, exceptionally 50 m) aquifers may exist in weathered rock with typical characteristics of shallow aquifers, i.e., short-term residence of water, enormous fluctuation of ground-water level, high vulnerability, and complicated capture.

Some authors (Knezek M., Krásny J., 1985) point out the high values of ground-water runoff especially in crystalline rocks of mountainous areas with precipitation exceeding 1,000 mm per year. According to these authors, the proportion of ground-water runoff from the mountainous crystalline rock is considerable when expressed as a long-term resource. From an area of 35,000 km^2 (accounting for 44.3 per cent of Bohemia and Moravia), ground water could provide 110 m^3 per second, i.e., 53.4 per cent of all natural ground water drawn in the Czech Socialist Republic.

However, the amount of ground water in crystalline rock that could be drawn for water supply systems is substantially less. Results of a regional investigation of the central Bohemian pluton, based on geophysical measurements, hydrochemical research, observation wells and pumping tests, and the evaluation of older wells have demonstrated a statistically significant relation between yield and the depth of a borehole. Higher occurrences of ground water in boreholes were recorded mainly from 20 to 40 metres below the surface. At greater depths, occurrences of ground water into the boreholes were rare. The yield of the wells also increases with the thickness of eluvial deposits and with the extent of fracturing of the rocks. However, the yield of boreholes and wells is low on the whole, usually about 1 l/s, and only exceptionally 3 l/s. Similar data were obtained during study of a part of the Brno granitic massif.

4. Hydrogeology of Cretaceous basin

Owing to its more than 15,000 square kilometres area and its ground-water resources of which more than 6 cubic metres per second are currently being used, the Czech Cretaceous Basin is the most important hydrogeologic unit in the Czech Massif. The average thickness of Cretaceous sediments (Cenomanian-Senonian) is 300 metres, reaching over 1,000 m in some areas. Aquifer systems are developed particularly in the western and eastern parts of the basin. The dominant tectonic directions are NW-SE and NE-SW.

In the western sector, the vertically delimited tectonic blocks are dominant for ground-water circulation. The main aquifer is situated in middle Turonian sandstones. The ground water is generally of the calcium-bicarbonate type and is potable.

Typical for the eastern sector of the basin is slight folding, with ground-water accumulations usually on the axial parts of the synclines. The main aquifers are located in the lower and middle Turonian, and many are artesian and flowing. Overflows from wells in excess of 10 litres per second are common, reaching exceptionally more than 100 litres per second. Springs yielding several hundred litres per second exist and are used for large water-supply purposes (e.g., city of Brno). The lower Turonian water is potable. The middle Turonian aquifer contains up to 50 mg/l of nitrates and locally more as the result of human activity.

In the central part of the basin are the thickest Cretaceous sediments but their permeability is generally low. No regional-scale aquifer exists. There are local occurrences of mineral water containing carbon dioxide in the Cenomanian.

A deterministic model based on a conceptual idea of the hydrologic system, and the Aquifer I model based on the method of finite elements and the simulation of steady-state and unsteady flow in a non-homogeneous anisotropic aquifer were used to calculate ground-water resources in the Czech Cretaceous Basin (SG Praha). Selected quantitative and qualitative characteristics are listed in table 2.

Among Paleozoic rocks, the most important hydrogeologically are the Devonian limestone and Permian and Carboniferous sandstone.

Devonian organo-detrital and mud limestones in the central part of the Czech Paleozoic basin (Barrandian) in Bohemia are karstified to a low degree, and are relatively limited over the erosion base of the Berounka River. The average specific runoff of ground water is low, about 3 $l/s/km^2$, and also the specific yield of investigation boreholes is only tenths of a litre per second per metre.

Of Paleozoic rocks in the Czech Massif, the most important hydrogeologically are the Devonian limestones north-east of Brno, referred to as the Moravian Karst. Hydrogeologic investigation confirmed the availability of about 150 l/s of ground water in the highly karstic limestone complex, covering some 100 km^2. However, the karst hydrogeologic system is vulnerable to human activity.

Table 2. Czechoslovakia, hydrogeology of Cretaceous Basin

Sector	Main aquifer system	Thickness and lithology	Ground-water level	Specific ground-water runoff ($l/s/km^2$)	Transmissivity (m^2/s)	Specific yield (l/s/m)	Quality
West	Lower Turonian	80 m, sandstone	70 m below surface, phreatic	5 to 9	10^{-3}	3 and more	Potable
	Cenomanian in part	40 m, mainly sandstone	40 to 60 m below surface, non-artesian	3 to 5	10^{-3} to 10^{-5}	Varies	Higher contents of iron and radioactive elements
East	Lower Turonian	60 m, sandstone and marlite	Frequent overflow	3 to 5	10^{-2} to 10^{-4}	10 to 0.1	Potable
	Middle Turonian	50 to 100 m, sandstone and marlite	Frequent overflow	3 to 7	10^{-2} to 10^{-4}	10 to 0.1	High nitrate content
Centre	Cenomanian locally	20 m, mainly sandstone	Overflow or non-artesian	0.5 to 1	10^{-5}	0.01	Mostly mineral water high CO_2 content or temperature

The Permian-Carboniferous formations are found in several basins (e.g., Plzen, Kladno-Rakovník, Inner Sudeten, Podkrkonosská, and Ostrava-Karvinná Basins) and in "furrows" (the Blanická Furrow, Orlice Perm, the Boskovice Furrow) in western, central and north-east Bohemia and in Moravia. Extensive occurrences of Permian-Carboniferous have also been found in the basement of the Czech Cretaceous Basin. The intensity of ground-water circulation generally depends on rock fissures and, to a lesser extent, their lithology. The main tectonic directions are NNE-SSW and E-W. The alternating sandstones and claystone reduce the possibility of recharge into deeper layers of the basins. The permeability of the sedimentary complex declines with depth. The highest values of transmissivity are usually related to depths from 100 to 120 metres (10^{-4} m^2/s), and in claystone at depths from 500 to 600 metres the transmissivity is very low (10^{-7} m^2/s). Also the quality of water varies with depth. In deep levels of Permian-Carboniferous basins with slow circulation, highly mineralized brines exist.

Ground-water flow is 2 to 5 $l/s/km^2$ and in areas with less precipitation is under 1 $l/s/km^2$. In most of the basins, however, the ground-water flow is affected by coal mining in the upper Carboniferous beds. Intensive mining activity at depths of several hundred metres requires draining of the overburden, which lowers the water table. In areas without coal deposits, for example the Orlice Perm, formed predominantly of arkose sandstone, conglomerate and siltstone, the transmissivity is 10^{-3} to 10^{-2} m^2/s. The ground-water runoff is 2 to 6 $l/s/km^2$, and the water is chiefly of the $Ca\text{-}HCO_3$ type.

Hydrogeological conditions in the Tertiary basins of the Czech Massif have been the subject of long-term study in connection with lignite, kaolin and mineral water in northern Bohemia, and ceramic clay and gravel sand in south Bohemia.

In the basins of northern Bohemia, e.g., the Sokolov Basin and the North Bohemian Basin, the hydrogeologic system has been deeply affected by long-term human activity, particularly lignite mining. Underground shafts were gradually replaced by giant open-pit mines requiring the removal of several tens of metres of overburden and draining of the pits. In some areas, the water table dropped 30 to 40 metres which had considerable ecological effect on the adjacent forested slopes of the Ore Mountains; it is now a subject of hydrogeologic investigation. Many factors complicate the problems of mining lignite while preserving the area's mineral waters. The Tertiary basins are related to a major tectonic structure and neovolcanic activity has made possible thermo-mineral springs, but a continuing flow of ground water is essential. In this respect, integrated investigations are required, with hydrogeology playing the leading role.

In sedimentary Tertiary basins, low-permeability argillaceous sediments predominate. Their considerable extent, particularly at the surface, substantially reduces the infiltration and recharge potential. In spite of this, in areas with highly fissured rocks important ground-water resources have been identified (for instance, "cypris" series of strata in the Cheb Basin in northern Bohemia). The south Bohemia's Budejovice and Trebon Basins are important reservoirs of ground water tapped for large water supply systems.

The main hydrogeologic data on some of the Tertiary basins are listed in table 3. Neovolcanic rocks (mostly basalt and pyroclastic rocks) comprise two large areas in northern Bohemia (the Doupovské Mountains and Ceské Stredohori). There are numerous isolated occurrences in the northern parts of Bohemia and Moravia. They are incorporated into the ground-water system of the Czech

Table 3. Czechoslovakia, hydrogeology of Tertiary basins of Czech Massif

Basin	Main aquifer system	Thickness and lithology	Specific ground-water runoff ($l/s/km^2$)	Transmissivity (m^2/s)	Specific capacity (l/s/m)	Quality
Cheb Basin (Northern Bohemia)	Quaternary, Neogene (upper part)	40 to 70 m, sand, clay	0.5	10^{-3} to 10^{-5}	0.1 to 4.0	Potable, high iron and manganese
	Neogene (lower part)	35 to 40 m, sand	0.5 to 1	10^{-3} to 10^{-4}	to 0.1	Potable, high iron and manganese
North Bohemian Basin	Miocene	20 m, sand	0.5 to 1	10^{-4} to 10^{-6}	0.5	Potable, high iron and manganese
South Bohemian Basins						
Budejevice Basin	Senonian	140 m, sandstone	2	10^{-4}	0.3	Potable, high iron and manganese
Trebon Basin, southern part	Senonian	150 m, sandstone	2 to 3	10^{-4}	0.2	Potable, high iron and manganese
Trebon Basin, northern part	Senonian	60 m, sandstone	2	10^{-3}	2 to 3	Potable, high iron and manganese

Cretaceous Basin, and in some cases the Tertiary basins. The system consists of a number of local aquifers drained by small temporary springs or into the major Cretaceous aquifers. The ground-water flow is not known exactly, but is considered generally low, in the range of 1 to 3 $l/s/km^2$. Neovolcanic rocks have not been considered in the regional investigations, but they should not be dismissed as hydrogeologically unimportant.

In the Quaternary deposits of fluvial and glaciofluvial origin, important shallow aquifers exist in the Czech Massif. Unfortunately, their hydrogeologic connection with surface water, which is polluted by intensive agricultural activity, frequently results in contaminating the shallow aquifers and rendering the water non-potable. In spite of this, ground water from glaciofluvial deposits, especially in the basins of the Elbe, Morava and Dyje rivers, is used to supply towns and for irrigation.

Eolian deposits (loess and loess sands) are of hydrogeologic importance only locally in Moravia. The lithology of eluvial deposits governs recharge of the system, but no larger resources of ground water occur in them.

The following features are typical of the fluvial deposits in the Czech Massif: uniform lithology including alternating sands and gravel above a rough-grained fraction at the base of the terrace; small variations in thickness, 8 to 12 metres on average exceptionally 20 metres or more in old river beds - similar values of permeability (10^{-3} to 10^{-4} m/s), transmissivity and specific yields (in low terraces, usually several litres per second per metre), a relatively low ground-water flow (2 to 3 $l/s/km^2$) and high nitrate content (more than 40 mg/l on average).

A regional hydrogeologic investigation of the deposits started toward the end of the 1950s. All important river confluences have been explored such as Elbe-Orlice, Elbe-Ohre, Morava-Dyje. An extensive ground-water quality monitoring system exists in the Elbe and Morava river basins and the impact of farming (especially fertilizers) on ground-water quality is being monitored.

Glaciofluvial and lake deposits are widespread mainly in northern Moravia, and reach thicknesses of several tens of metres. There are aquifers with good quality ground water, utilized for water supply systems.

From the geologic point of view, the Carpathian foredeep and the flysch belt in Moravia do not belong to the Czech Massif. Hydrogeologically most important in the Carpathian foredeep are Miocene basal clastic sediments. Confined aquifers exist in them with water resources of the order of several hundred litres per second. Ground water in the central parts of the foredeep has good quality, but near the fringes has been affected by human activity.

The greater part of eastern Moravia is underlain by flysch. Hydrogeologically, the area is quite heterogeneous owing to abrupt alternations of psammitic and pelitic sediments and profound tectonism which has fragmented the aquifers. However, in areas where sandstone predominates, conditions exist for large accumulations of ground water. In general, the ground-water flow decreases from north-east (5 $l/s/km^2$) to south-west (1 $l/s/km^2$), reflecting differing geomorphology and climate.

The Czech Massif abounds in <u>mineral waters</u> of varying chemical composition but with uniformly high content of carbon dioxide. The waters are genetically related to the European province of carbonate waters which include Eifel, northern Rhine, Harz and Bohemia-Silesia. The origin of mineral and thermal waters in the Czech Massif is primarily associated with the deep and regionally extensive tectonic lineaments in an area of strong Variscan tectonism.

The most important mineral water resources are located in north-western Bohemia and include Frantiskovy Lázne (Franzensbad), Mariánské Lázne (Marienbad), Karlovy Vary (Carlsbad), Jáchymov and Teplice, all used intensively as spas for medical purposes. The most popular gaseous water in Karlovy Vary is of the $Na-HCO_3-SO_4-Cl$ type with TDS of 6.4 g/l, temperature of 73° C and a yield of 35 l/s. The successful captation of the Karlovy Vary thermal water in inclined wells has been accomplished, with yields of the investigation boreholes up to 50 l/s. Balneologically valuable are the radioactive thermal waters in Jáchymov.

Cold mineral water of the $Na-Cl-HCO_3$ type with high contents of carbon dioxide is exploited from Cenomanian sandstones of the Czech Cretaceous Basin at the spa resort of Podebrady. Thermal waters at Jáanské Lázně are similar to the mineral waters of Silesia in Poland.

Mineral water from abundant springs and wells, with a low mineralization, usually alkaline and containing CO_2, is bottled and sold as table water.

Mineral waters in Moravia are mostly carbonated, with a shallow circulation and high CO_2 contents (e.g., Luhacovice, Karlova studánka). Waters of the sulphur type and various temperatures exist in the crystalline rock of Hruby Jeseník (circulating deeper than 1,000 m) and in flysch sediments (circulating deeper than 300 m). Balneologically important are resources of fossil marine water with a high content of iodine salts (e.g., Darkov).

5. Hydrogeology of the West Carpathians

The investigation of ground water in the West Carpathians has a rather short history when compared to that of mineral waters. The first data on mineral water in Slovakia were publiched by J. Wernher (1549), M. Szentiványi (1689) and M. Béla (1723).

Local problems concerning ground-water exploration were discussed in the early 1950s. Complex hydrogeologic investigation in the West Carpathians started in the 1960s. Quaternary deposits and Mesozoic carbonate sediments were considered the most important ground-water systems. Research activities were carried out by GUDS Bratislava, IGHP Zilina and VZ Bratislava, and were gradually extended to include other rock complexes. In the 1970s, hydrogeologic maps of the Slovak territory at the scale of 1:200,000 were published. Ground water is currently being explored with respect to its resources and protection. In 1984, a complex evaluation of ground-water resources in Slovakia was carried out (Hanzel and Kullman et al., 1984). The new investigation programme also includes the exploration of geothermal ground-water resources.

Ground-water resources in the Czechoslovak Carpathian system are associated with rocks ranging in age from pre-Mesozoic to Quaternary. The percentages of ground-water resources in the several geologic formations are tabulated below.

Formation	Per cent of overall resources
Pre-Mesozoic	6.1
Mesozoic sediments	31.9
Paleogene sediments of Inner Carpathians	3.1
Sediments of klippen and flysch belts	8.5
Neogene volcanics	8.9
Neogene sediments	1.8
Quaternary sediments, lowlands	27.1
Quaternary sediments, alluvial	12.6

From the point of view of ground-water availability to areal distribution and utilization for water supplies, the most important is ground water in Mesozoic sediments (32 per cent), Neogene volcanics (9 per cent) and, above all, in Quaternary sediments (40 per cent).

The Mesozoic (Triassic through Jurassic) mainly carbonate formations constitute a large part of the Core Mountains, and also the northern and southern margins of the Spissko-gemerské Ore Mountains. Ground-water flow is mainly related to the intensive tectonism. Middle and Upper Triassic limestone and dolomite, and some Jurassic carbonates, contain the best aquifers with an area of 3,280 square kilometres.

Geologic conditions gave rise to numerous, usually not large, fissure-karst structures. There are 79 hydrogeologic structures with areas ranging from several square kilometres to more than 100 km^2. Over a total area of 2,996 km^2, ground-water resources yield 27,000 to 31,000 l/s. Many springs exist and those with an average yield from 5 to 50 l/s predominate. Less common are springs with an average yield above 50 l/s. In spite of high annual precipitation (900 to 1,200 mm), the dense forest cover causes great evapotranspiration. The specific runoff of the ground water is not high, from 8.5 $l/s/km^2$ (500 m above sea level) to 14 to 17 $l/s/km^2$ (950 to 1,100 m). Carbonated water with TDS ranging from 0.1 to 1 g/l, most of it (56 per cent) with TDS within the 0.3 to 0.5 g/l range, is the dominating genetic type. Some of the ground water, circulating freely in open faults, contains up to 20 mg/l of carbon dioxide.

Neogene volcanic complexes over an area of 5,200 km^2 comprise the greatest part of Central Slovakia, and a part of East Slovakia (Slánské and Vihorlatské hills). Their hydrogeologic properties are determined chiefly by block structure. Neotectonic forces broke them into horsts and grabens separated by crushed zones that create favourable conditions for ground-water flow.

Effusive rocks are characterized by fissure permeability. The pore permeability of volcano-clastics composed of psephite and psammite fractions decreases with the increasing pelite component.

Specific yields of wells are from 0.5 to 4.0 l/s/m. In regions predisposed to intensive tectonism, the figures are 10 to 20 l/s/m and more (Central Slovakia).

The central part of the Stiavnické and Kremnické Hills is drained by mines with yields of 60 to 400 l/s of ground water.

Specific ground-water runoff from neovolcanic rocks varies between 2 and 5 $l/s/km^2$, dropping in some areas to less than 1 $l/s/km^2$.

The composition of neovolcanic rocks (mainly andesite, rhyolite, their volcano-clastics, and basalt) adds certain silicates to the ground water. The most common type of ground water is the Ca-Mg HCO_3 type with TDS ranging from 50 to 800 mg/l.

Quaternary deposits of varied origin, thickness and area fill a dense and long system of river valleys, intermontane depressions and extensive lowlands. Fluvial gravels and sands predominate; eolian, proluvial, glacial and other deposits are less common.

The Quaternary deposits occur the most in the Danube, East Slovakian and Záhorská lowlands, and cover more than 5,600 km^2. Their thickness is 10 to 50 m, reaching up to 400 m in the Danube lowland. The permeability of the sandy gravel and sand deposits is 10^{-4} m/s to 10^{-2} m/s. The specific yield of wells is 2 to 15 l/s/m, amounting to 10 to 190 l/s/m in the Danube lowland. More than half the ground-water resources in Quaternary deposits are in the Danube lowland which is one of the most significant ground-water reservoirs in Czechoslovakia.

Fluvial deposits in intermontane depressions, especially in the basins of the Váh, Nitra, Hron, Ipel, Slaná, Poprad, Hornád, Torysa, Bodva, Topla, Ondava and Laborec rivers cover 3,600 km^2 with a thickness ranging from 2 to 65 m, most commonly from 8 to 13 m. Their permeability is 10^{-4} to 10^{-3} m/s, dropping to 10^{-5} to 10^{-6} m/s locally. The specific yield of wells is 0.01 to 30 l/s/m.

In the Tatra Mountains and their foreland, glacial deposits cover about 400 km^2 and range in thickness from 25 to 80 m, locally above 400 m. Their permeability is 10^{-3} to 10^{-4} m/s. Glaciofluvial deposits have a lower permeability 10^{-7} m/s. Moraine deposits, however, show a permeability of up to 10 m/s. Specific yields of wells are 0.08 to 3 l/s/m. The average specific ground-water runoff is 8 to 12 $l/s/km^2$. Ground water is usually recharged by infiltration from surface streams. The TDS of this water ranges from 100 to 900 mg/l, and in case of contamination up to 1.2 to 1.4 g/l. The Ca-Mg-HCO_3 type predominates. In fluvial deposits of large rivers intermediate and mixed types are found.

Mineral waters occur in all of the tectonic zones - the Inner Carpathians, and the klippen and flysch belts. Most mineral waters are found in the Inner Carpathians belt in Triassic carbonates and dolomites of the sedimentary envelope and nappes. The maximum yield of some localities amounts to 50 l/s at a temperature of 70° C. The water is of the Ca-Mg-HCO_3 and Ca-Mg-SO_4 types, and their intermediates, with a TDS of up to 5 g/l. The most valuable are mineral waters in Piestany, Trencianské Teplice, Sliac and Dudince. In the Inner Carpathian belt, mineral water is also found in Neogene sediments (sand, sandstone, conglomerate) of the Vienna basin, the Danube basin, and the East Slovakia and South Slovakia basins. The maximum yield of wells is 24 l/s, water temperature being 92° C. The water is of the Na-HCO_3 and Na-Cl types, with intermediates, and TDS of up to 35 g/l.

The most important is the central depression of the Danube basin which includes the localities of Dunajská Streda, Topolniky, Calovo and Gabcíkovo. The klippen and flysch belts are mainly associated with cold mineral water containing carbon dioxide. The yields of most of the localities are generally very low (0.01 l/s). The water is of the Ca-Mg-HCO_3, Na-HCO_3 and Na-Cl types with intermediates and TDS of up to 1 to 2 g/l, exceptionally 10 to 40 g/l. In the klippen belt, Nimnica is the most important locality, and in the flysch belt, those of Luhacovice, Oravská Polhora, Bardejov and Cígelka, producing the most popular mineral waters. Most of the above mineral waters are used in spas for curative purposes.

D. Exploration and development

Investigation methods in Czechoslovakia are based on extensive and systematic hydrogeologic mapping, original hydrogeologic investigations, measurements of spring discharge, surface-flow discharge, and of ground-water fluctuation and quality.

1. Geophysics

Geophysical methods include geoelectrics, gravimetry, thermometry and seismics, and also gasometry for mineral waters, all of them widely applied. Thermometry and the method of specific conductivity yield valuable results in determining hidden discharges of ground water into surface streams. It is particularly Slovakia where the method of evaluating ground-water discharge curves is being used (Kullman, 1977, 1980, 1983).

2. Data banks

Two types of data banks had been established in the country long before the process of computerization started. One of them is Geofond, a federal bank containing all the data and results of geologic and hydrogeologic investigation and prospecting, information on boreholes and wells, description of geologic profiles, results of pumping tests, qualitative characteristics of ground water and related data. Informing the Geofond about the results of boreholes and wells testing is obligatory. The Hydrofond data bank includes all results of the observations of the meteorological, hydrologic and hydrogeologic network. Historical data are collected as well. Since the two banks were established a long time ago, the data, originally stored in traditional archives are now being converted to machine-readable form suitable for computer analysis. Calculations of ground-water resources are also collected in a specialized data bank.

3. Assessment

From the hydrogeologic point of view, the state territory is divided into separate regions according to the Map of Hydrogeological Zoning of Czechoslovakia. The main characteristics of the systems in all important regions are known from long-term investigations. Ground-water assessment is provided at different levels, referred to as categories. Requirements as to the precision of ground-water resources calculations increase with the level of category. Knowledge of the geologic structure and the hydrogeologic system, including boundary conditions obtained from geophysical investigation, mapping, investigation boreholes and pumping tests, the monitoring of the qualitative and quantitative characteristics of ground and surface water and precipitation, and the determination of the principles of regional hydrogeologic protection - all these are among the main features for determining the level of category of ground-water resources (there are four categories altogether).

Traditionally, the water balance approach is used to calculate ground-water resources and safe yield. At least two different methods are required to calculate the safe yield. The most frequently used include the separation of the ground-water components of the hydrograph, Kille's method based on the construction of the duration curve from monthly minimum discharges, and analysis of the duration curve of mean daily discharges. Calculation of ground-water resources based on conventional hydrodynamic methods is also performed commonly.

Pumping tests of up to a year in several boreholes simultaneously are carried out for the calculation of water resources. In some fissure-karst structures with variable spring yield, the use of ground water is optimized by the seasonal exploitation of resources accumulated in structures.

4. Control

Ground water is considered a natural resource, and calculations of its reserves must, therefore, be registered and approved by the State Commission for Resources Classification (Ground Water Subcommission). Each regional commission issues reports on the quantity, quality and category of ground-water resources. Results of the continuing investigations help to make more accurate the data on the quantity and category of ground-water resources. Approval of the calculations of ground-water resources prior to exploitation is obligatory.

5. Mathematical models

Mathematical models are used to calculate the safe yield and exploitable resources, and for analysis of quality. For calculating exploitable resources, the Aquifem model is used, and for special problems, Trescott's three-dimensional model.

Numerous water quality models have been tested and are used to study contaminant transport in aquifers.

The programs used for stochastic modelling are based on mathematical statistics, theory of probability and theory of errors. Computer programs for calculating the statistical characteristics of the data taken and programs for correlation and regression analyses are applied in regional hydrogeologic investigation. Deterministic models based on physically defined relations of mass water transport in a porous environment are employed in calculations of the quantity of water used for water supply systems.

6. Monitoring

A standard monitoring network was established by the Hydrogeological Department of the Hydrometeorological Institute and a network was established temporarily for special hydrogeologic projects. This network is operated partly by the Hydrometeorological Institute and partly by other organizations engaged in ground-water investigation.

A permanent monitoring system for springs and ground-water levels was set in the 1930s. In Bohemia, 1,334 boreholes located in the main aquifers, 559 springs and 590 sounds in hydrogeologic profiles are currently monitored. Readings are taken at least once a week. In pilot boreholes continuous monitoring goes on with the use of recording gauges. Since 1982, the quality of ground water has been monitored in selected wells and springs, generally two to four times a year.

In experimental areas, special monitoring systems and data acquisition systems have been established for the purposes of research.

7. Ground-water development

Ground water is mostly developed by means of such conventional technologies as captation of springs and drilling of vertical wells. In the case of karstic springs under favourable geologic conditions, horizontal galleries are also employed. Inclined wells have been used with success for mineral water captation.

The combined exploitation of ground water and springs in karst includes the captation of springs in periods of high yield, captation by siphon, and pumping by submersible pumps in extremely dry periods when extensive drawdown of ground water is inevitable. The protection of ground water from contamination in areas of fissure and fissure-karst permeability is a special problem. The greatest attention is paid to the protection of open fissure zones draining ground water over long distances.

Hydrogeologic conditions of the aquifer, requirements of well yield, water quality, and other factors determine the drilling method and well casing to be used. Cable-tool percussion, hydraulic rotary drilling, jet drilling, hydraulic-percussion and air-rotary drilling are commonly employed. Different drilling fluids are used, but mud mixes are generally preferred for test wells. An artificial gravel pack is installed if necessary. Stainless steel pipes and Johnson type stainless steel screens are the most frequently employed materials for well casing and screens.

Wooden, glass and earthenware pipes, and anti-corrosive pipes for mineral water, are used for well casing. Bored or slotted steel pipes are also employed, particularly for observation wells. Polypropylene material screen with a diameter of up to 617 mm are installed, too, featuring a high percentage of perforation (up to 41 per cent), resistivity against incrustation, and no threat to health.

Numerous corrosion problems exist from carbonic and sulphuric acids and incrustation problems from calcium bicarbonate especially in the case of mineral water. Incrustation is removed by acids, by boring, or by exchanging the pipes. Incrustation inhibitors are also applied, forming thin coats and preventing corrosion of the pipes.

The cost of wells depends largely on their diameter, technology of drilling, casing and screens. The cost per metre of a well (to include drilling, casing, screens development and completion, pumping tests, chemical and biological analysis and hydrogeological evaluation) is usually some 3,000 to 4,000 Czechoslovak koruna.

8. Pollution

In Czechoslovakia, intensive industrial and agricultural activity and high density of population are a source of ground-water contamination. The State Water Inspection as a specialized control office for water protection registers, for example in the Czech Socialist Republic, an average of 200 cases a year of water pollution from oil and petroleum products, of which more than 30 per cent affect ground water. Oil and gasoline which escapes during storage or transport accounts for about half the cases. Reports on oil spills and pollution of aquifers were presented by Czechoslovak specialists at the International Symposium of the International Association of Hydrogeologists held in Prague in 1978. This symposium considered only the problem of pollution from oil and petroleum products.

Ground-water pollution caused by agricultural activity has been systematically recorded since the mid-1970s. Most cases (about 50 a year) occurred because of the leakage of animal waste into the ground-water system. An increasing content of nitrates from fertilizers in ground water has been recognized, particularly in shallow alluvial aquifers along the middle Elbe and Morava rivers. Intensive monitoring is now in operation in the above areas to study the extent and mechanism of nitrate pollution in the unsaturated and saturated zones. Research into the impact of fertilizers on ground-water quality under field conditions have been under way for several years. The movement and interaction of nitrates within the vegetation cover, soil, rock and ground-water system are being studied under natural conditions.

The increasing number of ground-water pollution cases during the period 1970-1980 led the Government to establish a long-term ground-water protection system. The repair or replacement of oil tanks, containers, waste lagoons and oil pipelines is urgently needed. Monitoring wells have been bored near known pollution sources and ground-water quality regularly monitored. Boreholes are cased in order to be ready for emergency use in removing the contaminating medium. If ground-water contamination occurs in the future, the responsible polluters may use the monitoring wells, but will have to reimburse the Government for the costs incurred.

E. Training, research and international co-operation

1. National and parastatal agencies

The Faculty of Natural Sciences at Charles University, Prague, has the longest tradition in the country of a systematic education of hydrogeologists. The Department of Hydrogeology and Engineering Geology was founded in 1950. The first Czechoslovak professor of hydrogeology was Otta Hynie, author of a multi-volume treatise on general hydrogeology and regional hydrogeology of Czechoslovakia. The first specialists in hydrogeologic sciences graduated in 1954. A Department of Hydrogeology was later founded at the Comenius University in Bratislava. Technical universities in Prague, Brno, Ostrava, Bratislava and Kosice also train specialists in ground-water problems, mining-hydrogeology and other fields. There are specialized secondary schools for educating intermediate-level technicians in geology.

Large hydrogeologic units work within the Czech and Slovak Geological Departments investigating and developing ground water, solving ground-water protection problems and preparing hydrogeologic maps, and at the Ministry of Forest and Water Management monitoring ground water and planning for use of ground water in various supply systems.

International hydrological courses, covering also hydrogeology for experts from developing countries, are regularly held in co-operation with UNESCO. There is close co-operation with the International Association of Hydrogeologists and the International Association for Hydrological Sciences. Czechoslovak hydrogeologists participate in the International Hydrogeological Programme of UNESCO. Several international symposia concerned with ground water and mineral water protection were organized, or are being prepared in Czechoslovakia.

The most important companies engaged in hydrogeologic activities include Stavebni geologie, Prague; Vodní zdroje, Prague; Geotest, Brno; Vodné zdroje, Bratislava; and IGHP (Inzenyrsko-geologicky a hydrogeologicky pruzkum), Zilina.

There are hydrogeology departments at the Central Geological Institute in Prague and the Dionyz Stur Geological Institute in Bratislava.

2. Technical assistance

Under the umbrella of the Czech and Slovak Geological Offices and the Ministry of Forest and Water Management, there are organizations entrusted with various types of technical assistance to developing countries. Hydrogeologic surveys including mapping, drilling of investigation boreholes and water wells, and consultancy services are carried out in several countries of Africa, South America and Asia, operating through bilateral commercial contracts and intergovernmental co-operation and agreements. Technical and scientific assistance through agencies of the United Nations has emphasized educational projects. In particular, ground-water projects with Czechoslovak participation have been implemented in Bolivia, Cuba, Egypt, Ethiopia, Islamic Republic of Iran, Iraq, Jordan, Mongolia, Nigeria, Venezuela, Zambia and Zimbabwe. Co-operative arrangements exist with many European hydrogeologic institutions, including BRGM in France, VITUKI in Hungary, various hydrogeologic organizations in Poland, VSEGINGEO in the Soviet Union, and the British Geological Survey in Great Britain.

References

A. Czech Massif

Balek, J. A Year-round water balance model, its Application and its Future. IAHS Publ. No. 115, 1975.

__________. A field monitoring programme for the assessment of the quantity and quality of ground water resources in the Czech Cretaceous Basin. Int. Symp. on Methods and Instrumentation for the Investigation of Groundwater Systems, Nordwijkerhout, Netherlands, 1983.

Hynie, Otta. Hydrogeology of Czechoslovakia, vols. 1 and 2, Czechoslovak Academy of Sciences, Praha.

Regional Geology of Czechoslovakia, Atlas of 1:1,000,000 maps, Central Geological Institute, Praha, 1967.

Skorepa, J. et al. Occurrence of nitrogen materials in ground water in the territory of the middle Elbe River in Bohemia. Memoires IAH, vol. 1, Praha, 1982.

Skorepa, J. and J. Vrba. Hydrogeological investigation of crystalline rocks of the Czech Massif. Memoires IAH, USA, 1985.

Svoboda, J. et al. Regional geology of Czechoslovakia. vols. 1 and 2. Czechoslovak Academy of Sciences, Praha, 1964.

Vrba, J. Hydrogeological zoning of Czechoslovakia. Praha, 1966.

B. Czechoslovak Carpathian System

Bela, M. Hungariae novae et antiquae prodromus (1723). (In G. Rebro, 1983, Alfa, Bratislava.)

Buday, T. et al. Regional geology of Czechoslovakia, volume II: West Carpathians, 2. Praha, 1967.

Franko, O., S. Gazda and M. Michalícek. Origin and classification of mineral waters of the West Carpathians. Dionyz Stúr Geological Institute, Bratislava, 1975.

Franko, O. and M. Kolárová, Map of mineral waters in Czechoslovakia, 1:500,000. Dionyz Stúr Geological Institute, Bratislava, 1983.

Hanzel, V., S. Gazda and I. Vaskovsky. Hydrogeology of the southern part of the High Tatras Mountains and of their foreland, West Carpathians. Hydrogeology and Engineering Geology, series 5, Dionyz Stúr Geological Institute, Bratislava, 1984.

Hanzel, V., E. Kullman et al. Prognoses for ground water in Slovakia, Regional geology. Dionyz Stúr Geological Institute, Bratislava, 1984.

Hynie, Otta. Hydrogeology of Czechoslovakia, volume II: Mineral Waters. Czechoslovak Academy of Sciences, Praha, 1963.

Krásny, J., M. Knezek, A. Subova, A. Danková, M. Matuska and V. Hanzel. Ground water runoff on the territory of Czechoslovakia. Czech Hydrometeorological Institute, MHP UNESCO, Praha, 1982.

Kullman, E. Hydrological and hydrodynamical evaluation of ground water in fissure and fissure/karst environment. Geological Work, Reports 67. Dionyz Stúr Geological Institute, Bratislava, 1977.

__________. L'évaluation du régime des eaux souterraines dans les roches carbonatiques du Mésozoique des Carpates Occidentales par les sourbes de tarissement des sources. West Carpathians. Hydrogeological and Engineering Geology, series 3, Dionyz Stúr Geological Institute, Bratislava, 1984.

__________. Problems and possibilities of increasing usability of karst water. Proceedings, Fourth Slovak Geological Conference. Dum techniky CSVTS. Bratislava, 1981.

__________. Ground water régime with turbulent flow in fissure/karst rock environment. Geological Work, Reports 79. Dionyz Stúr Geological Institute, Bratislava, 1983.

__________. Search for, and possibilities of captating fissure/karst water covertly flowing into surface water. Proceedings, Eighth National Hydrogeological Conference. Dionyz Stúr Geological Institute, Bratislava, 1984.

__________. Captage d'une source karstique par forages horizontaux. Hydrogéologie des terrains karstiques. International Contribution to Hydrogeology, vol. 1, Hannover, 1984.

__________. Etude en vue du captage et de l'exploitation les plus favorables de la source karstique de Jergaly (Great Fatra, Czechoslovakia). Hydrogéologie des terrains karstiques. International Contribution to Hydrogeology, vol. 1, Hannover, 1984.

Mahel, M. Mineral springs in Slovakia in respect to their geological structure. State Geological Institute, vol. 27, Bratislava, 1952.

IV. FINLAND

Area: 337,010 km^2 Population: 4.9 million

A. General

The Republic of Finland opens on the Baltic Sea by its shores on the Gulf of Finland and the Gulf of Bothnia (figure 17). The country is mainly a peneplain, with only a minor area belonging to the Scandinavian mountain range, the highest peak on the Finnish side being 1,328 m. In the south of Finland, a few slopes reach 200 m and in the north, 500 m. The mean elevation of the country is 152 m. Despite the general flatness, the topography is markedly undulating in most parts of the country. Finland has more than 60,000 lakes which cover 9 per cent of the area.

Most of the rivers flow into the gulfs of the Baltic Sea, a few flow to the Arctic Sea and the White Sea. Eight river catchment areas cover 60 per cent of the area of Finland. The largest river is the Vuoksi, which drains through Lake Ladoga in the Soviet Union into the Gulf of Finland. Its catchment area at the border is 61,000 km^2 and the mean discharge 550 m^3/s. Six other rivers discharge more than 200 m^3/s. The larger catchment areas discharge 8-11 $l/km^2/s$, being higher in the north than in the south.

Finland lies within the north temperate zone, in the transitional zone between maritime and continental climates. Rapid weather changes are typical, depending on the prevailing wind direction.

The mean annual temperature is 5.5° C in the south-west of Finland, 0° C somewhat north of the Arctic Circle, and the coldest areas register about -1° C. The July mean temperature ranges from 18° C in the south-east to 14° C in northernmost Finland and, respectively, in February from -4° C in the south-west to -14° C in the north-west.

The mean annual precipitation is from 700 mm in the south to 400 mm in the north. In the north, about half the precipitation is in the form of snow. The land is snow-covered in the south-western islands for 100 days per year and in the north-west for 205 days per year on the average. Permafrost does not occur except in some peat hummocks in the north.

B. Geology

The rocks of Finland are either very old or very young. They include Precambrian, mainly Archean, crystalline basement and Quaternary, mainly Holocene, unconsolidated sediments. Finland's territory is a central part of the Baltic Shield, 2,800 to 1,550 million years old. It consists predominantly of plutonic and metamorphic rocks, largely granite, schist and gneiss.

The continental Quaternary ice sheet covered the entire region, and its movements swept the old peneplain clean, removed the loose material and eroded the rock surface by an average of 7 m. The material was transported and partly deposited inside the region as glacial drift including nonstratified till and stratified sand and gravel. The crust was warped under the weight of the ice for more than 100 m and most of the region was covered by water during the succeeding lake and sea stages of the Baltic Sea.

Figure 17. Finland, geographic setting

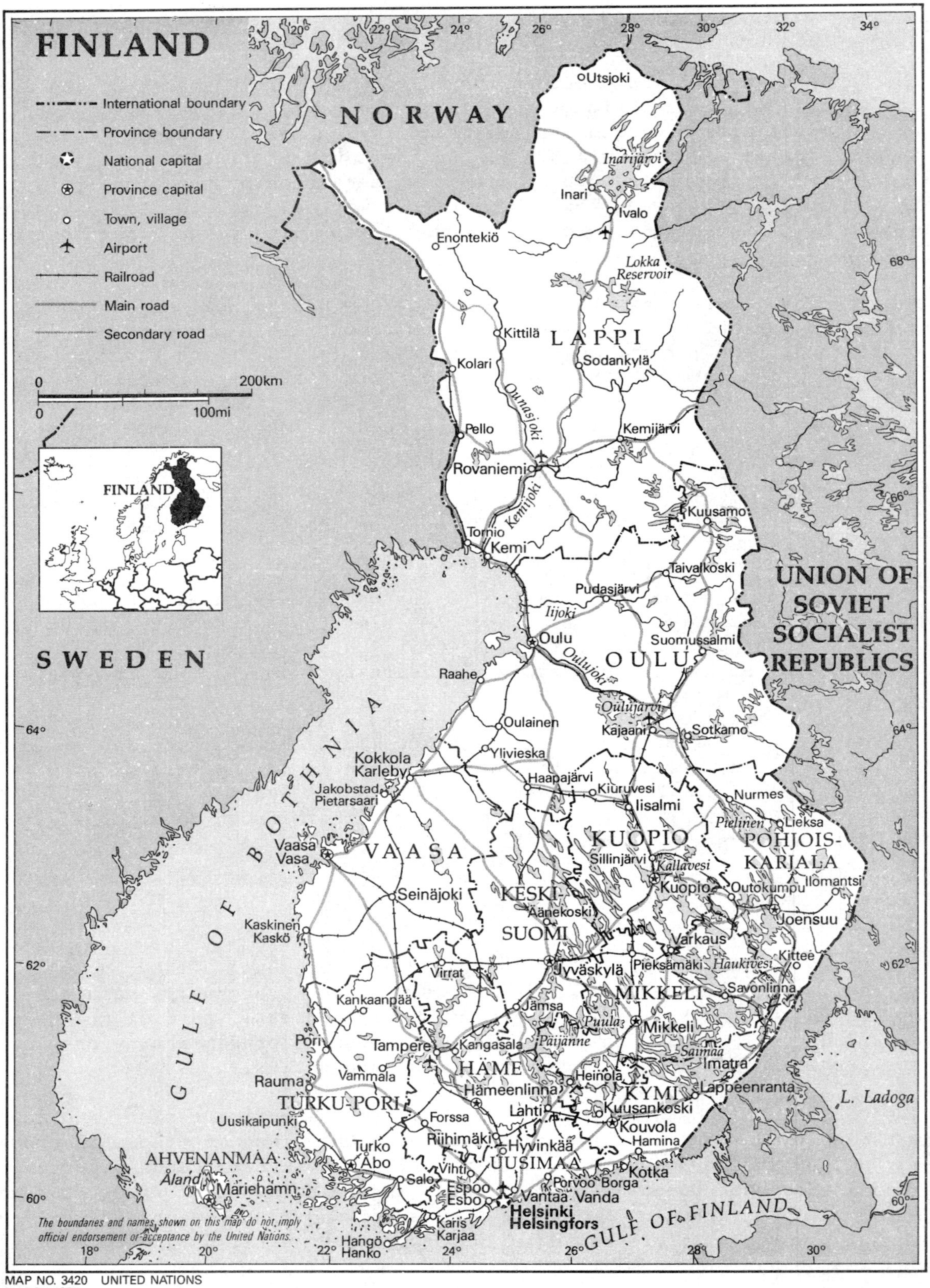

MAP NO. 3420 UNITED NATIONS
MARCH 1987

The deglaciation caused uplift of the crust and the subaquatic areas gradually rose above the surface. The uplift process still continues and is most rapid in the northern parts of the Gulf of Bothnia, being at present 9 mm per year. Waterlaid sediments are met with in the subaquatic areas, consisting of clay and other fine-grained fractions.

The youngest deposits, peat, cover 32 per cent of the land area. The thickness of the Quaternary deposits averages only about 7 m and barely 100 m anywhere.

C. Hydrogeology

1. Historical

In the 1800s no significant difficulties were met in providing water for domestic supply as the population was scattered, towns and villages were small and ground water was sufficiently available from shallow wells in the rather thin deposits covering the basement and the springs discharging from them. In addition, domestic water was available in Finland from surface-water courses, and also from the tens of thousands of lakes. Only on islands surrounded by the sea did supply in sufficient quantity cause difficulties. There, in places, the supply was drawn from wells drilled into the basement. Accordingly, it is natural that ground-water research lacks long traditions in Finland.

Hydrogeologic investigations became current in connection with the organization of the water supplies of the major cities, Helsinki and Turku, at the turn of the century, as the few separate wells could no longer satisfy the demands of the growing population. The municipality of Helsinki carried out large and thorough ground-water investigations during the first two decades of the present century in the eskers of the surrounding countryside and even as far away as the I Salpausselkä terminal moraine. The investigations were possibly more extensive than any executed there since. The results proved that the glaciofluvial deposits might yield 1,000-5,000 m^3 per day of ground water and in exceptional cases in excess of 10,000 m^3 per day. It was considered that the requirements of Helsinki for water could not be satisfied by ground water at acceptable costs, so a surface-water plant was established. By contrast, Turku got its ground-water plant in 1902, with a yield of 3,000 m^3 per day. Despite the good geologic location of many towns, enabling them to exploit sufficient quantities of ground water, the example of Helsinki had a negative influence on hydrogeologic research. During the following decades, surface-water plants were favoured. Only a few hydrogeologic investigations were made and mostly by people whose main field was other than hydrogeology.

After the Second World War, as industry expanded, surface-water supplies began to be exposed to pollution and the population to become increasingly concentrated, attention was again turned to ground-water research. In the 1950s and 1960s, several engineering companies were established to design and construct ground-water plants for densely populated communities and for industry. Around the same time scientific hydrogeologic research was started at, e.g., the Engineering Department of the Board of Agriculture and at the Geological Survey of Finland; and in 1970, the National Board of Waters was founded to assess the exploitable ground-water resources. Ground-water research was also conducted in the private sector. By now, there is a good general grasp of the quality and quantity of Finland's ground-water resources, but there is still a lack of detailed information.

2. Principal aquifers

The larger water supplies mainly tap the aquifers of the glaciofluvial and morainal deposits. The aquifers of the other Quaternary deposits and the bedrock are still mainly exploited for local consumption only (figures 18 and 19).

The eskers are glaciofluvial deposits in the drainage channels of the retreating ice of the Quaternary glaciation either in tunnels near the ice front or at their mouths. They consist of sorted, rounded rock material which is coarsest in the central bottom parts, containing even large boulders, mantled by gravel and sand strata. The eskers are mainly deposited on bedrock, only rarely on till.

By and large, they do not conform to the underlying topography but commonly follow fault lines that in many cases are eroded depressions. The eskers lack continuity in places and the topography forms ground-water divides, and in valley bottoms the esker aquifers discharge into surface-water courses. Thereby, related aquifer units seldom exceed 10 km in length. They are mostly a few hundred metres wide and a few tens of metres thick; few are more than 50 m thick. The lower parts of the esker slopes are mostly overlain by clay in subaquatic areas, which elevates the water table.

In the eskers as well as in the ice-marginal formations in southern Finland, about 60 per cent of the annual precipitation seeps, under favourable conditions, into ground water, even more in the north. With the average recharge in the aquifer, this equals nearly 1,000 m^3 per day per square kilometre. The rate of recharge is nearly constant in the whole country, as the higher precipitation in the south and smaller evapotranspiration in the north nearly balance each other. The highest yielding esker aquifer is the Säkylä Esker in south-western Finland, yielding 40,000 m^3 per day from a catchment area of 56 km^2.

The morainal formations, especially the two Salpausselkäs, are large aquifers. Their structure is complicated by alternating layers of coarse and fine sorted material and beds of till. The largest morainal aquifer known in Finland is Kukonharju-Hatsina in the I Salpausselkä near the town of Lahti, yielding about 40,000 m^3 per day from an area of 55 km^2. In certain other areas, successive lowly permeable strata restrict the ground-water flow to such an extent that exploitation is not possible. At several such locations, there are eskers on the ice-proximal side in connection with the marginal formation partly imbedded in it. These form drainage channels connecting the permeable water-yielding strata and many have high yields.

Glacial till, the most widespread sediment in Finland, has in general a low permeability owing to its high clay content. Wells sunk into it yield but a few cubic metres per day. Higher yields are obtained when even thin, sorted or leached strata are imbedded; however, in such cases, the yields are sufficient only for individual farms or households. In some moraines, the till may be sorted to such an extent that even a few hundred cubic metres per day can be exploited.

Figure 18

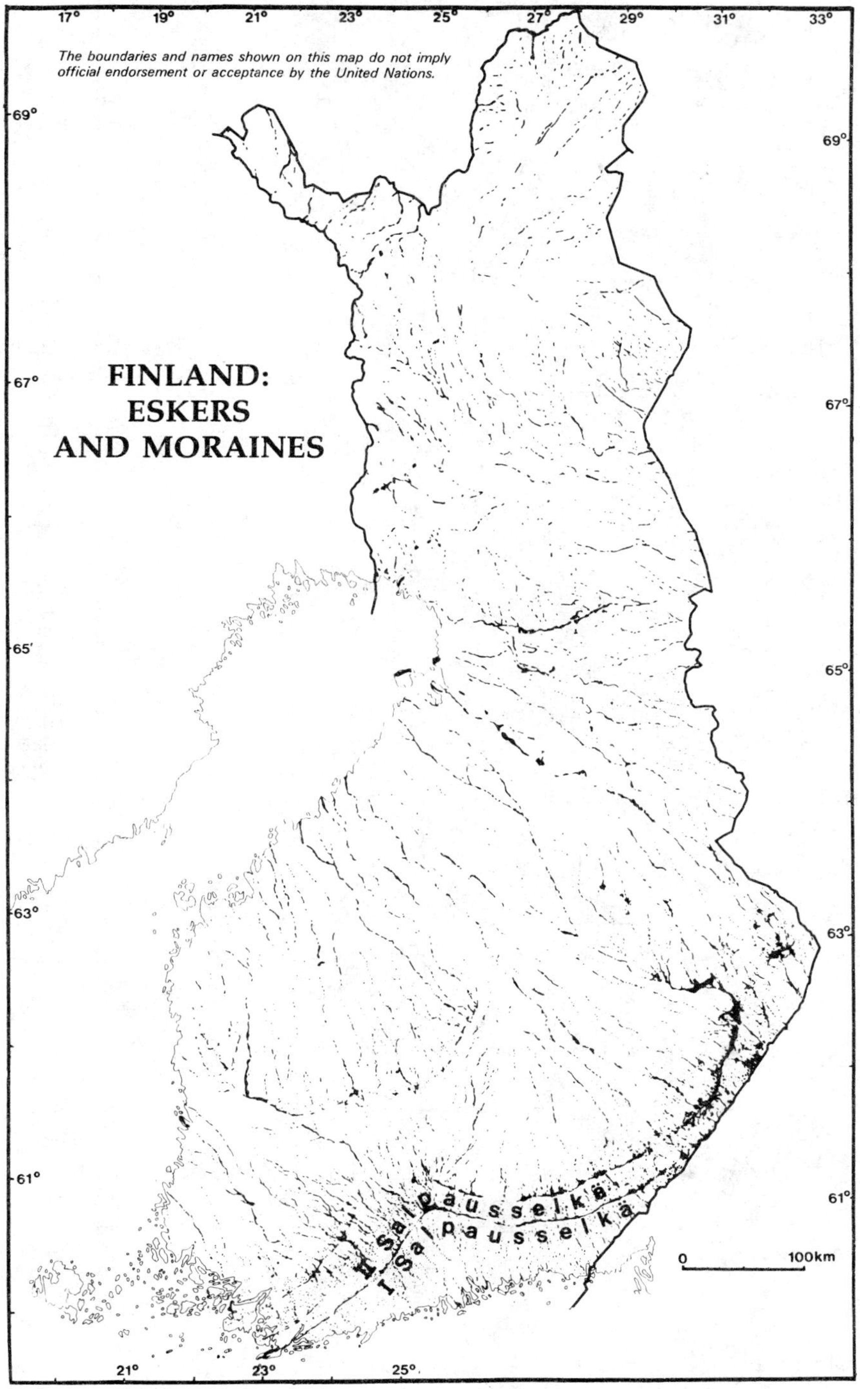

3546.10x

Source: Geological Survey of Finland

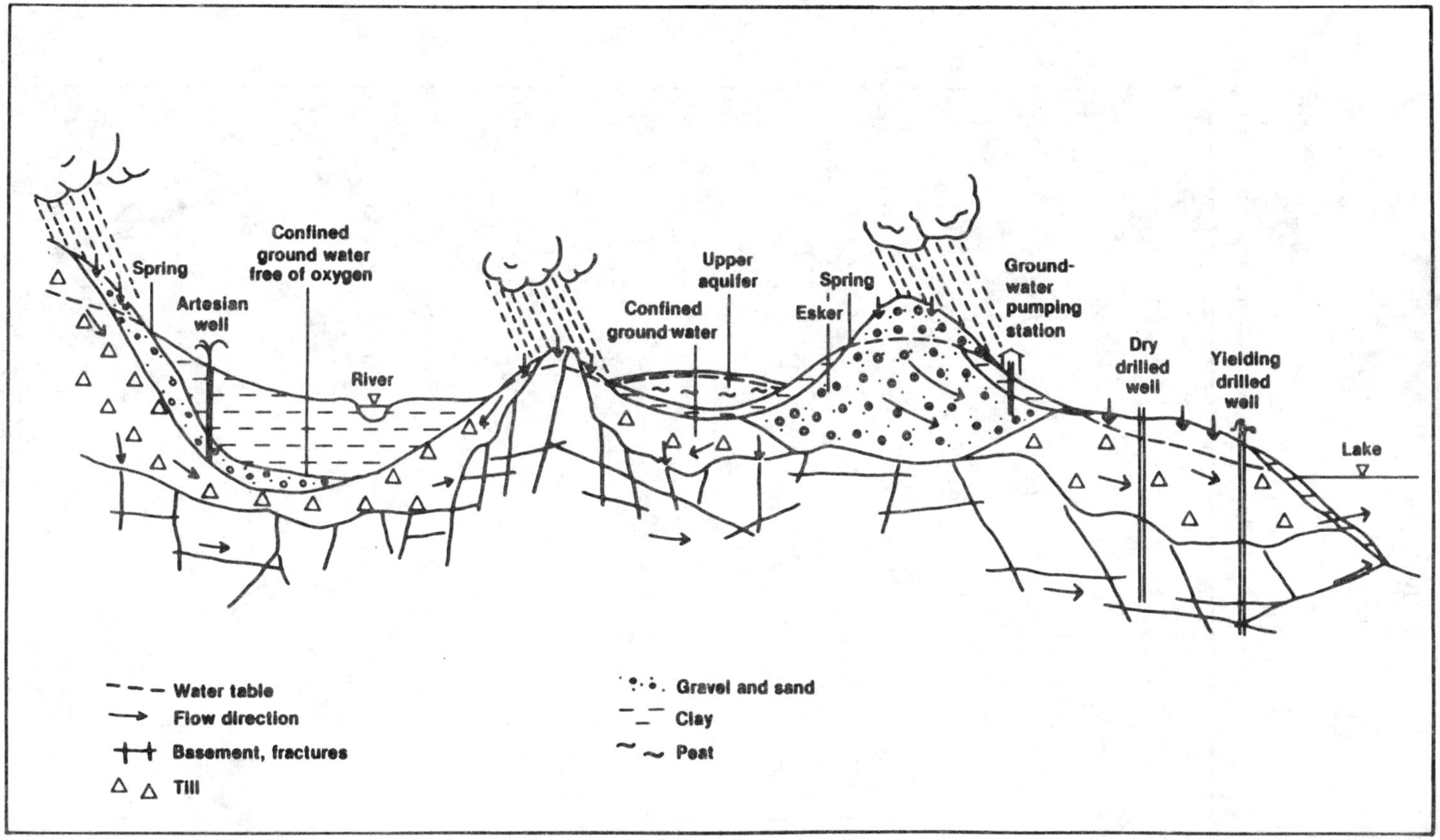

Figure 19. Ground-water flow in glaciated terrane

The crystalline Precambrian basement is nearly impermeable and the rock conducts water only as a result of fracturing and tectonic faulting. The type of rock is not the main factor determining the yield of a well but rather its penetration by open fractures. Fracturing varies greatly over short distances and two closely situated wells may discharge extremely different quantities of water. The openness and density of the fractures decrease with depth so that most of the yield of a well is generally reached at a depth of some 50 m, and drilling further to a depth of 100 m usually gives only a slight increase and drilling still deeper only exceptionally improves the yield appreciably.

The average yield of wells drilled into the basement is a cubic metre per hour, some reach 10-20 m^3/h and, in exceptional cases, nearly 100 m^3/h. The highest average yield is reached in the rapakivi granites, being 1.5-2 m^3/h; also, most schists yield more than the average. The thickness of the overlying surficial sediments influences the yield of wells; extensive permeable sediments have the effect of increasing the discharges most.

3. Ground-water quality

In chemical composition, Finnish ground water is mostly bicarbonate, with calcium as the main cation (table 4). Their electric conductivity, which involves the total dissolved electrolyte concentration, is in the range of 5-30 μS/m. The ground waters of Quaternary deposits, also including the most exploited aquifers with the highest yields, are generally soft and acid. Their hardness is mostly in the range of 1-5° dH and they contain free carbon dioxide, which often requires neutralization before piping. Ground water in the basement is, on the average, richer in electrolytes (15-40 μS/m) than those of Quaternary deposits. Also its pH is close to neutral or slightly alkaline.

Table 4. Finland, chemistry of ground-water types

	Sand and gravel		Till		Basement	
Chemical aspect	Average	Median	Average	Median	Average	Median
pH	6.4	6.4	6.4	6.4	6.7	6.7
El. cond. in μS/m, 25° C	20.4	14.9	23.8	19.2	35.0	24.3
HCO_3 mg/l	50.0	34.1	62.2	47.1	97.5	73.7
NO_3 mg/l	13.9	3.8	15.1	6.9	11.8	2.3
Cl mg/l	15.3	6.7	17.1	9.2	32.6	12.5
SO_4 mg/l	17.2	10.0	17.7	13.0	18.5	11.2
Ca mg/l	17.8	13.3	20.9	17.1	26.2	19.6
Mg mg/l	4.4	2.9	5.1	3.7	7.8	5.6
Na mg/l	8.9	5.4	10.1	6.9	25.1	8.9
K mg/l	7.8	3.0	8.9	4.4	7.1	3.2

Source: Geological Survey of Finland.

In coastal areas covered by the _Littorina_ stage of the Baltic Sea, the ground water contains on the average more electrolytes than in areas farther inland. Occasionally, in aquifers confined by clay beds chloride concentration exceeds the bicarbonate concentration. The geographic location affects the quality of the ground water in that the electrolyte concentration diminishes from the coastal areas towards the north and east.

The influence of the chemical composition of the basement is mainly overwhelmed by other, more dominant factors. Only the exceptionally high fluoride concentration of the rapakivi granites can be noticed in both the Quaternary and basement aquifers. Owing to the reducing conditions in many aquifers, the iron and manganese concentrations are so high that treatment is required before domestic consumption.

D. Investigation techniques

1. Surveys

Investigations are generally initiated by means of remote sensing surveys during which aerial photo interpretation is used, with both monochromatic and polychromatic photographic material in black and white or colour versions. In addition, the national base map at scale of 1:20,000 is well adapted for remote sensing investigation.

Field investigations usually begin with a site reconnaissance by a geologist or hydrogeologist. The next phase is geophysical prospecting, using both seismic refraction and, to a lesser extent, electric resistivity sounding.

After the surficial investigation, either light or medium to heavy drilling is undertaken, depending on the type of borehole. Small-diameter holes are drilled for reconnaissance and observation of piezometric levels (by means of 32 or 50 mm piping). Larger boreholes are drilled for aquifer test and collection of water samples. In investigations where information about the quality of the water is needed, plastic pipes 50 mm in diameter are used.

Geophysical and geochemical readings are then generally taken, of which measurements of temperature, dissolved oxygen, pH, electric conductivity and redox potential (Eh) are the most important. The movement of ground water is also measured from the observation holes either by the point dilution method or with radioactive tracers.

Long-term pumping tests assess water quality and its stability. Mathematical hydraulic calculations are applied to determine the hydraulic parameters of the aquifers.

Sophisticated mathematical and analog models are seldom practical because of Finland's special geologic features which make such techniques very difficult to apply.

The Geological Survey of Finland has carried out hydrogeochemical mapping of the ground waters of the country. During the years 1978-1982 ground-water samples were collected, on the basis of two samples per site, from wells and springs in different aquifers in the surficial deposits and from wells drilled in the basement rocks. Data concerning wells or springs, the geologic environment, prevailing climatic conditions and possible man-made influence were recorded and a large-scale physico-chemical analysis prepared. The information has been computerized and organized for ultimate publication.

The National Board of Waters keeps a register of water-intake plants having more than 200 consumers. The records contain reports on the hydrogeologic investigations made to establish the plants as well as follow-up information on discharge and water quality.

The need for national archives of drilled wells is recognized, but as of 1988 no law had been passed requiring contractors to provide well data. The National Board of Waters was due to establish a central register of water data of a number of State and other organizations in order to centralize the available information. As of 1988 the register awaited necessary financing.

Mapping of the Quaternary deposits on a scale of 1:20,000 (in the north at 1:50,000) was initiated in 1977 jointly by the National Board of Survey and the Geological Survey of Finland. Compilation is scheduled to cover the entire country in 30 years. The map and its explanatory text will include a broad range of information on hydrogeology, including aquifer characteristics, piezometry, the discharge of springs, ground-water plants and the physico-chemical quality of ground water.

2. Monitoring and modelling

The National Board of Waters has made an inventory of ground-water areas important for public water supply. It includes the areas that are used or will probably soon be used to satisfy public demand or meet the needs of the food industry. Information was collected on boundaries, areas, yields and the exploration of important aquifers as well as on plants and areas of hazardous ground-water resources. The inventory includes more than a thousand localities in an area of 2,200 km^2 and a total estimated yield of 1.7 million m^3 per day; some 500 of the localities being exploited. All aquifers yielding more than 250 m^3 per day are included.

The National Board of Waters established a ground-water monitoring network in 1974. It includes 54 stations in different climatic and geologic environments. The following data are registered at the stations: precipitation, water equivalent of snow cover, chemical composition of the precipitation, soil moisture, frost, evapotranspiration, surface runoff, piezometry of the ground water, quantity and physico-chemical quality of the ground water.

The data obtained have a considerable use for basic research and such practical applications as regional and seasonal balance assessments and forecasts.

The Quaternary aquifers are, as compared with many other regions, small and irregular in shape and have complex hydraulic connections with other surficial deposits and the basement. The basement aquifers are governed by unevenly distributed fracturing and faulting and complicated connections with surficial deposits and surface-water courses. Therefore, the practical application of mathematical models requires so many assumptions, in most instances, that their reliability is in question. To a certain extent mathematical models are used in ground-water exploration as, for example, in connection with investigations for the final disposal of nuclear waste in the basement rocks. However, much more work is necessary before practical and universal models can be applied.

3. Ground-water development

By the end of 1983, there were 794 public water supply plants serving 3.79 million people (78 per cent of the population). The daily consumption averaged 1.06 million cubic metres during the year. Of this, 47 per cent was ground water or 0.5 million cubic metres per day. Of total consumption, 15 per cent was industrial. Most industrial enterprises are supplied by their own plants and their consumption is not included in the foregoing figures. The daily per capita consumption was 279 litres.

In Finland, three main types of wells are built: dug wells, tube wells and borehole wells. The diameters of the dug wells range from 1 to 3 m. Most are made of prefabricated rings and are less than 10 metres deep. The cost of a dug well ranges from Fmk 10,000 to Fmk 150,000, depending on size and circumstances.

Two methods are used to build tube wells: (1) the applied Johnson method, i.e., a naturally developed tube well, and (2) the artificially gravel-packed well. In both methods, a casing, mainly of plastic and 400 mm in diameter, is used. Well screens differ, depending on the construction method, so that in naturally developed tube wells Flux-screens are used, resembling the continuous-slot Johnson well screen. The well screens are made in Finland. In an artificially gravel-packed well, the screens are usually of the wooden-pipe type.

Tube wells range from 6 to 40 m deep and 70-80 per cent are naturally developed. The average cost of a tube well is Fmk 100,000.

Borehole wells average 160 mm in diameter. Casing is necessary only where there are loose soils, and no casing or screens are installed in hard-rock boreholes. The casing is set in the rock surface by cement grouting. The casing used in these wells is also usually of plastic but exceptionally steel is used. The diameter is generally 200 mm and its cost approximately Fmk 150 per metre.

In dug wells, standard centrifugal pumps and annular water pumps are used; in boreholes and tube wells the pumps are of the submersible type. The capacity of the pumps is chosen so as to correspond to the delivery lift needed and the highest continuous yield. The maximum allowable yield is determined by a pumping test.

4. Problems

The aquifers are mainly phreatic and shallow. The buffering capacity of the soils is low, owing to the small concentrations of calcium carbonate. As a result, the ground water is highly sensitive to pollution; but, as the population density is relatively low, there are fewer sources of pollution than in most other European countries.

The influence of acid rains on ground water have not been observed as diminishing pH, but increasing concentrations of dissolved electrolytes have been registered. It is assumed that if the buffering capacity is exceeded, the consequences will be rapid and dramatic. The acidification of some small forest lakes gives warning of this.

Agricultural fertilization and other human activity show up in places as excessive nitrate concentrations. Many roads traverse the eskers and the transportation of oil and other hazardous products creates considerable risk.

Excessive exploitation has in several cases lowered the water table and stirred stagnant bodies of ground water into motion. As a result, iron and manganese concentrations and sometimes even salinity have increased in the discharged water. This is often the case as, traditionally, the ground-water plants have been situated on the flanks of eskers, many of which are clay covered. There the ground water is often stagnant and reducing conditions prevail. Modern drilling techniques enable wells to be sunk into the coarse central parts of the eskers with the main ground-water flow.

Sea-water intrusion is not a major problem in Finland. In the coastal aquifers connected with the sea, the discharge has to be limited to prevent the water table from declining below sea level. In drilled wells along the coast and on the islands, the depth and discharge have to be kept regulated as the water is drawn from fresh water floating on saline water. It should also be noted that the salinity of the sea surrounding Finland is low, having an electrolyte concentration less than a tenth of that in the world oceans.

Brackish or saline water is occasionally met with in deep drilled wells mainly in areas less than 100 km from the coast. Such water is old and occurs in sequences of the basement not involved in the ground-water flow.

Public water plants must obtain a permit stating the maximum allowed exploitation, which must not exceed the calculated yield of the aquifer. Overdraft often causes deterioration of water quality before excessive decline of the water table. The yield of some aquifers can be greatly improved by artificial recharge.

Human activity in urban areas tends to lower the water table. Also, uplift of the earth's crust has the same effect in coastal areas. The damage caused by lowering of the water table is quite limited and the situation is under control. The principal aquifers occur in coarse sorted material not susceptible to subsidence.

E. Training, research and international co-operation

Courses in hydrogeology are given by the Universities of Helsinki, Turku and Oulu. Sciences involved in hydrogeologic research, such as geophysics, remote sensing and chemistry of water, are studied at several institutions of higher education. Water engineering is taught at the Helsinki University of Technology and other technical institutions, including the Technical Research Centre of Finland.

Practical and scientific ground-water research is conducted by the National Board of Waters, the Geological Survey of Finland, the Agricultural Research Centre and various other institutions dealing with agriculture and forestry. The nuclear power companies and a project under the Geological Survey of Finland are doing sophisticated hydrogeologic research for the final disposal of nuclear wastes in bedrock.

Consulting engineering companies dealing with ground water pursue, to some extent, scientific hydrogeologic research both as projects with outside financing and also on their own.

Finland's co-operation with developing countries in the field of ground-water engineering has been limited by financial constraints. Development co-operation was started in the early 1970s in the form of expert assistance. At that stage, Ethiopia and the United Republic of Tanzania were the chief beneficiaries. Since then, assistance to Ethiopia has been discontinued and continued with the United Republic of Tanzania only on the basis of bilateral aid. Multilateral expert assistance has not been given to any appreciable extent in the domain of ground-water engineering.

Development co-operation projects have been implemented in the field of ground-water engineering since 1972 when a project was launched in the United Republic of Tanzania to develop the water supply system of the Mtwara and Lindi regions. A similar project was started in 1980 in the province of Harispattuwa in Sri Lanka. In the Western Province of Kenya, development of a rural water-supply system, based largely on ground water, was initiated in 1981.

In addition, Finland has co-operated with developing countries through four post-graduate courses dealing with water supply, including ground water, organized by the Helsinki University of Technology and the Tampere University of Technology. Participants have come from Ethiopia, the United Republic of Tanzania, Kenya, Sudan and Zambia.

References

International Hydrogeological Map of Europe. Map sheets D 2, Haparanda, and D 3, Stockholm, with explanatory notes. UNESCO, Paris, n.d.

Map: Pre-Quaternary rocks of Finland, 1:100,000. Compiled by Simonen, A. Geol. Surv. Finland, 1980.

Map: Quaternary deposits of Finland, 1:1,000,000. Editors, R. Kujansuu and J. Niemelä. Geol. Surv. Finland, 1980.

Hydrological yearbook 1980. Publications of the Water Research Institute 53. Helsinki, 1983.

Yhdyskuntien vedenhankinnalle tärkeät pohjavesialueet, English Summary: Groundwater Areas Important for Public Water Supply. Vesihallitus - National Board of Waters, Finland, Tiedotus - Report 109. Helsinki, 1976.

Vesilaitosten veden laatu vuonna 1980, Water quality in water utilities in 1980. Vesihallitus - National Board of Waters, Finland, Tiedotus - Report 226. Helsinki, 1983.

Vesihuoltolaitokset 31.12.1983, Water supply and sewer systems 31.12.1983. Vesihallitus - National Board of Waters, Finland, Tiedotus - Report 249. Helsinki, 1984.

Hyyppä, J. Chemical composition of ground water in the bedrock of Finland. Geol. Survey of Finland, Nuclear Waste Commission of Finnish Power Companies, Report YJT-84-10, Helsinki, 1984.

Lahermo, P. Chemical composition of ground and surface waters in Finnish Lapland. Bull. Comm. Geol. Finlande 242, 1970.

Mälkki, E. Ground-water flow velocity as an indicator of permeability and internal structure of eskers. Publications of the Water Research Institute 32. Helsinki, 1979.

Rönkä, E. Drilled wells and ground water in the Precambrian crystalline bedrock of Finland. Publications of the Water Research Institute 53. Helsinki, 1983.

Simonen, A. The Precambrian of Finland. Geol. Surv. Finland, Bull. 304, 1980.

V. GREECE

Area: 131,990 km^2 | Population: 10 million

A. General

The Hellenic Republic is largely surrounded by seas. On the south and west is the Ionian Sea and on the east the Aegean Sea. To the north is a border of 1,200 km with Albania, Yugoslavia, Bulgaria and Turkey. There are more than a thousand Greek islands of which only fifty are larger than 40 km^2. Accordingly, Greece has about 15,000 km of coastline.

Three fourths of Greece is mountainous or hilly. The mean elevation of the country is about 500 metres. The main mountain chain is the Pindus (2,638 m at Smolika); the highest mountain is Olympus (2,917 m). All the geomorphologic features have a common trend or direction. Nearly all the mountain chains of Greece show a clear north-west to south-east orientation. Like trends are evident in the coastlines, rivers and alignment of islands.

The climate, mostly of Mediterranean type, is strongly influenced by the Pindus Mountains. Rainfall in the western part is fivefold higher than in the east (figure 20). Long periods of sunshine is one of the main features of the climate. Wind directions vary greatly in space and time. Usually the winds blow from the north, but north-east to north-west directions are common in summer.

Four climatic types prevail: (1) Alpine in mountainous areas, (2) an intermediate climate between the European and Mediterranean types, which predominate over northern and central Greece, (3) Mediterranean over western Greece and islands of the Ionian Sea, and (4) a mild climate over the Aegean Sea, Attica and the rest of the country.

Lakes and rivers are more abundant in the western and northern part of the country, due to the higher rainfall levels in these areas. The rivers Evros, Ardas, Strymon, Nestos and Axios enter the country from the north, while the Aoos River runs from Greece to Albania. The Acheloos River has the greatest volume and the Aliakmon River is the longest (297 km).

Based on morphologic features, the country has been divided into 14 water districts, each crossed by one or more rivers.

Figure 20

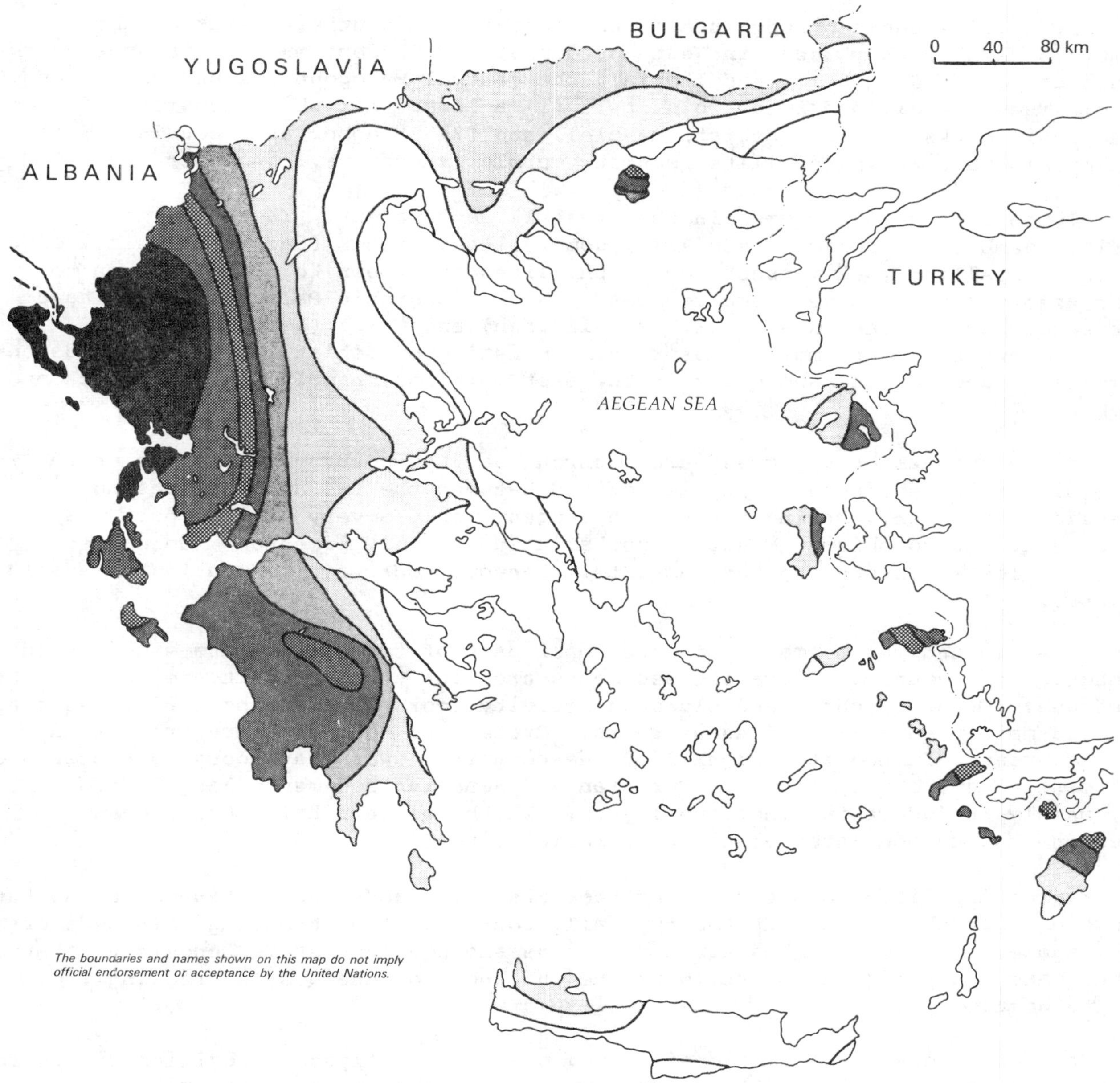

GREECE:
MEAN ANNUAL PRECIPITATION

Over 1200 mm	700 to 800 mm
1100 to 1200 mm	600 to 700 mm
1000 to 1100 mm	500 to 600 mm
900 to 1000 mm	400 to 500 mm
800 to 900 mm	300 to 400 mm

3546.11x

B. Geology

1. Stratigraphy

Crystalline rocks occupy much of the eastern and southern parts of the country. Mainly, they occur in West Macedonia, the Rhodope massif, islands of the Aegean Sea (including Samos and Ikaria), the central Peloponnesus and the island of Crete. They are divided into two units: (1) a lower unit which consists of highly metamorphic rocks (gneiss, schist, marble), and (2) an upper unit consisting of semi-metamorphic rocks (phyllite, schist, marble, sandstone, conglomerate).

Plutonic rocks are common in the crystalline basement. These are mostly granite, gabbro and peridotite. Also, some volcanic rocks (dacite andesite) exist in the crystalline rock masses. Upper phyllites are known to be of Permian to lower Mesozoic age. Modest occurrences of non-metamorphic Paleozoic rocks have been recognized on the islands of Kos (Silurian) and Chios (Devonian). Carboniferous rocks are more widespread, consisting of schist, phyllite, sandstone, graywacke, conglomerate and occur in the east mainland, namely Lokris, Attica and Lakonia.

Mesozoic rocks occupy great areas throughout the country. They lie on the Paleozoic rocks; a disconformity is noticed between the two ages reflecting a Triassic regression. During the Alpine orogeny, only a very limited part was affected by metamorphism. Mesozoic rocks include mainly limestone, dolomite, chert, shale, sandstone and conglomerate. Igneous rocks are common in the Mesozoic terranes.

Triassic sediments (mainly in the upper part of the series) are encountered throughout the country. Jurassic sediments are, for the most part, represented by limestone. Great volcanic and plutonic activity took place during the Jurassic and produced mainly subsilicic igneous rocks. Cretaceous sediments are present in several areas. In the eastern half of the country, Lower Cretaceous is represented by limestone and cherty schist. Turonian and Senonian sediments largely prevail. They mainly include thin limestone layers. In the western half, Cretaceous limestone has common intercalations of radiolarite.

Flysch deposition started during the Late Cretaceous and continued during the Cenozoic. Flysch is abundant and typically consists of alternating thin beds of clay, shale, marl, marly limestone, sandstone and conglomerate. Deposition started in the east and progressed west, with the thickness of the beds decreasing from east to west.

The Cenozoic age, as mentioned above, opened with flysch deposition except for the very western part where limestone deposition continued into the Paleocene, Eocene, Oligocene and Miocene ages. At the same time, the deposition of molasse formations took place in the eastern part of the country, and especially in two main troughs. The first trends nearly east-west and is located along the south part of the Rhodope massif; the second lies along the east side of the Pindus Mountains in a north-west to south-east direction.

Sedimentation continued all over the country, with limestone, flysch or molasse, up to the middle Miocene. During the late Alpine phase, fault tectonics caused depressions, some closed, represented today by the Neogene basins. In some, lignite is present.

The Quaternary sediments are mostly unconsolidated and are present as diluvium and alluvium in basins, and along the coast as terra rossa and on terraces. Quaternary deposits 100 metres thick are found in several basins where they constitute significant aquifers.

A simplified geologic map of the country is presented in figure 21.

2. Tectonics

Hellenic mountain chains (with the Pindus Mountains as their axis), belong to the Dinarides and, as already mentioned, show a NW-SE orientation. The same trend is also demonstrated by the crystalline massifs, with a second arc inside the first, although the formations of the second one had been folded before the Alpine orogeny which initially shaped the arc.

It is obvious that although the main effect was caused by the Alpine orogeny, previous similar uplifts before this (Hercynian, Caledonian) also contributed to the present structure.

Broad basins, in the form of geosynclines, were established on the old crystalline basement, during the late Paleozoic and early Mesozoic ages. In the geosynclines, the whole constituting the Hellenic Alpine system, the Mesozoic sediments were deposited, from Triassic to Late Cretaceous (eastern part of the country) and up to Eocene (the western part).

This process was followed by Alpine folding, started during the Late Cretaceous. The folding began to the east and progressively moved westward covering all. Folding activity signified the end of the Hellenic geosyncline and gave rise to the geotectonic zoning. The Mesozoic sediments affected by the Alpine orogeny exhibit such common features of stratigraphy and tectonics as to allow classification into recognizable geotectonic zones of the Hellenic Alpine geosyncline (figure 22).

The zones in sequence from west to east are: Paxos, Ionian, Gavrovo, Pindus, Beotian flysch, Ossa unit, Parnassus, Pelagonic, Axios, Circum Rhodope belt, Serbian-Macedonian and Rhodope massifs and the Attic cycladic massif. The Rhodope, Axios, Pelagonic and subpelagonic zones are called the inner zones because they occupy the inner parts of the mountain chains surrounding the Rhodope massif and were affected by more or less similar tectonic events and magmatic intrusions during the Mesozoic. The rest of the zones are called the outer zones and generally surround the first group.

The Paxos Zone is characterized by tectonics affecting rocks of low flexibility which produced wide and faulted anticlines, while a greater plasticity and mobility is demonstrated by the succeeding Ionian Zone owing to the presence of gypsum. The structure is one of a sequence of synclines and anticlines overthrusting one another from east to west.

The basement of the Gavrovo Zone moved only very slightly towards the west during emergence of the zone. A principal feature of the Gavrovo Zone is its wide and strongly faulted anticlines.

Figure 21

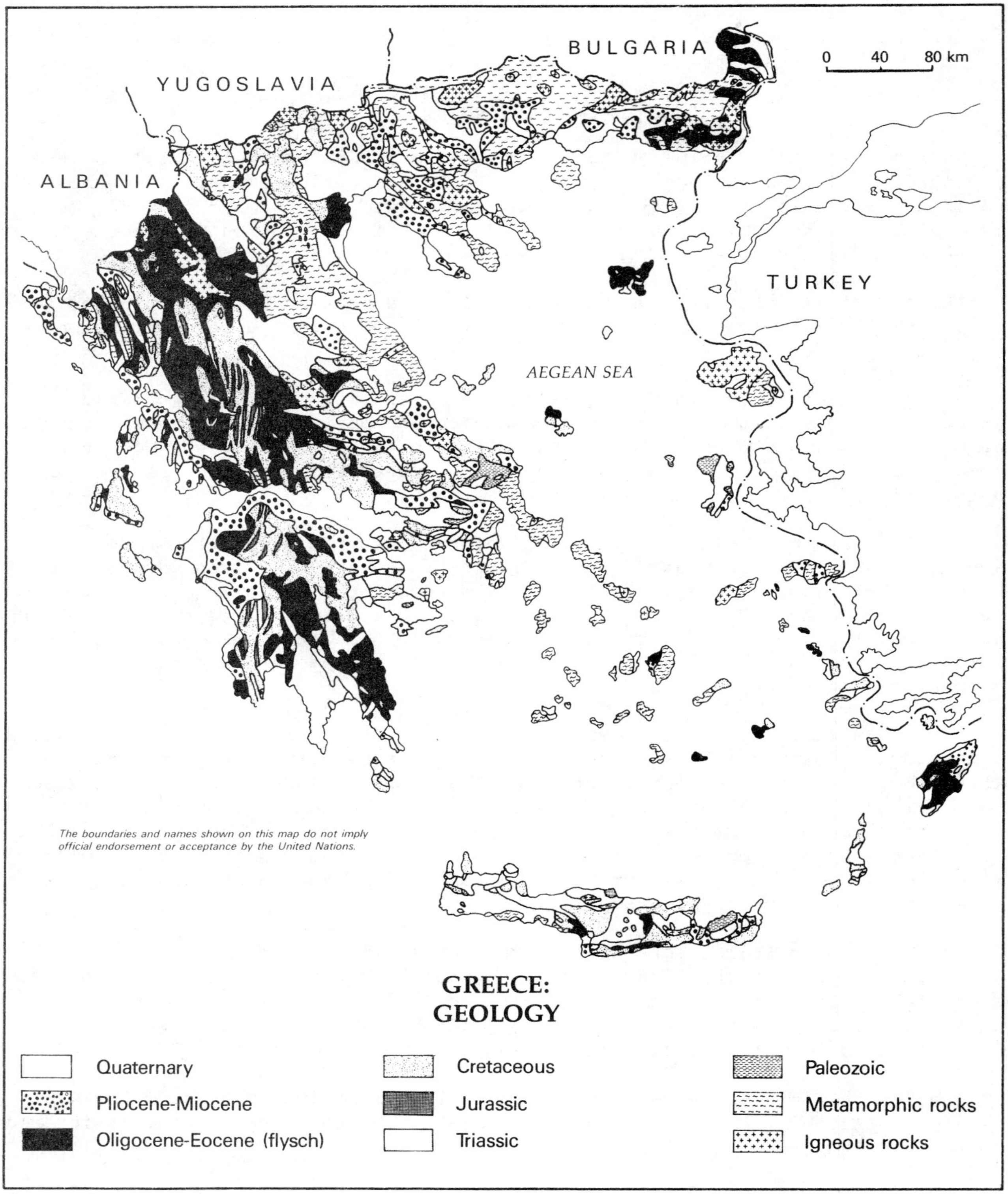

3546.12x

Figure 22

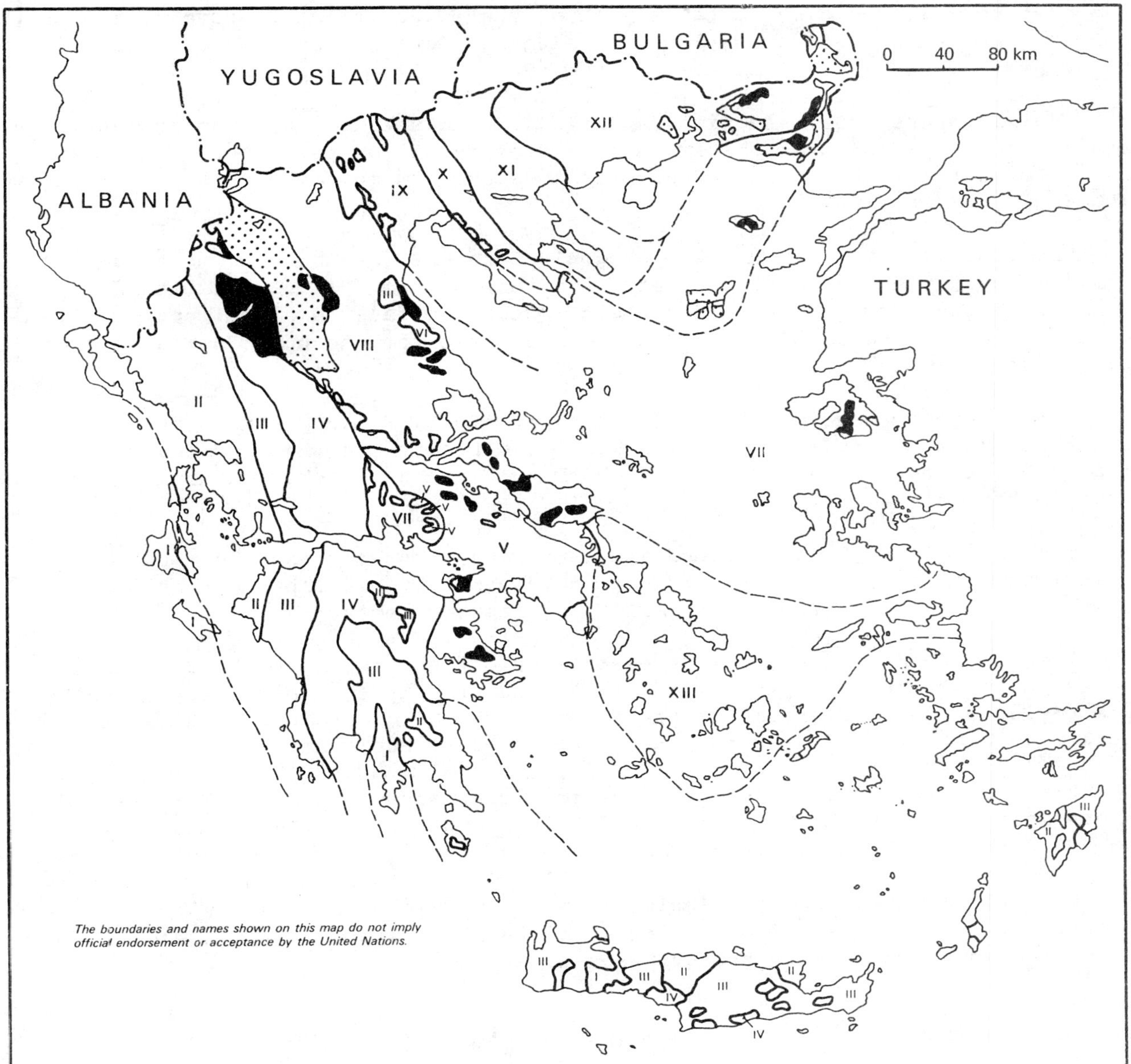

GREECE: GEOTECTONIC ZONES

I	Paxos Zone	VI	Ossa Unit	XI	Serbian-Macedonian Massif
II	Ionian Zone	VII	Parnassus Zone	XII	Rhodope Massif
III	Gavrovo Zone	VIII	Pelagonic Zone	XIII	Attic Cycladic Massif
IV	Pindus Zone	IX	Axios Zone	(dotted)	Molasse
V	Beotian Flysch	X	Circum Rhodope Belt	(black)	Ophiolites (Sub-Pelogonic)

3546.13x

The Pindus Zone shows very plastic tectonics, reflecting the limited thickness and many alternations of its sediments as well as a lack of basement, which had been submerged by stresses related to the tectonic activity taking place to the east. All the sediments were folded so as to display a sequence of overthrusts from east to west and also on the flysch of the neighbouring Gavrovo Zone. In the same way the ophiolites of the late Subpelagonic Zone were thrust over its flysch and nappes.

The Parnassus Zone tectonics resulted in alternating synclines and anticlines with tectonic grabens and horsts. A main feature of the zone is strong faulting traceable to the great thickness of its limestone and dolomite units and relative lithologic homogeneity.

The rigid basement of the Subpelagonic Zone and the thick limestones and dolomites occurring in the lower part, are responsible for the low plasticity and mobility of the zone, which caused the formation of faulted anticlines and synclines, while increased flexibility is observed for the upper units owing to the ophiolites and thin shales of the zone.

The Pelagonic Zone was a stable block of crust, already shaped by the Hercynian orogeny. Evidences of its limited mobility are the wide folds which show great radii of curvature and many of which are faulted. Autochthonous ophiolites occur in the zone.

The Axios Zone shows intermediate tectonics, characterized both by wide anticlines, mainly in the basement, and a sequence of overlapping plates (from east to west) formed due to the inflexibility of its sediments.

The Rhodope massif appears as a stable block, shaped by tectonic events earlier than Alpine. Completion of the Alpine orogeny (mid-Miocene) was followed by a period of less compressing stress but which saw great verticle faulting resulting in horsts and grabens containing Pliocene sediments.

Finally, Pleistocene epeirogeny caused broad submergence resulting in the modern morphology of the Greek mainland. These movements also led to formation of the islands in the Aegean Sea.

Eustatic movement due to glacial advance and recession resulted in a variety of effects, such as the different levels of karstification or appearance of Recent marine sediments well above present sea level.

C. Ground water

1. Management

Water resources management is handled by government organization. The existing legislation is fragmentary and in many aspects insufficient. Nevertheless, several decrees specify the way in which water should be managed and set priorities for the various uses of water, in order to serve society most beneficially.

On the subject of water management, Greek law applicable to ground water covers nearly all cases related to exploitation, exploration and use of water for agricultural purposes. Under the law for agricultural water, a distinction is made between public and private water. As regards irrigation water the law specifies that the Minister of Agriculture regulate its use and management according to the type of cultivation and to the extent, duration and frequency of the irrigation cycles. The same law provides for possible expropriation of any private water surplus to the needs of the land owner. In fact, any private water could be taken over by the State.

In cases of conflicting uses of water and more specifically in the case of water for power and irrigation, a responsible committee decides, provided that the maximum discharge does not exceed 200 litres per second or the installed capacity is not greater than 25 horsepower (18,650 W).

Another reference to water management is made in decree 439/1945 under which restrictions are imposed on excavation of irrigation wells and pumping of ground water, in order to secure a dependable supply for cities, towns and villages.

The quantities of water authorized to be pumped from the existing wells and the duration of these restrictions are decided upon jointly by the Ministers of Agriculture and of Public Works.

Decree 560/1948 provides for the emergency supply of water to residential areas. In such cases the municipal authorities can take over all surplus private water and distribute it according to the specific needs of the population.

Finally, a later decree provides for the case of surplus municipal water that can be taken over by another municipality in cases of severe shortage of the second.

2. Exploration and development

Hydrogeologic investigation is principally carried out by the Institute of Geology and Mineral Exploration (IGME). This institute is responsible for ground-water exploration as an aspect of scientific research, for advising governmental bodies, municipalities and local authorities, in solving a variety of ground-water problems, and responding to public and private questions regarding water use.

A significant number of hydrogeologic projects have been carried out since 1950 and large areas of the country have been thoroughly investigated by the IGME. Specialized programmes are under way on karst hydrology (of great importance for the country), to inventory of all kinds of springs (thermal, mineral, karstic) and in general to monitor its ground-water resources through the use of tracers and related technology.

As a result a goodly amount of hydrologic information is collected, organized and used by IGME. This information is at the disposal of all public authorities, institutions and organizations. Information included in the present paper is based mostly on data from IGME.

Apart from IGME, the Ministry of Agriculture also acts as a service engaged in such ground-water activities as exploration, development and exploitation. The Ministry of Agriculture and the Ministry of Interior, share the task of primary planning and design of ground-water development for agricultural and domestic uses. The development of water resources for hydropower is entrusted to the Public Power Corporation.

The Hydraulic Works Service of the Ministry for Public Works is responsible for the design and implementation of public hydraulic programmes and related ground-water investigations. Finally, many private firms act in the field of ground-water exploration and development on behalf of both private enterprise and public authorities.

Ground-water development is mainly carried out by the private sector. In case of domestic use, this is a task reserved for municipalities and local authorities. In case of thermal water, the exclusive authority is the Greek Tourist Organization. Water supply systems of larger cities including project design, exploitation, treatment, distribution and management are the responsibility of special public utility firms.

3. Methods and techniques

The hydrogeologic information available deals mostly with karst hydrogeology, karstic and other spring hydrology, quantity and quality of ground water, monitoring and ground-water hydrology. The lithologic complexity of the country, an effect of its complex geologic structure, has frustrated efforts to obtain comprehensive, accurate and useful hydrologic data country-wide.

The most widely used methods and techniques to estimate ground-water parameters are, more or less, the following: the piezometric level is determined by measuring the depth of water in wells, provided that elevation and position are known with accuracy. Although a few automatic water level recorders are in use, such measurements are usually made manually, with the density of measurements depending on water fluctuation. Discharge of pumped wells is usually measured volumetrically; flowmeters are used on streams and springs. In order to assess transmissivity and the storage coefficient, long-duration pumping tests of boreholes and wells are systematically included. The development of ground-water resources includes abstraction from boreholes, large-diameter wells and springs.

Borehole depth shows great variation, usually in ranges from 100 m to 200 m, although ever-deeper boreholes are needed to exploit some aquifers (mainly karstic). Regular sampling is taken at selected sites to determine ground-water quality. Continuous monitoring and recording of quality is obtained with electronic instruments at a few sites.

In most cases infiltration of water is estimated indirectly when computing the basic balance equation of the hydrologic cycle. Because the infiltration value is related to precipitation, surface runoff and evapotranspiration, the accuracy obtained is dependent on the accuracy of each of the many components of the equation.

Lysimeters for direct estimation of infiltration are used in a few cases only. This method is, in fact, applicable only in measuring flow through porous media. For the rest, represented by fissure flow systems (karstic) common throughout the country, direct infiltration estimation might introduce serious error and has little practical application.

Evapotranspiration estimates usually introduce the most serious error in water balance calculations. Of course, this element is closely related to climate, soil and vegetation, all of which can show local variations. Attempts have been made to establish methodologies valid in the varied country conditions. Accordingly, empirical formulae are usually utilized, although evaporation pans to measure evaporation losses are in occasional use. Surface runoff is measured in most cases.

Standard gauges are widely used to measure rainfall and are read daily or weekly. A few recording gauges of the tipping bucket type are in use. About 830 rain gauges are monitored regularly by different government services, providing a coverage density of one gauge per 150 km^2, approximately. Permanently installed and in use for a long time, these gauges produce rainfall information which, while perhaps not optimal, is sufficient for reliable evaluation and calculation.

4. The aquifer systems

From the hydrolithologic point of view, four units may be distinguished for the country (figure 23).

1. All carbonate formations (pure limestone or dolomitic and marly) have generally high permeability,

2. Neogene and Quaternary deposits are the principal porous media and are characterized by locally high to locally low permeability,

3. Igneous rocks show locally medium permeability, and finally

4. Basement rocks (mainly schistose members) and flysch are practically impermeable.

Intergranular and fissure flow developed in the porous media and carbonates, respectively, represent the two prevailing aquifer systems. Apart from carbonates, other rocks such as igneous rocks, if fractured, also contribute to the fissure aquifer system.

The following relations exist between aquifer systems and surface waters: the porous aquifers of river basins exchange water with rivers to a great extent. In winter time surplus river waters feed the underground strata, while during summer, rivers drain water from the aquifers. The main surface water reservoirs, both natural and artificial, are used for water supply or hydropower. Inflow to and outflow from these storage reservoirs are fully controlled by man, hence, their effect on the aquifers cannot be considered as resulting from natural conditions.

5. Aquifers in porous media

Aquifers in porous media are mostly developed in Quaternary deposits of the lowlands and in Neogene basins. These occur mainly in the eastern half of the country, which has the more extensive lowlands. The lowlands are filled by post-Alpine sediments which consist, for the most part, of gravel, pebbles, sand, silt and marl, all belonging to rather moderate granulometric phases. Coherence of the members of these aquifers ranges from medium to low. Where the aquifers are in carbonate environment, or near their origin, increased coherence is observed locally.

Figure 23

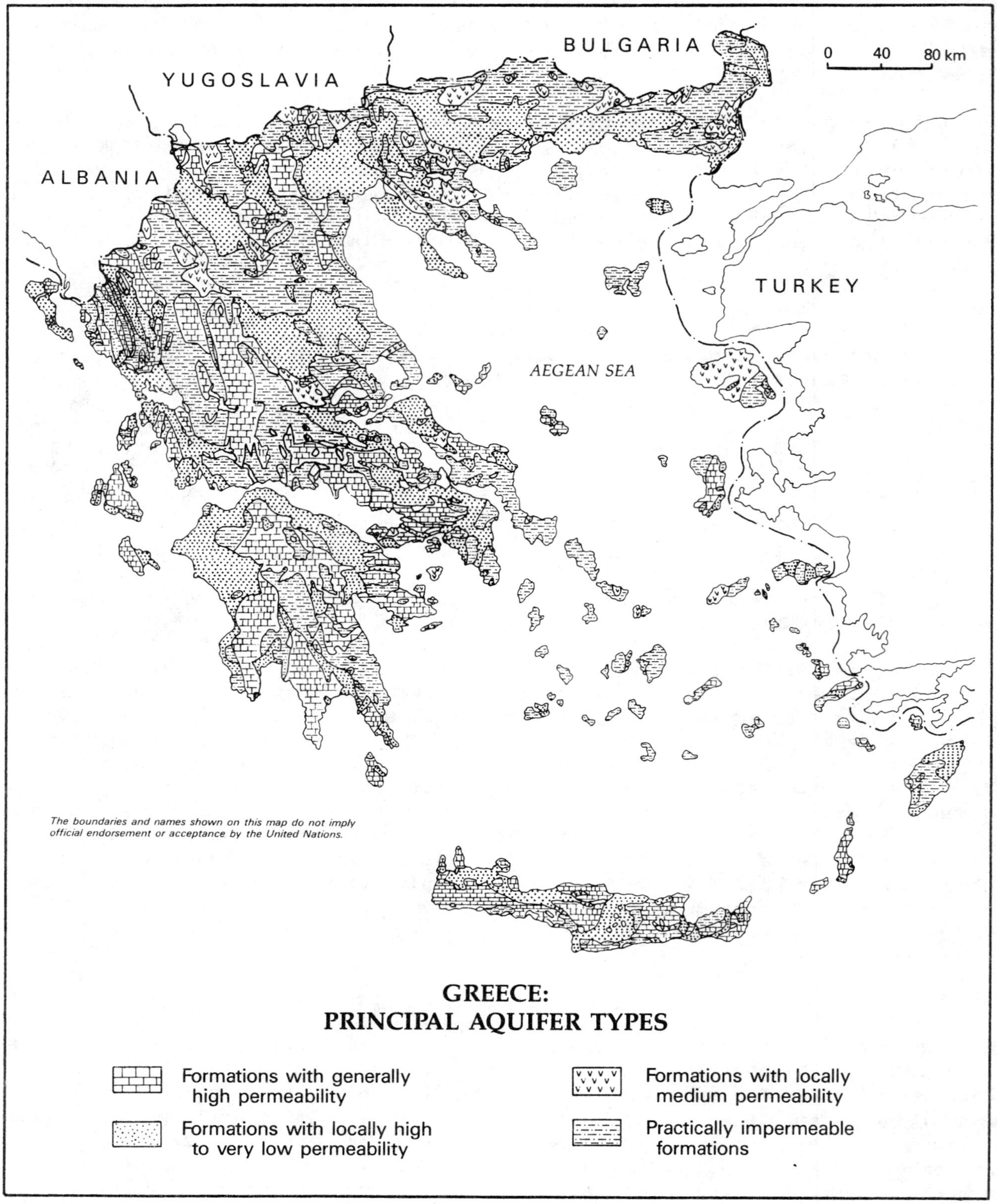

3546.14x

A thickness of a hundred metres for the system is identified locally. Average thickness increases greatly where Neogene sediments underlie Quaternary formations. Impermeable silt and clay layers, which occur mainly in the deeper parts, may create confined or semiconfined conditions. In most cases, boreholes drilled in such formations deliver substantial yields of water. In 90 per cent of the porous media, the specific capacity ranges from 3 to 5 $m^3/h/m$, and the transmissivity ranges from 10^{-2} to 10^{-4} m^2/s.

The storativity of the porous aquifer systems, for the most part is less than 10 per cent, where unconfined conditions exist. The storativity of confined members usually averages between 10^{-3} and 10^{-4}. The water table mostly shows a rather limited annual fluctuation, less than 3 metres, and is generally encountered at shallow depths, usually less than 20 metres. Aquifers in porous media were, in many cases, quickly exhausted as they were the first to be developed.

During the period 1950-1960, about 2,500 boreholes were drilled in such aquifers, mainly to meet increasing agricultural and industrial demands. Furthermore, the number of operating boreholes and wells increases continuously. Overdevelopment of some aquifers caused a permanent lowering of water tables, or in other cases, where there is hydrologic continuity with the sea the water quality deteriorated due to mixing with the salty sea water.

6. Aquifers in fractured rocks

Rocks in which fracture-flow aquifers exist, occupy about 35 per cent of the country. These aquifers are formed in carbonate rocks (including metamorphic, non-metamorphic, pure, marly, and dolomitic). Fracture-flow aquifers predominate in the south-west half of the country, but are common elsewhere. On the mainland and Peloponnesus they are found in hydrologic continuity with the sea except where the carbonate masses are isolated in an impermeable environment.

Karst is found both lower and higher than present sea level for such reasons are epeirogeny and eustasy. Karstic aquifer systems in Greece are ruled by the degree of fracturing, the enlargement of fractures due to solution and faulting. Fracturing is a principal feature. The thick carbonate strata show the greater fracturing effect, the enlargement depending directly on the calcium carbonate content of the rock.

Carbonate rocks formed in deep-sea environments are rather thin-bedded and marly and are poorly fractured, whereas carbonates of shallow deposition are generally well fractured. As a rule, deep zones overthrust the shallow ones. The distribution of carbonates in space, is well explained by the tectonic pattern, described earlier as a sequence of overthrusts from east to west of the different geotectonic zones. The major tectonic effects also contributed to local dipping of the carbonates and also defined the orientation of their main fracture systems. The role of faults seems to be significant in providing the hydraulic conditions for fissure-flow aquifers.

Storativity and transmissivity reflect the secondary fracture network and depend directly on the size and number of fissures, fractures and solution forms. Transmissivity values are around 15 X 10^{-3} m^2/s and the average yield of 90 per cent of all boreholes is 100 m^3/h.

In many cases, salt contamination problems due to hydraulic continuity of the sea with the aquifers enforced the imposition of low abstraction rates for boreholes, which could otherwise be operated under higher degrees of discharge. The specific capacity of boreholes is less than 15 m^3/h/m in only 15 per cent of the cases; in more than 60 per cent of the boreholes the specific capacity ranges from 15 to 200 m^3/h/m, below the critical yielding capacity. For the most part, boreholes drilled in the system average 100 metres in depth. Owing to overdevelopment, however, aquifers as deep as 300 metres or more are being tapped. Annual fluctuation of the water table varies between 1 and 10 metres depending on the local tectonics and hydrogeology.

Karstic aquifers are of great significance for the country, since they represent the more extensive and, for hydrogeologic reasons, the better replenished as well. Hence, great effort is made to develop these aquifers in all their potential. Thus, karstic springs discharging aquifers protected from sea intrusion are sought, also karstic aquifers free of contamination problems, provided that elevation, topography and other factors do not raise costs too high. Brackish and submarine springs discharging karstic aquifers in hydraulic continuity with the sea can be exploited if care is taken to abstract fresh water before it can mix with the sea.

Annual replenishment of karstic aquifer systems is calculated to be 12×10^9 m^3 of water, as product of the areal extent of the system multiplied by 0.25 m, which is taken as average infiltration. Assuming that annually 7.5×10^8 m^3 per year is needed for domestic use, 60×10^8 m^3 per year for agriculture, and 6×10^8 m^3 per year for industry, then a total water supply of approximately 7.5×10^9 m^3 per year is necessary, which is less than annual replenishment of the system; hence the importance of the karstic system for the country. Nevertheless, increasing development activities will lead to the need for more water, possibly in excess of this annual replenishment. If such water deficits occur, artificial recharge should be adequate to meet the needs.

D. Water districts

Boundaries of the 14 water districts of Greece are shown in figure 24, to which figure the number in parentheses in the following descriptions refer.

1. Western Peloponnesus

Impermeable rocks occupy the 30 per cent of the Western Peloponnesus (7,120 km^2) water district (01), the rest, 70 per cent being divided onto porous aquifers (18 per cent), fissure aquifers (34 per cent) and mixed flow aquifers (18 per cent). The annual replenishment of porous, fissure and mixed flow aquifers is estimated to be $1,610 \times 10^6$ m^3, 200×10^6 m^3 and 72×10^6 m^3, respectively, for a total of $1,882 \times 10^6$ m^3 per year. An amount of 340×10^6 m^3 per year is discharged through 60 springs, 52 of them karstic. The rest of the ground-water demand is covered by means of boreholes. Total abstraction is 607×10^6 m^3 per year of which 25×10^6 m^3 per year, 560×10^6 m^3 per year and 22×10^6 m^3 per year are consumed by the urban, agricultural and industrial sectors respectively. Additional potential resources are estimated to be about 650×10^6 m^3 per year.

Figure 24

GREECE:
WATER DISTRICTS

1	Western Peloponessus	8	Thessaly
2	Northern Peloponessus	9	Western Macedonia
3	Eastern Peloponessus	10	Central Macedonia
4	Western Mainland	11	Eastern Macedonia
5	Epirus (M. Pindos)	12	Thrace (M. Rhodope)
6	Attica	13	Crete
7	Eastern Mainland	14	Aegean Islands

3546.15x

2. Northern Peloponnesus

Impermeable formations cover less than 5 per cent of the Northern Peloponnesus (02) (5,946 km^2). About 46 per cent of the area is occupied by porous media formations, which show a permeability ranging from 10^{-3} to 10^{-4} m/s. Infiltrated water is estimated to be about 380 X 10^6 m^3 per year, assuming that the infiltration rate is 15 per cent of precipitation. Fissure aquifers are developed in the rest of the district area. An amount of 1,360 X 10^6 m^3 infiltrates the fissured aquifer system. An amount of 367 X 10^6 m^3 per year is discharged through 24 karstic springs. The water demand for irrigation in the lowlands is estimated to be 460 X 10^6 m^3 per year, of which 340 X 10^6 m^3 is from surface water, the rest from aquifers. Domestic needs require 35 X 10^6 m^3 annually and industry consumes about 68 X 10^6 m^3 per year. Taking in account that total abstraction is 223 X 10^6 m^3, the water resources are adequate.

3. Eastern Peloponnesus

Impermeable rocks occupy 34 per cent of the Eastern Peloponnesus (03), 12 per cent is reserved for deposits with intergranular flow, the latter showing an average thickness of 300 m. The infiltration coefficient is about 15 per cent of precipitation. Hence, the amount of infiltrated water is about 80 X 10^6 m^3 per year. Half of the district area is underlain by carbonates. An amount of 1,680 X 10^6 m^3 of water infiltrates the karstic aquifer system annually. A great quantity of water is discharged from springs, most of them the karstic type, estimated to be about 850 X 10^6 m^3 per year. In addition, important submarine springs exist along the eastern coast. A total of 770 X 10^6 m^3 per year is used for irrigation, abstracted from boreholes and springs, and 15 X 10^6 m^3 is used for livestock. For the water supply of urban areas, 20 X 10^6 m^3 of water is used annually, abstracted from springs and boreholes. About 25 X 10^6 m^3 of water is consumed by industry. Thus, a total abstraction of 830 X 10^6 m^3 takes place annually, the potential resources being about 960 X 10^6 m^3 per year.

4. Western Mainland

An aquifer of low potential and at a depth of 4 to 15 metres, is developed in the Western Mainland (04) lowlands. Free karstic aquifers feed coastal and submarine springs, which are found along the coast. In mountainous areas, a deep karstic aquifer system discharges through springs with an average yield of 100 l/s or more, although karstic springs with yields as high as 5,000 m^3/h are also common. Two geotectonic zones occur in the district. In the Pindus Zone, vertical flow prevails whereas in the low karst of the Ionian Zone considerable horizontal flow occurs, leading to development of a water table with a level at the base of the karst. Maximum spring discharges occur in the Ionian Zone, ranging from 40 to 14,000 m^3/h. In the Pindus Zone, discharges range from 40 to 5,000 m^3/h.

5. Epirus

The areal extent of the Epirus (05) district is 8,693 km^2. Three types of aquifer are found: (1) Porous, (2) Fissure and (3) Mixed flow. These comprise 10 per cent, 26 per cent and 6 per cent of the district, respectively, the rest being occupied by impermeable flysch. The porous aquifers are found in Neogene and

Recent deposits of river basins, mainly located in the southern and western part. They occur also in the Ioannina basin and in the Konitsa plain to the north. In the Konitsa basin, an unconfined karst aquifer may exist below the sediments; deep circulation and a geothermal field are also probable. Fissure-flow aquifers are developed in various units of the Ionian and Pindus zones, found nearly throughout the district. Deterioration of water quality occurs locally, due to gypsum.

6. Attica

Attica (06) is the most populous part of the country. The different aquifer units of Attica are discussed separately owing to the complex structure, geomorphology, local conditions and problems. In the Quaternary and Neogene deposits of Athens, Elefsis, Megara and Mesogeia basins, porous aquifers occur, with the following hydrogeologic conditions:

In the Athens basin, ground-water levels occur at an average maximum elevation of 450 m. The hydraulic gradient ranges from 0.06 to 0.01. The maximum depth to water table is nearly 20 m. Maximum annual water-level fluctuation is less than 2 m, over 95 per cent of the area. The average critical yield of boreholes and large diameter wells is 25 m^3/h. Transmissivity varies between 1.5×10^{-4} m^2/s and 5×10^{-3} m^2/s, and abstractions are 2.2×10^6 m^3 per year.

In the Elfsis basin, the water table is found at a maximum elevation of 4 m. The mean annual fluctuation is 1.1 m and the hydraulic gradient ranges from 0.001 to 0.0002. The critical yield of large-diameter wells shows great variations ranging from 5 m^3/h to 70 m^3/h. Transmissivity values are determined to be from 2×10^{-4} m^2/s and 3.5×10^{-3} m^2/s, and abstractions amount to 8.7×10^6 m^3 annually.

In the Megara basin, semi-confined aquifers are developed in the north-west part and phreatic aquifers in the south-east part. Mean annual fluctuation of the water table is locally found at about 1.8 m. Overpumping, especially in the north-west part, causes a serious decline of the piezometric surface, i.e., about 1.5 m per year.

In the Mesogeia basin, ground water is locally found under semi-confined conditions in the central area. Water levels are 10 m to 220 m above sea level, with annual fluctuation being about 1.2 m. Boreholes and large-diameter wells usually yield between 10 m^3/h and 40 m^3/h. Transmissivity ranges from 2×10^{-4} m^2/s to 9×10^{-4} m^2/s, and abstractions are 22×10^6 m^3 per year.

In the northern graben areas (Marathon), a phreatic water table discharges to the sea. Depth to the water table ranges from 5 m to 90 m in the north, and from 5 m to 130 m in the south. The mean annual water-level fluctuation is 1.6 m. Critical yields of boreholes and wells vary from 5 m^3/h to 50 m^3/h. Transmissivity is between 6×10^{-4} m^2/s and 1.4×10^{-3} m^2/h. Water abstractions amount to 6.5×10^6 m^3 per year.

Impermeable substratum and nature of the fissure network are the two main factors governing the hydrogeology of karstic aquifers, which can be divided to those developed in West and East Attica separately.

In West Attica, the drainage conforms to the major anticlinal axis of the Parnis and Kithairón mountains. Two thirds of the area drains to the north, the rest to the south mostly through large fault systems.

In West Attica, circulation of water towards the saturation zone, which is at about sea level, is rapid due to the high slope gradient of the impermeable substratum. Furthermore, a mixing trend with sea water is pronounced.

During the dry season, a water table forms in the saturation zone at about 10 m above sea level. Depths to it range from 200 m to 300 m. The mean annual water table fluctuation is 2 m, the yields of boreholes range from 30 m^3/h to 150 m^3/h, and transmissivity values range from 1.5 X 10^{-2} m^2/s to 5 X 10^{-3} m^2/s. Finally, water abstractions come out to be 6 X 10^6 m^3 per year.

In East Attica, the principal karstic aquifer is the so-called lower marble. The water table is at an average maximum elevation of 5 m in the dry season, and has a mean annual fluctuation of 5 m. Depths to the saturation zone are between 100 m and 230 m. In the upper marble, the aquifer is at depths less than 150 m, depending on the position of their impermeable substratum. Transmissivity is 2 X 10^{-3} m^2/s to 1 X 10^{-2} m^2/s and boreholes yields are from 10 m^3/h to 80 m^3/h. Total water abstractions from all marble aquifers are about 13 X 10^6 m^3 per year. The annual replenishment of the porous system of Attica is 53 X 10^6 m^3. The ground-water deficit of the system is 12.4 X 10^6 m^3 per year, since annual withdrawal is 65.4 X 10^6 m^3. Not all the basins described share this deficit equally. The Elefsis, Megara and Marathon areas are the most affected, due to overpumping. Because the annual rate of increase in abstractions is 1.8 X 10^6 m^3, most of the porous aquifers would very soon be fully depleted or made irreversibly saline by the sea unless remedial steps are taken. The fissure aquifer system of Attica produces 136 X 10^6 m^3 per year as surplus water, given that the annual replenishment of the system is 156 X 10^6 m^3, while the abstractions are 20 X 10^6 m^3 per year. The greater Athens area is adequately supplied from lakes and artificial reservoirs of surface water so that the abstractions cited above are for the rest of Attica, including domestic use, agriculture and industry.

7. Eastern Mainland

The Eastern Mainland (07) covers 11,688 km^2. Porous rocks occupy 20 per cent of the area and multilayered aquifers are found between the surface and 300 m. Transmissivity ranges from 10^{-3} to 10^{-6} m^2/s, and boreholes drilled in the porous rocks yield between 5 m^3/h and 150 m^3/h. Carbonate rocks occupy 27 per cent of the area and their water tables are virtually at sea level. Transmissivity is mostly in the range of 10^{-2} m^2/s to 10^{-4} m^2/s. Boreholes in the karstic systems yield from 80 m^3/h to 140 m^3/h. Two boreholes in the south-western part of the district yield as much as 500 m^3/h. Total annual abstractions are estimated to be from 80 X 10^6 m^3 to 100 X 10^6 m^3. The volume of water taken from the basin of the Sperchios River alone is as high as 70 X 10^6 m^3 per year.

8. Thessaly

The Thessaly area (08) covers 14,000 km^2. Aquicludes cover 30 per cent of the area and the remaining 70 per cent has a mix of pore (38 per cent), fissure (11 per cent) and mixed-flow (21 per cent) aquifers. The permeability of the porous aquifer system ranges from 1 X 10^{-4} m/s to 1 X 10^{-3} m/s. The annual infiltration to the aquifer is 850 X 10^6 m^3 per year. The amount of water annually infiltrating the karstic and mixed-flow aquifers is 1,185 X 10^6 m^3 and 114 X 10^6 m^3, respectively. The annual replenishment is 2,149 X 10^6 m^3 resulting in a water surplus of 1,000 X 10^6 m^3, given that abstractions are 1,149 X 10^6 m^3 per year. Of the water drawn each year, agriculture uses 1,050 X 10^6 m^3 per year, industry 49 X 10^6 m^3 per year and domestic use 50 X 10^6 m^3 per year.

9. Western Macedonia

Western Macedonia (09) covers 13,033 km^2. Porous aquifer systems prevail in 29 per cent of the district, namely, in the basins of Florina (north), Ptolemais-Amideo (central) and along the eastern part of the district. Fissure aquifer systems occur in 21 per cent of the district, and in the south-western 15 per cent, mixed-flow aquifers prevail. Aquicludes, mainly crystalline, occupy 35 per cent of the district.

The following conditions predominate in each of the main hydrologic units of this district. In the central part, the Vernon and Askion mountains, a low-potential aquifer discharges mostly through small springs. Towards the south of the central area is an extended karstic aquifer whose replenishment is 242 X 10^6 m^3 of water annually and abstractions 6 X 10^6 m^3 per year. Towards the eastern part of the district two significant hydrologic units are related to Mount Vermion. The first, north of the mountains, is a karstic aquifer replenished by 90 X 10^6 m^3 per year and flowing into Lake Nisi and the Kara Kayia springs. The water flow is principally used for hydropower. The second is holokarst in the Klidion area. Discharge is to Lake Vegoritis. Annual replenishment is 350 X 10^6 m^3 per year and abstraction is limited to 30 X 10^6 m^3 annually. In the northern part of the district, the Florina basin, a significant production of natural gases (CO_2) occurs. Hence, all ground water of the basin is acid. The porous aquifers of this basin have low transmissivity, an average of 5 X 10^{-2} m/s. The annual replenishment is about 22 X 10^6 m^3 and abstraction 4 X 10^6 m^3 annually.

In the Ptolemais-Amideo basin (central part of the district) the following aquifers are distinguished. In the Amideo area are significant porous aquifers. The average transmissivity is 2 X 10^{-3} m^2/s and the average annual fluctuation of water level is 3-4 m. The amount of ground water abstracted annually does not exceed 1 X 10^6 m^3 and potential resources amounting to 19 X 10^6 m^3 per year. A new hydroelectric power plant will require 15 X 10^6 m^3 per year. In the Ptolemais area, a confined aquifer has transmissivity values ranging from 4 X 10^{-4} m^2/s to 5 X 10^{-2}/s. Annual replenishment is about 6 X 10^6 m^3. But abstractions for nearby villages and industries amount to 10 X 10^6 m^3 per year. This deficit results in a drawndown of 0.8 m per year.

Finally, in the eastern part, namely, the Sarighiol basin, a centrally located aquifer has transmissivity ranging from 5 X 10^{-2} m^2/s to 4 X 10^{-4} m^2/s. Annual replenishment is 10 X 10^6 m^3. Potential resources are not more than 2 X 10^6 m^3 per year.

10. Central Macedonia

The district of Central Macedonia (10) covers 10,987 km^2. It is divided geomorphologically into the Axios Valley and Chalkidiki Peninsula. In the western part lowlands occupy 36 per cent of the area, mostly along rivers. Porous lowland aquifers are found in the basins of Thessaloniki-Giannitsa and of Almopia. Fissure-flow aquifers cover 7 per cent of the district. They are mainly located in the Chalkidiki and Almopia areas and around Mount Paikon. The area has few mixed-flow aquifers.

Two types of aquifer are distinguished in the Thessaloniki-Giannitsa basin. First are thick and extensive volcanic tuffs in the northern part. Their replenishment is about 21 X 10^6 m^3 of water per year. Second are confined and semi-confined aquifers, the result of impermeable clay intercalations in Quaternary alluvium. Confined conditions prevail in the beds of ancient Lake Giannitsa and in the southern part of the basin. Three overlapping artesian aquifers exist. The first is 5 m thick at 30 m depth; the second is 10-12 m thick at 90-110 m depth; and the third is 15 m thick at 150 m depth. Boreholes 150 m deep, with unconfined flow, yield 100-120 m^3 per day. In this basin also are three unconfined aquifers. The first is at 20-30 m depth, is 3-4 m thick and its water level fluctuates 0.5 m. The second is at 40-50 m depth, is 5 m thick and its water level fluctuates 0.7 m. The third is at 100 m depth, is 5 m thick and its water level fluctuates 1.5 m. The third is the most important; boreholes to it yield up to 170 m^3 per day.

The main aquifer of Mount Paikon discharges through a spring located on the south slope. Its annual is about 90 X 10^6 m^3. This is one of the principal water sources for the city of Thessaloniki. The basin of Almopia is a separate hydrologic unit. Significant springs flow at the margin of the basin and in its flat area. Some are thermal. The total annual discharge is about 50 X 10^6 m^3. Other aquifers here, both confined and unconfined, provide the area with virtually unlimited supplies of water. Finally, total annual abstractions from the basins of Thessaloniki-Giannitsa and of Almopia are about 200 X 10^6 m^3 of water.

11. Eastern Macedonia

The Strymon river basin constitutes nearly the entire Eastern Macedonian water district (01), which covers 7,416 km^2. The district is 45 per cent lowlands. Of the district's aquifers, 34 per cent are porous, 8 per cent fissured and 1 per cent mixed. Its principal aquifers are in the basins of Serres and Drama. Annual abstraction from them are up to 150 X 10^6 m^3 in the Serres basin and 200 X 10^6 m^3 from Drama (exclusive of springs). Irrigation, applied May through October, accounts for 60-70 per cent of the demand. Potential annual resources are 103 X 10^6 m^3 for Serres and 95 X 10^6 m^3 for Drama.

12. Thrace

The Thrace district (12) covers 11,054 km^2 and has two major rivers, the Evros and Nestos. The district is about 45 per cent lowlands. Of the district's aquifers, 30 per cent are porous, 7 per cent fissured and 10 per cent mixed. Karstic aquifers in marbles of the metamorphic system yield up to 190 m^3/h to springs. Two significant karstic aquifers exist near Mount Lekani. Their annual capacities are 6 X 10^6 m^3 and 38 X 10^6 m^3 of water, respectively. Replenishment of the fissured aquifer system is about 223 X 10^6 m^3 of water annually. The principal porous-flow aquifers are in the Tertiary basins of Orestiada (north), Alexandroupoli (south) and Xanthi (central). Others are in the plains along the Evros and Nestos rivers and in the Nestos delta. In the Nestos River area, evaporation, surface runoff and infiltration coefficient were 60, 18, and 22 per cent of precipitation, respectively, in a year when precipitation was 658 mm. In the Evros River area, evaporation, surface runoff and infiltration were 70, 23, and 7 per cent of precipitation, respectively, in a year when precipitation was 755 mm.

Finally, in all lowlands (4,971 km^2) of the district, the following values were calculated for components of the hydrologic balance equation: evaporation, 2,283 X 10^6 m^3; surface runoff, 720 X 10^6 m^3; and infiltration, 509 X 10^6 m^3 of water.

13. Crete

The island of Crete (13) which covers 11,200 km^2 is mountainous and has widespread carbonate rocks of high transmissivity. Storativity is also high, due to the pronounced karstic nature of the fractured rocks. Carbonate rocks, therefore, constitute the major water reservoir of the island. Karstic aquifers feed high-yielding springs located at various altitudes; even low-yielding springs discharge 1-4 m^3/s. Some are submarine or brackish. Better hydrogeological conditions and greater rainfall in the western part of the island gives it a greater water-resource potential.

In Quaternary deposits, mainly in the plains along the northern coast and the Messara and Ierapetra areas to the south, are shallow water tables with ample water potential to satisfy local demands. Near the sea, overpumping has resulted in salt-water intrusion of the aquifer. Although total water abstractions are not well known, additional resources are to be expected within the karstic system which has not yet been thoroughly exploited. The alluvial aquifers, however, have already been overdeveloped to serve primarily the tourist industry and agriculture. Further development of the karstic system is necessary if already evident needs for water, particularly during the dry season, are to be met.

14. Islands of the Aegean Sea

Two major groups of islands, the Cyclades and Dodecanese, and many individual islands constitute the Aegean Islands district (14). More than 1,000 islands and islets are scattered in the Aegean Sea; some of the better known include Lesbos (1,630 km^2), Rhodes (1,398 km^2), Chios (842 km^2), Lemnos and Samos (476 km^2) and Naxos (428 km^2). The main hydrogeologic feature of the islands is the limited occurrence of aquifers. Carbonate rocks, if contained within impermeable barriers, yield to springs at various altitudes. Carbonates open to the sea constitute aquifers much of whose water is lost. Open karstic aquifers in continuity with the sea and feeding coastal brackish and submarine springs have been identified in the islands of Chios, Syros, Samos, Ikaria, Calymnos, Sifnos, Antiparos, Paros, Naxos, Sikinos, Amorgos, Astypalea, Symi, Tilos and Carpathos. Bodies of fractured crystalline rocks yielding water under pressure due to marble intercalations exist in the islands of Lesbos, Kea, Kythnos, Serifos, Sifnos, Andros, Tenos, Mykonos, Naxos and Ios.

Porous aquifers occur in the narrow plains of some islands, significantly Lemnos, Lesbos, Chios, Samos and Rhodes. Porous aquifers are best exploited by wells and are drawn upon for domestic use and agriculture. Some have been intruded by sea water because of overpumping. In general, the water resources of islands are fully developed and exploited, and with the influx of summer visitors, the need for more water becomes increasingly apparent. Finally, owing to volcanism and active tectonics of the islands, there are many thermal springs.

References

Angelier, J. Tectonic evolution of the Hellenic Arc since the late Miocene. Tectonophysics, 1978.

Burton, J. D. and N. Papakis. Handbook of karst hydrogeology with special reference to the carbonate aquifers of the Mediterranean region. FAO, 1963.

Christodoulou, Th., G. Kallergis, A. Morfis, N. Kourmoulis and Ch. Papaspyropoulos. Hydrogeological investigation in the River Axios basin. IGME, Hydrol. and Hydrogeol. Investigations. Athens, 1975.

Dounas, A., N. Kakavas and N. Tassios. Hydrogeological investigation in the River Sperchios basin. IGME, Hydrol. and Hydrogeol. Investigations. Athens, 1978.

Dounas, A., G. Kallergis, A. Morfis and M. Pagounis. Hydrogeological investigation of karstic brackish springs of Ag. Apostoli area (Kalamos-Attica). IGME, Hydrol. and Hydrogeol. Investigations, 31. Athens, 1980.

Food and Agriculture Organization of the United Nations/SF2. Karst groundwater investigation-Greece. I.G.S.R.

International Association of Hydrogeologists (IAH). Karst hydrogeology. Proceedings of the 12th Intern. Congress, n.d.

Kakavas, N. Inventory of karstic springs in Greece. vol. 6: East Mainland. IGME, Hydrol. and Hydrogeol. Investigations. Athens, 1984.

__________. Results of borehole and pumping research in various aquifers of the River Sperchios Valley. IGME, Hydrol. and Hydrogeol. Investigations. Athens, 1985.

Kakavas, N. and G. Gioni-Stavropoulou. Hydrogeological investigation in the area of Peristeri-Attica Municipality. IGME, Hydrol. and Hydrogeol. Investigations. Athens, 1978.

Kallergis, G., A. Morfis, Ch. Papaspyropoulos and Th. Christodoulou. Hydrogeological investigation in Western Thessaly. IGME, Hydrol. and Hydrogeol. Investigations 8. Athens, 1973.

Kallergis, G. and St. Skayias. Inventory of karstic springs in Greece. vol. 3: Thessaly. IGME, Hydrol. and Hydrogeol. Investigations. Athens, 1980.

Knithakis, M. Inventory of karstic springs in Greece. vol. 4. IGME, Hydrol. and Hydrogeol. Investigations. Athens, 1983.

Kounis, G. D., A. C. Vitoriou, G. C. Gioni and D. C. Dimitropoulos. Evaluation of Groundwater Resources in Greece. Final Report for E.E.C. IGME, Athens, 1983.

Kourmoulis, N. Inventory of karstic springs in Greece. vol. 5: West Mainland. IGME, Hydrol. and Hydrogeol. Investigations. Athens, 1984.

Maratos, G. N. Geology of Greece. vol. 1. Athens, 1972.

Mariolakos, H. Geology of Greece. Laboratory and Museum of Geology and Paleontology Editions, University of Athens, A No. 13, Athens, 1976.

Orphanos, G. and G. Sfetsos. Hydrogeological investigation in Kammena Vourla. IGME, Hydrol. and Hydrogeol. Investigations. Athens, 1975.

Pagounis, M. and G. Gioni-Stavropoulou. Policies and legislation on the protection and management of groundwater in Greece and elsewhere. Proceedings of groundwater protection seminar of the United Nations. Athens, 1983.

Papaconstantinou, A. Hydrogeological study of Florina basin. Athens, 1975.

__________. Hydrogeological investigation of Vermion mass karst. Thessaloniki, 1983.

Skayias, St. Inventory of karstic springs of Greece. vol. 1: Peloponnesus, Zakinthos, Kefalinia. IGME, Hydrol. and Hydrogeol. Investigations. Athens, 1978.

Vergis, E. Hydrogeological investigation in the River Nestos basin. IGME, Athens, 1980.

__________. Hydrogeological investigation in the River Evros basin. IGME, Athens, 1981.

__________. Hydrogeological investigation in the area between N. Karvali and Paradissos. IGME, Hydrol. and Hydrogeol. Investigations. Athens, 1985.

__________. Inventory of karstic springs of Greece. vol. 7: Thrace. IGME, Hydrol. and Hydrogeol. Investigations. Athens, 1986.

VI. HUNGARY

Area: 93,030 km^2 Population: 10.6 million

A. General

The Hungarian People's Republic, in the centre of Europe, is in the central, deepest part of the Carpathian basin. All the country belongs to the Danube River system. The greatest part is covered with plains; not more than 0.8 per cent lies at altitudes higher than 500 m and only 32 per cent is above 200 m (figure 25).

The eastern and south-eastern half of the country is occupied by the Great Hungarian Plain, a sediment-filled basin of the Tertiary Pannonian Sea, with elevations mostly between 85 and 100 m. The Kisalföld Plains in the north-west part of the country and those of the Dráva River in the south-west are smaller.

The area west of the Danube, called Transdanubia, is of higher relief, but even the highest peaks are below 900 m. The central range, crossing the country in SW-NE direction, reaches the greatest altitude, 1,015 m in the Mátra Mountains.

The three main climatic zones of Europe, namely the Mediterranean, the Atlantic and the continental zones, overlap in Hungary, the characteristics of each being detectable, but with the continental influence predominating. This is reflected in cold winters and dry, hot summers with frequent droughts. The annual mean temperature in the country ranges from 7° to 12° C, the thermal sum varying from 2,900 to 3,300 degree (C) days.

In a manner typical of continental climate, the temperatures often fluctuate unpredictably within wide extremes. Although the normal monthly mean temperature in January is only between -1° and -2° C, severe winters are common with temperatures remaining below -5°, or even -10° C for several weeks. Thick ice covers are then developed on the rivers and lakes, which on breakup cause disastrous floods.

The normal average precipitation is 600 mm per year, but the distribution in time is highly variable (figure 26). In some years the annual precipitation may be as little as 300 mm, resulting in disastrous droughts. At the other extreme, as much as 1,200 mm may fall in a year, with inundations due to undrained runoff causing loss of crops over several hundred thousand hectares. The seasonal distribution of the total annual precipitation is also unfavourable, since often a considerable part thereof falls outside the growing season. Over considerable parts of the country potential evapotranspiration values are higher than the average precipitation, characteristic of a semi-arid climate (figure 27).

Owing to its varied topographic features, the country collects waters rushing down from the surrounding higher parts of the Carpathian Basin. Detailed maps made during the first half of the nineteenth century show 2.3 million hectares of marsh area and 1.6 million hectares of periodically inundated meadows used as pasture only (figure 28).

The basin is drained by three major rivers, namely the Danube, the Tisza and the Dráva. The two latter discharge into the Danube outside the country.

Figure 25

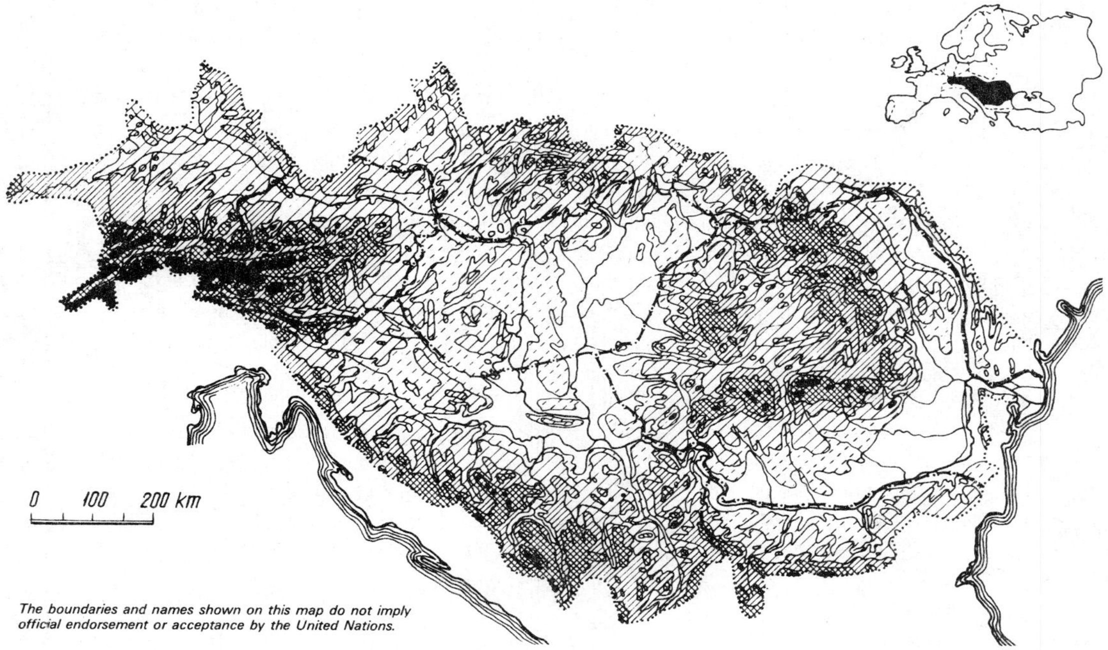

HUNGARY:
TOPOGRAPHY OF DANUBE BASIN

Over 2000 m
1000 to 2000 m
500 to 1000 m
200 to 500 m
100 to 200 m
Under 100 m

3546.16x

Figure 26

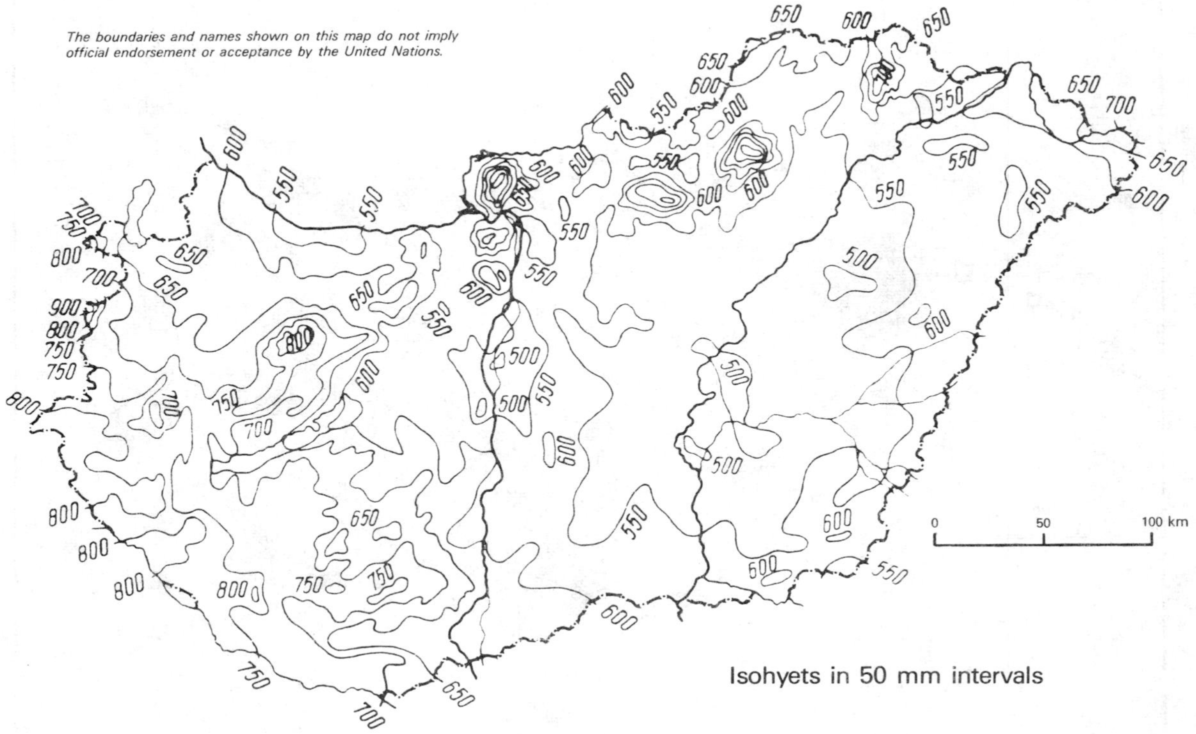

HUNGARY: RAINFALL

3546.17x

Figure 27

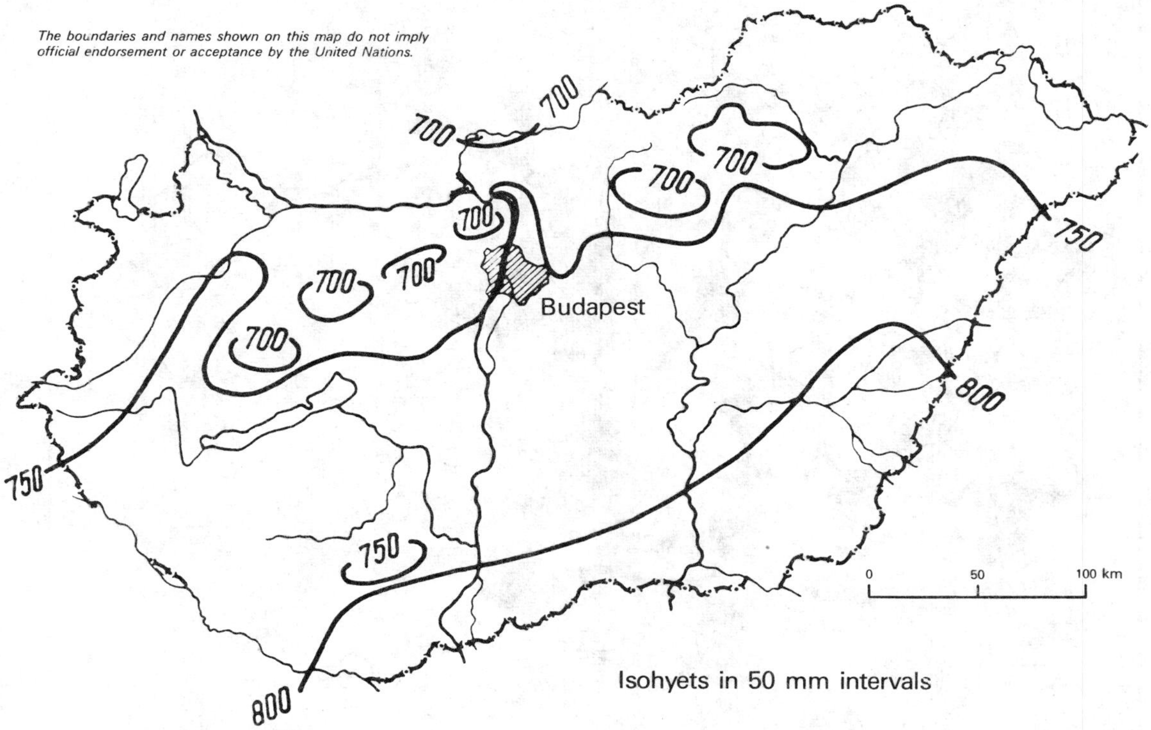

HUNGARY:
POTENTIAL EVAPOTRANSPIRATION

3546.18x

Figure 28

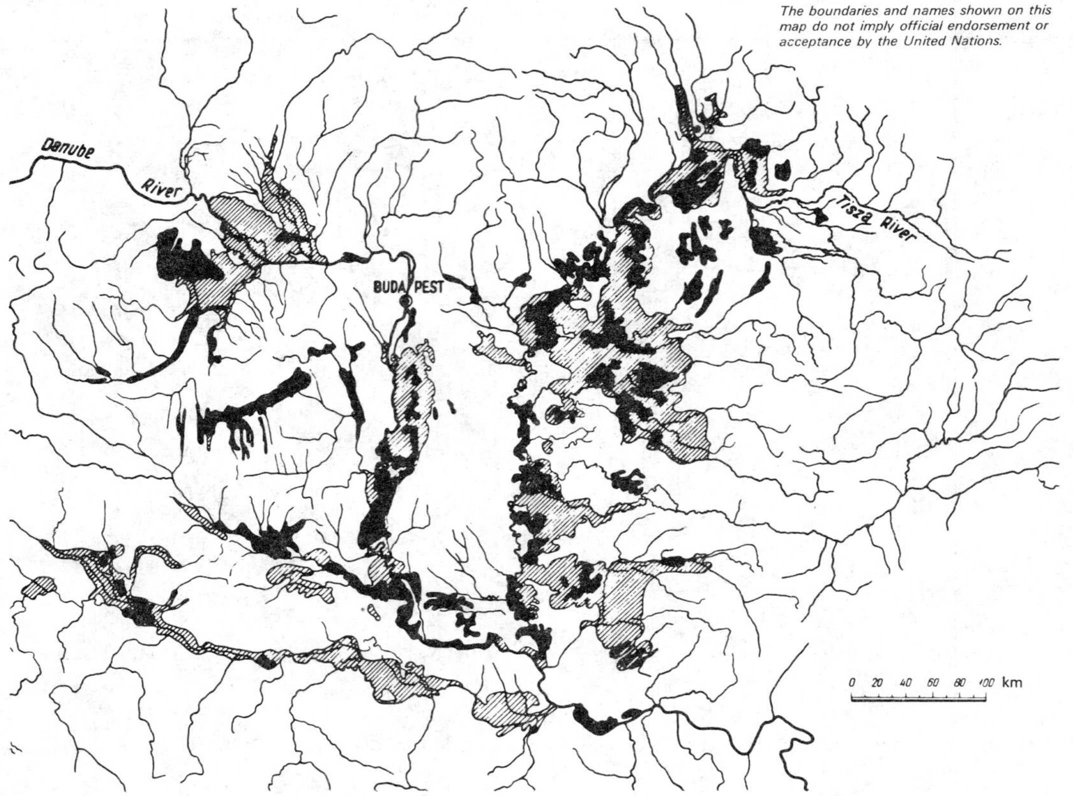

HUNGARY:
FLOODING IN CARPATHIAN BASIN, EARLY 1800's

Area inundated periodically — Perennial marsh

3546.19x

All major streams originate in neighbouring countries. The Danube and its western or right-bank tributaries originate in the Federal Republic of Germany, in Austria and in Yugoslavia while the northern right-bank tributaries of the Danube, further on the Tisza River and her tributaries enter the country from Czechoslovakia, the Soviet Union and Romania lying within the Carpathian Basin (figure 29).

The river pattern in the plains is very sparse due to the low slope of the rivers. The average slope over the Hungarian section of the Danube is 6 cm/km, whereas over the Tisza River it is only 2 cm/km, while the slope over the largest tributary of the Tisza, the Hármas-Körös, is hardly noticeable.

In the historical ages in the Carpathian Basin, owing to the abundance of surface waters, ground water was utilized only in the form of natural springs with high capacities; this is suggested by the remnants of Roman baths in this region. Although flood control dikes had been built as early as the Middle Ages, the most significant development in this respect was achieved in the eighteenth and nineteenth centuries as the result of extended flood and river control works in the Tisza River basin.

Owing to the effect of river regulation and drainage works, surface waters were confined into the river beds or below ground level, thus resulting in the development of near-surface water level in the lowland regions.

In this new situation, large numbers of dug wells were constructed, also as the result of deterioration of the quality of surface waters. Because the quality of the water of dug wells was not satisfactory either, the exploration and exploitation of deeper aquifers by means of drilled wells began, with the drilling of artesian wells by the hundreds, then by the thousands.

Vilmos Zsigmondy was among the first in Europe to have constructed artesian wells and his work was characterized also by the construction of the first well aimed at the exploration of the thermal waters of Budapest, near the thermal springs. Later, about 6 km from the springline, drilling operations explored a thermal aquifer at a depth of 970.4 m yielding water of 79° C temperature. Ever since, this well has yielded thermal water.

As early as 1866 the establishment of artesian wells was based on the hydrogeologic considerations of József Szabó who recognized the relation between cold and thermal karstic water and developed the first schematic model of geothermal flow in karstic media.

In earlier years, the centres of hydrogeologic research in Hungary were the University of Technology, Budapest, and the Hungarian Geological Institute.

Later the centre of research was shifted to the Hungarian Geological Institute, where under E. R. Schmidt, the first hydrogeologic atlas was prepared, illustrating the dynamics of subsurface flow in the confined artesian aquifers of the Hungarian Great Plain.

With establishment of the Research Centre for Water Resources Development (VITUKI) research into subsurface water resources was renewed in the form of a special research unit which became the primary intellectual centre of Hungarian subsurface water research.

Figure 29

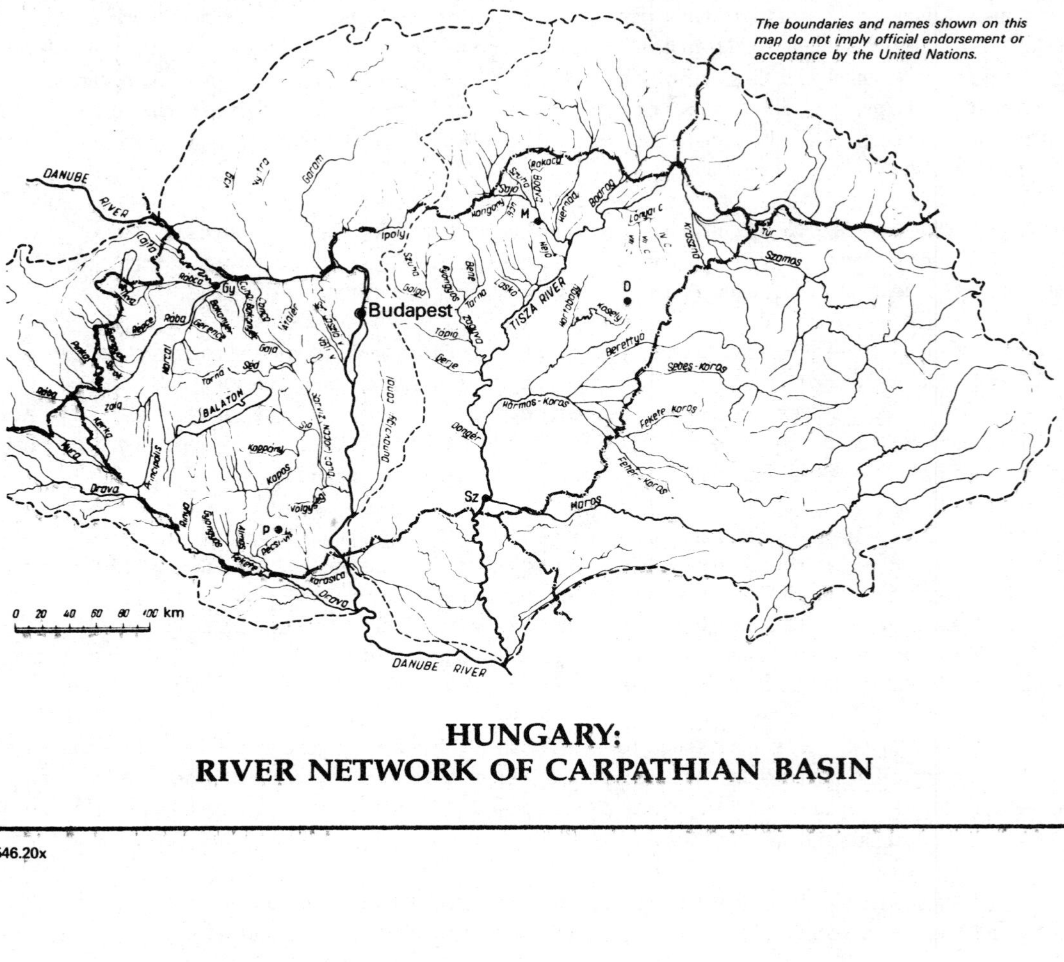

HUNGARY:
RIVER NETWORK OF CARPATHIAN BASIN

3546.20x

B. Hydrogeology

1. Principal aquifers

The mountain ranges surrounding the Carpathian Basin consist of ancient, folded, consolidated and metamorphic formations. Similarly, the bedrock of the basin consists of ancient formations (figure 30).

Streams flowing from the Carpathian Mountains have, ever since the Pliocene, deposited within the basin loose and deep beds of gravel, sand and clay. In some places, such deposits are more than 2,000 metres thick. The bedrock is covered by several hundred metres of marly clay within which are a series of confined artesian aquifers of several hundred metres total thickness. After the Carpathian Mountains had emerged the area was covered with alluvial deposits and gradually converted into a fresh-water lake which eventually filled with sandy deposits providing a series of excellent aquifers (figure 31).

Washout of debris and sediments from the Alps and Carpathian Mountains have continued, creating gravel cones along the piedmont, i.e., along the border of the basin, with alluvial gravel fields along the river itself. Finally, in the west the Pannonian Basin is divided into two parts by a transverse chain of mountains consisting of carbonate-karstic formations 3,000 m thick, also providing a means of communication with the confined aquifers of the basement complex.

Owing to the above outlined hydrogeologic situation, the aquifers of Hungary are closely interconnected and consist of several sub-systems which are hydrodynamically balanced, having fairly different character in respect to their exposure to contamination, that is, the degree of natural protection varies with the local hydrogeologic situation (figure 32).

2. Subsystems

The following five subsystems can be distinguished:

1. The <u>karstic ranges</u> in the western part of the country, to which the extended karstic thermal aquifers of the basement complex are also connected. Under natural conditions the geothermal flow systems, controlled by the infiltrating cold karstic water from the surface, are exposed to pollution only in the zone of infiltration (figure 33).

2. <u>Sediment cones and bank-infiltration systems</u>. Sandy gravel formations along the rivers are of 15-20 m depth on average and, exceptionally, up to 300 m. Owing to the close communication of these formations with the river water, there is a natural exchange of water between the parent river and the aquifer. Generally the gravel aquifer is open to the surface with no obstacle to vertical infiltration.

3. There are aquifers of quite varied character, consisting of Holocene <u>flood silts, sediment cones along the foot of mountains and valley bottom deposits</u>. Their near-surface zone is usually open or confined with a thin impermeable cover and the transitional upper zone of Quaternary deposits is generally 10 m deep with a maximum depth of 30 m.

Figure 30

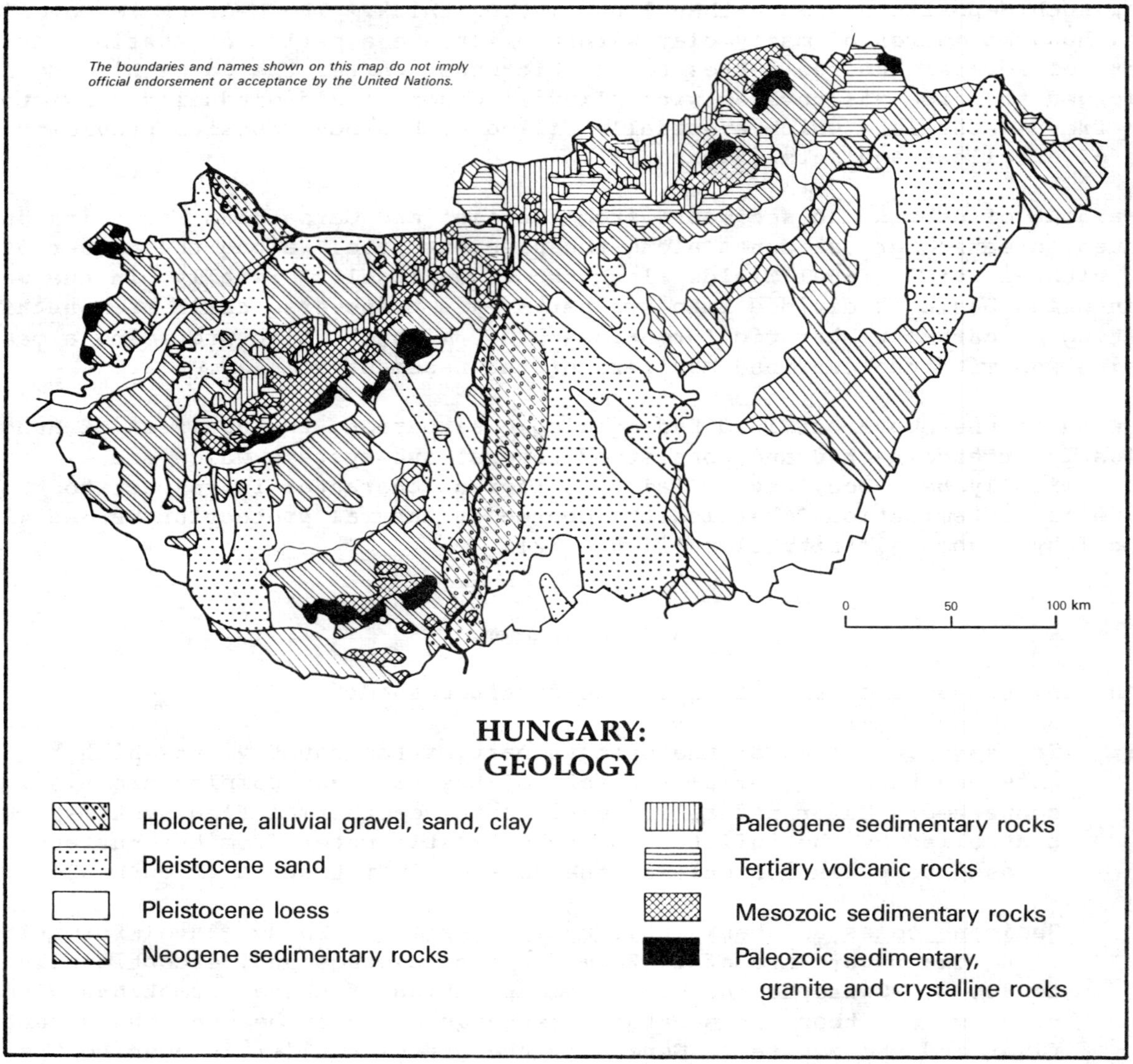

3546.21x

Figure 31

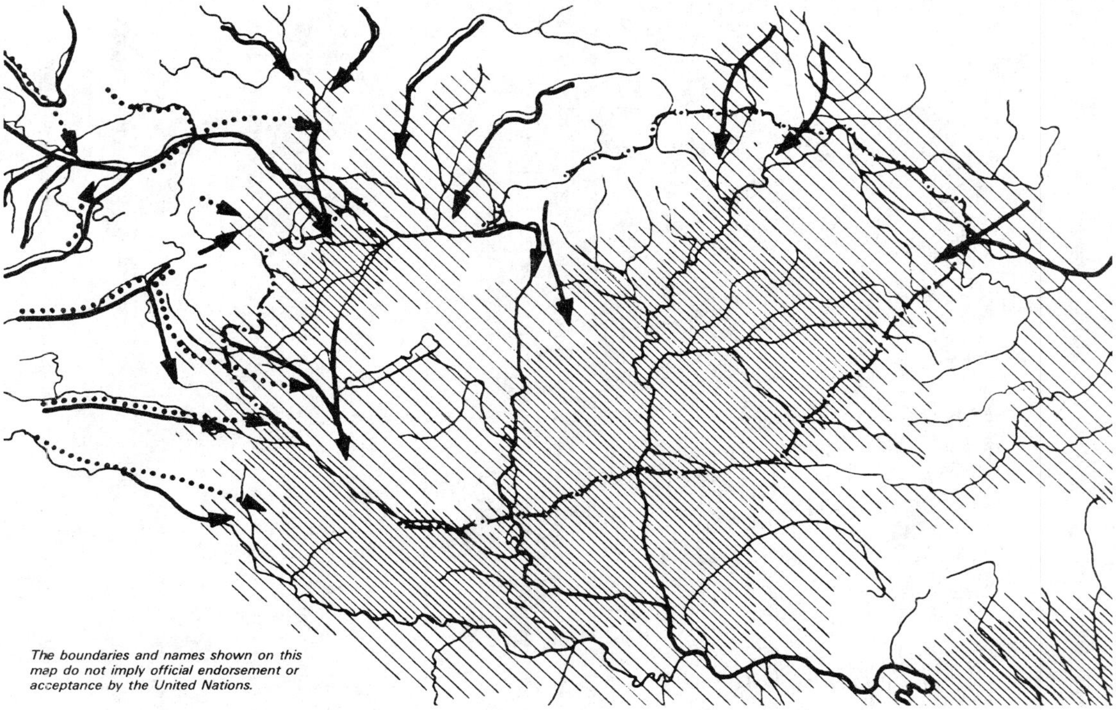

HUNGARY:
PLIOCENE - PLEISTOCENE ALLUVIAL DEPOSITS

Upper Pliocene lake

Lower Pliocene lake

Direction of ancient river

3546.22x

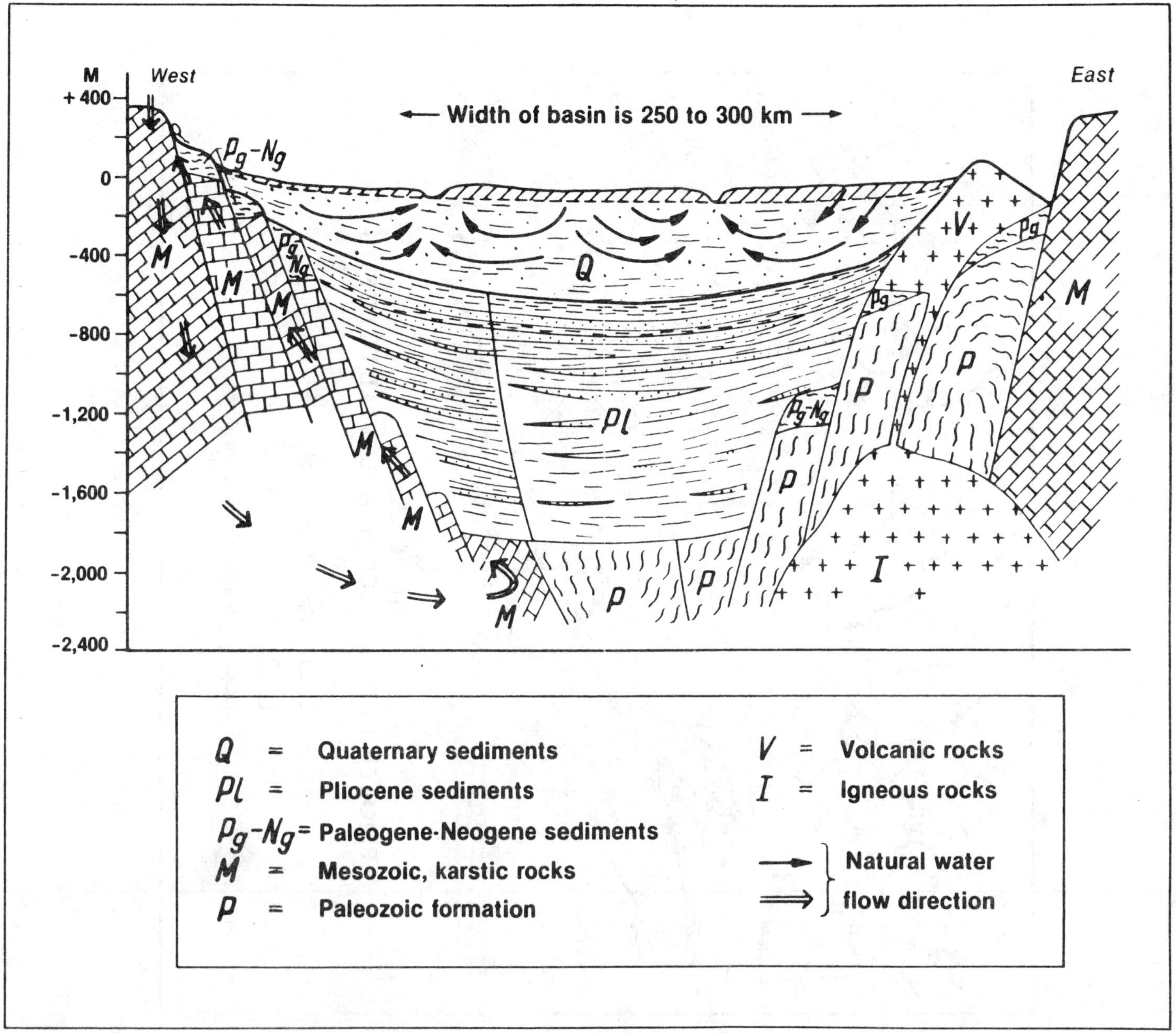

Figure 32. Ground-water flow in the Hungarian Basin

Figure 33

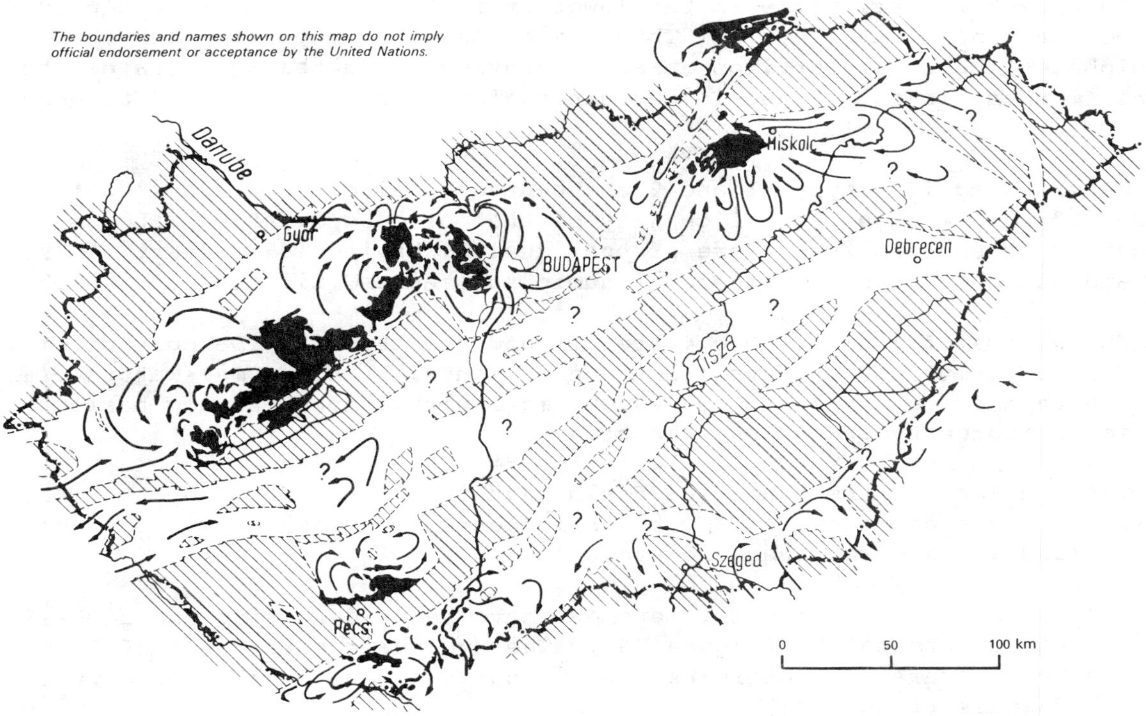

HUNGARY:
MESOZOIC KARST AND DIRECTIONS OF FLOW

Open karst

Subsurface karst

Miscellaneous formations

Natural flow direction

3546.23x

4. Quaternary alluvial aquifers of layered structure with lenses. Some alluvial aquifers are characterized by a natural water circulation extending to considerable depths and mainly with indirect communication to the surface. On the basis of head differences of the artesian aquifers infiltration zones and upward flow zones can be well distinguished.

5. The upper Pannonian aquifer system, only with distinct indirect surface connections, yielding cold water along the basin's border and thermal water in the middle.

Most of the waters in the aquifers of Hungary are of calcium-magnesian bicarbonate character, while in the lower lying Pannonian aquifers sodium bicarbonate conditions prevail. This basic character may differ by the occurrence of sulphates, chlorides or iron at some locations. In the aquifers of the southern plains, methane is common and in certain regions concentrations of arsenic can be found.

Water of aquifer types 1 and 2 above have total dissolved solids (TDS) content of 400-500 mg/l with a general quality of that of good drinking waters. Most of the waters of aquifer type 3 are of poor quality having TDS content of 2,000-3,000 mg/l and with high concentrations of humic substances, nitrates and sulphates.

The waters of aquifer type 4 are of good quality with TDS content of 600-800 mg/l, characterized by methane and trace elements. The thermal waters of aquifer type 5 have a TDS content of 1,000-3,000 mg/l with a much varying basic character that is dominated by sodium bicarbonates.

The aquifers of the Hungarian Basin contain waters not older than 40,000 years. In areas of downward flow the salt content is less, while in those of upward flows it is more, than the average.

The paths of downward water movement have been identified on the basis of absolute age determination (figure 34). The hydraulic parameters of Quaternary aquifers were accurately determined on the basis of the data derived from thousands of drilled wells (figure 35).

In the deeper part of the sedimentary basins the aquifers contain different kinds of thermal water. The salt concentration of these thermal waters varies within wide ranges, that is, from 1,000 mg/l to 25,000 mg/l. The possibilities of thermal water development are well known in Hungary (figure 36).

Figure 34

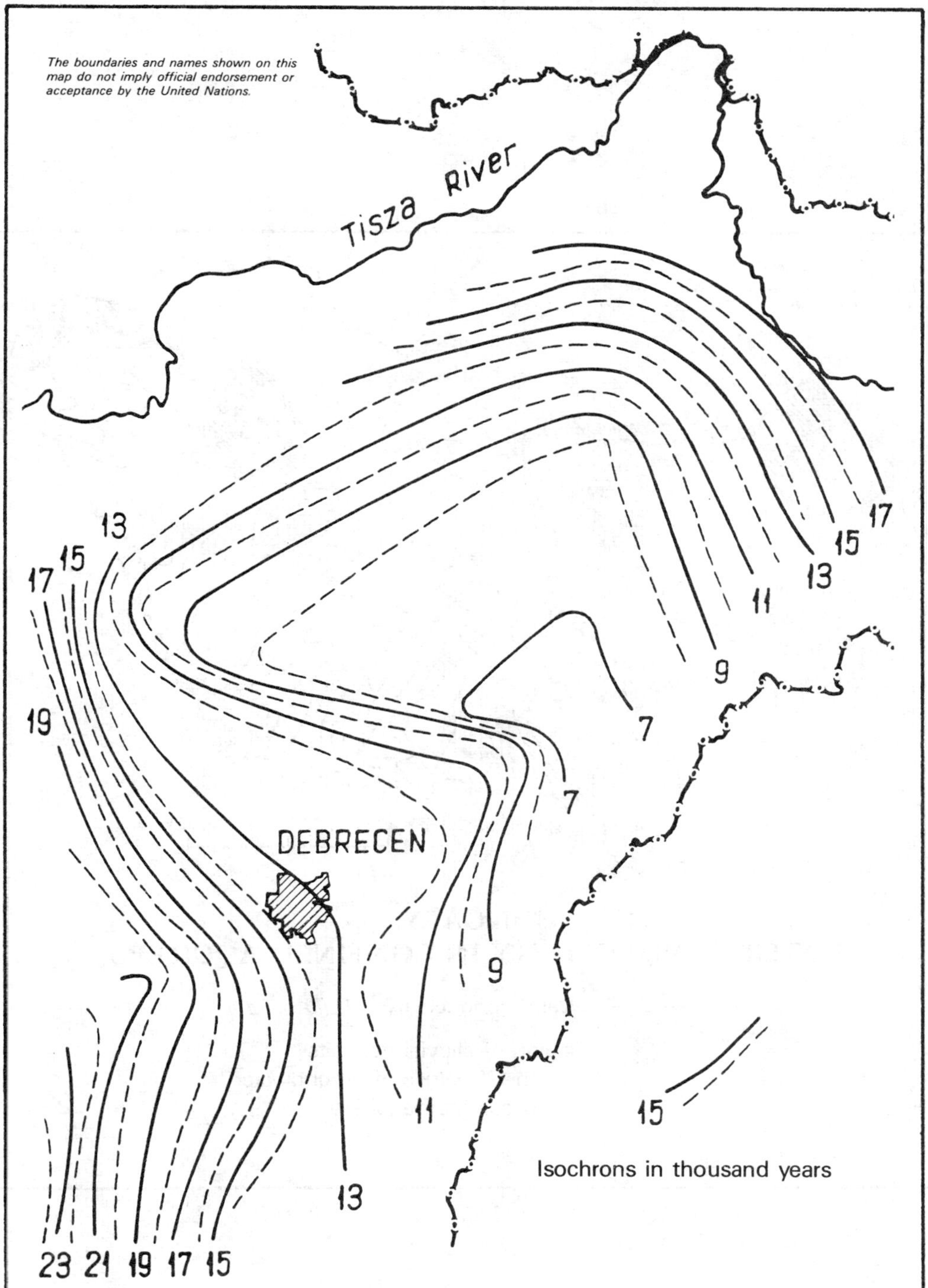

HUNGARY:
AGE OF GROUND WATER IN QUATERNARY AQUIFERS

3546.24x

Figure 35

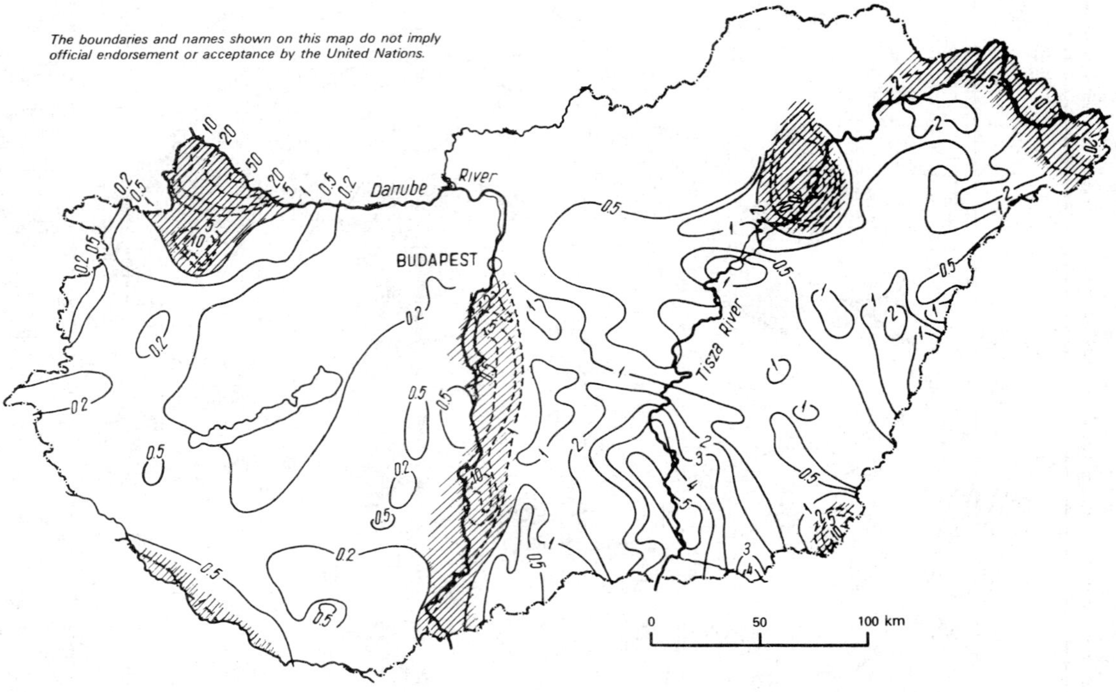

HUNGARY:
WATER CONDUCTIVITY IN CONFINED AQUIFERS

— 0.5 — Water conductivity *(T) (10^3 $m^2 \cdot d^{-1}$)*

Region of alluvial deposits with 'T' Values of groundwater and confined aquifers

3546.25x

Figure 36

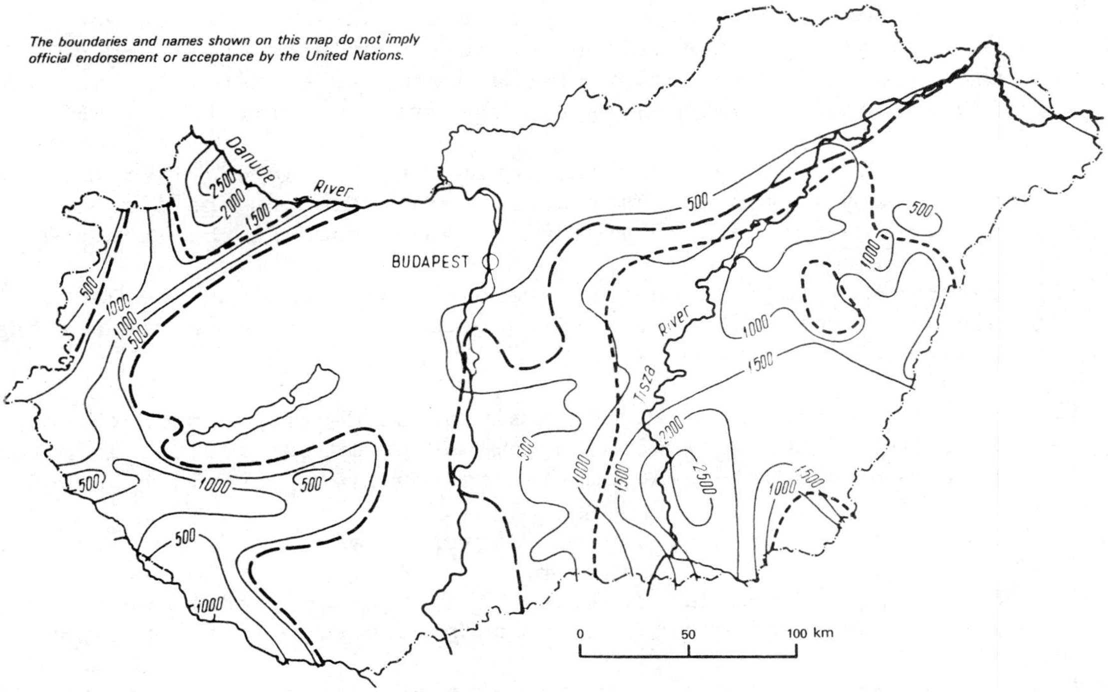

HUNGARY:
THERMAL WATER RESOURCES

—500— Depth of Upper Pannonian Basement in metres

Regions suitable for exploration of thermal-water of temperature:

— — — 35 to 60°C

- - - - - 35 to 100°C

3546.26x

C. Ground-water development

1. Data collection

The collection of certain data for each drilled production well is mandatory. They include:

1. A disturbed sample from each penetrated layer and from each 5-m depth; the geologic profile of the borehole should be developed on the basis of macroscopic investigation of these samples on site; the samples are later investigated in laboratory and the profile corrected;

2. Undisturbed samples from the aquifer to be tapped; from the overlying and underlying layers and from layers prescribed in the licence given by the relevant water authority; the equipment used is the standard core-barrel;

3. Geophysical logging in the borehole: electric and radioactive logging is mandatory for each drilled production well of depth greater than 50 m together with flow-metering;

4. Minimum 72 hours pumping test aiming at development of the capacity-function of the well with determination of the relationship between discharge and drawdown, carried out mainly by air lift method;

5. Sampling and chemical-bacteriological analysis of the water developed;

6. Sampling and chemical analysis of the gas (methane) contained in the water; sampling carried out through direct or indirect separator;

7. Establishment of the recovery curve after production is stopped and measurement of the head at regular intervals through a period of two hours;

8. Measurement of the pressure head at aquifer depth in wells deeper than 500 m; and

9. Thermo-logging of wells producing water above 30° C temperature.

In the preparatory stage of larger water intakes, and in the case of areal investigation, the interference of wells is evaluated through exploration wells sited in groups. The variations in chemical composition of water is evaluated through serial sampling and chemical analysis of the water during long-term pumping tests with the purpose of determining the proper water treatment.

A file is compiled for each drilled well which contains all the data and results obtained during the construction and pumping test of the well, together with its technical data. Such files for 60,000 wells are stored at the Research Centre for Water Resources Development (VITUKI), at the Institute for Water Resources Management and at the Hungarian Geological Institute. The twelve District Water Authorities hold the documentation from their area of reference. The documents are in traditional form, part on microcards and part in computerized form. The proportion of the two latter forms varies from place to place. A comprehensive computerization programme has been launched in VITUKI. The essential data of this central inventory are regularly published in printed form. Ten volumes of the series had been published as of 1988.

The results of areal exploratory works and those of the preparatory works of the water intake plants are summarized in preliminary studies and reports. Abstracts of the documents compiled by various institutions, are routinely prepared by the Institute for Water Resources Management. The abstracts are stored at this institute and the district water authorities.

The results of operational measurements and observations are mailed directly to the relevant District Water Authority; after review they are sent to VITUKI for evaluation and publication annually.

Data from the observation networks of deep and shallow ground waters and of karstic waters are collected by the relevant District Water Authorities and sent to VITUKI for processing and evaluation. The same system prevails in the case of a pilot quality observation network. Computerization of the system is under way.

The public health organization takes and analyses a large number of samples from production wells in the framework of its supervisory function. The results are collected partly by the District Public Health Authorities and partly by the National Institute for Public Health. A computerized central data bank of common use is being developed by public health and water organizations.

2. Resources assessment

For national and regional evaluation the exploitable resources of shallow ground water of porous aquifers is assessed in terms of the discharge which can be exploited as a difference between infiltration and evapotranspiration leading to a specific index of potential expressed in cubic metres per day per square kilometre. Water resources in alluvial deposits along the rivers are assessed by calculating the average transmissivity of the alluvium and, on the basis of an assumed gallery at a given distance from the river bed, leading to the specific index of potential. Water resources in karst are assessed from rainfall infiltration upon the uncovered rock surface on the basis of various measurements (e.g., readings in caves; relationships between rainfall and discharge or distribution of karstic springs) leading to the index of potential. Water resources in deep aquifers are evaluated from the vertical permeability of the geologic structures measured at operating water intakes leading again to a specific index of potential belonging to a selected uniform regional depression, say, 10 or 20 metres below the natural head prevailing in the area. Also, the exploitable highest capacity of the concentrated water intakes is given. Through these assessments, the total ground-water resources of the country are estimated to be 14.5 million cubic metres per day. Values for the various regions are used in the long-term planning of water management and in decisions at the regional level.

For local assessments, the data from exploration for different purposes (geologic survey, oil exploration) are collected, evaluated and supplemented through such exploratory works as geophysical survey, sounding, logging, exploratory drilling, pumping tests and interference studies. Evaluation of the results suggests the following questions: What amount of water can be exploited at a given site? What will the drawdown be in a well, or the regional depression? What will be the time-distribution of the depressions?

Mathematical modeling was applied experimentally in the late 1960s. Now it is used routinely in evaluating the useful ground-water resources of a given region, or of future water intake works under prescribed operational circumstances and constraints; and also to forecast the effects of future water exploitation on the environment, in water supply and mining.

Regional and local models are in use and the first steps taken to forecast water quality, its distribution in time and space in the environments of given or future water development. A model has been designed for the study of thermal water resources on a national scale; there are models for the description of heat movement in thermal water exploitation when production and injection wells of 1,000-1,400 m depth are used.

3. Monitoring networks

Operational measurements and observations are carried out at more than 300 water intake plants. The average monthly production of wells and the dynamic water level in them is recorded. Lacking sufficient observation wells, production wells are used to observe the static level in them during a one-hour break from production and then calculating the theoretical static level with a standard function. The number of water intake plants under observation grows regularly.

Observation networks are in operation around only the largest five water intake plants of the country with the purpose of tracing the cone of depression around the intakes. Most of the observation wells are abandoned production wells. The total number of intakes in the public water supply works is 1,200, from the smallest ones of a few hundred cubic metres per day capacity to water intake areas for the capital with a capacity of several hundred thousand cubic metres per day for each.

The deep ground-water areas of the lowlands are covered with an observation network of more than 500 wells tracing the regional depression developed by the high number of smaller and larger intakes.

The central area of the country is built up of Mesozoic carbonate rocks containing a large amount of karstic water. The area is heavily effected by industrial and mining activity. These effects have been monitored by a network of more than 300 observation wells for more than 15 years.

Data characterizing the régime of shallow ground water have been recorded by a network of 1,700 observation wells throughout the country for 30 years.

4. Well casing and pumps

Dug wells are commonly used in rural areas for the supply of single households, sometimes in animal breeding. They number about a million; fewer and fewer are used as public water supply is increased, and because of increasing contamination by nitrates in fertilizer. The walls of such wells are made of bricks, stones or prefabricated concrete and are generally 0.6 to 1.2 m in diameter. The number of drilled wells amount to about 60,000. They tap the upper 500 m of lowland alluvium for the main purpose of providing drinking water. For thermal water, about 800 wells have been drilled to depths of 1,000-2,000 m, both in loose clastic sediments and in carbonate rocks. Steel tubes of 90-600 mm diameter are used for casing. The structure is usually the traditional two- or three-part casing; the uppermost section is the largest diameter down to 20-40 m depth to exclude contaminated ground water; the second part extends down to the last aquitard above the aquifer; the third part is the screened section of smallest diameter, usually 90-160 mm. All the casings are cemented except the screened one. Their capacity ranges from a few tens of litres per minute to 1,000 l/min. Shallow ground water is drawn from alluvium by high-capacity, large-diameter tube wells constructed with double gravel screen around a production tube of 400-600 mm asbestos-cement casing. Their capacity ranges from 1 to 3 m^3 per minute.

The largest output is delivered by the radial wells used mainly by the Metropolitan Water Works of Budapest. A shaft 2.5-4.5 m in diameter penetrates the overlying layers of low permeability down to the aquifer at a depth of 20-40 m. Here radial screen tubes are driven out 30-50 m. Yields range from 3 to 8 m^3 per minute. Natural springs are tapped by means of various concrete and reinforced concrete structures. In some rare cases, water produced as a byproduct of mining is utilized in great quantities. Water is produced usually with shafts built by traditional means or by drilled shafts of various dimensions.

Pumps used in the production of water are usually electrically driven submersible pumps of various types. Well field operations may be automatically controlled.

5. Contamination and protection

The preparation of water intake works is normally based on hydrogeologic and technical experience. In comparing different potential alternatives, economic analysis has been done in a few cases; the vulnerability of the water resources to be exploited has been hardly taken into consideration, even in the case of large regional water supply systems. Lacking monitoring systems, the operation of intake works tapping the aquifer system has not been controlled and, not surprisingly, operational problems have arisen.

An advantageous hydrogeologic situation has saved Hungary from serious exploitation problems. Drawdown of 30 to 50 m has been experienced at a few large plants and this has led to some misunderstandings in connection with the future availability of water resources. But after proper study it was clear that exploitation may continue without serious problems, or at least be maintained at the same level, with proper management of the intake system.

Lack of strict monitoring and controls has led to serious contamination of some ground water. Some shallow ground water displays nitrate contamination as high as several hundred milligrams per litre mainly in populated areas of rural regions and their surroundings. The steady increase of nitrate content has been experienced in some bank-infiltration alluvial water production plants because of uncontrolled human activity in their source areas. At a few plants, permissible nitrate levels can be maintained only by mixing with other, purer, waters. Some water intake plants had to be shut down because of the high level of contamination. Some plans for future water intake areas had to be abandoned because of high nitrate contamination and their potential water resources no longer included for meeting future needs.

The main sources of contamination are uncontrolled storage and use of fertilizer in agriculture, a low degree of canalization and waste water treatment;

uncontrolled waste disposal by industry, farmers and the populace in general. The establishment of protection zones has been prescribed by legislation; but uncertainties in defining the geologic parameters of protection needed and because of legal delays the major proportion of such zones has not, as of 1988, been established. For about 400 to 500 water production plants a comprehensive system of protection zones is needed, but as of 1988 fewer than 20 had been established.

The cost of establishing protection zones is perhaps the greatest problem. Realistic determination of compensation for lost income from the interdictions prescribed in protection zones makes a nice dimensioning of the zones essential. Exploration, calculations and observations are needed and the compensation must be paid. In addition, the water supply enterprises are not always aware of the need for protection. All the necessary protection zones can probably not be established in less than 10 years, meanwhile losing a considerable amount of otherwise exploitable ground water.

D. Training, research and international co-operation

1. National and parastatal agencies

Ground-water resources are handled by the District Water Authorities at the regional level and by the National Water Authority (NWA) on a national scale. Regional exploration and country-wide assessments are carried out at the expense of the latter organization. Local explorations are sponsored by the user. Licences to explore and develop water sources are granted by the twelve District Water Authorities.

Monitoring of ground-water equality and the environmental effects of ground-water exploitation is done by the Research Centre for Water Resources Development (VITUKI). Here is placed the central inventory of drilled wells and ground-water quality data. VITUKI co-ordinates and supervises the operation of national observation networks operated by the District Water Authorities. The Institute for Water Resources Management is the central organ of NWA in the preparation of decisions made on the national level.

Three main Government consulting and engineering firms are involved in exploration either on a regional scale or for preparation of local and regional water supply works. Drilling is carried out by the Enterprise for Water Exploration (VIKUV) under the supervision of NWA, by the Geological Exploration and Drilling Enterprise (OFKFV) supervised by the National Geological Authority; the number of other smaller organizations involved in exploration is more than 40.

Also the geologic organization is interested in regional and local exploration. Its two main institutions are the Hungarian Geological Institute and the Eötvös Loránd State Institute for Geophysics. The planning and execution of exploration are co-ordinated with those of the water management organization.

Geologists and hydrogeologists are educated at the Eötvös Loránd University of Sciences, Budapest, at the József Attila University of Sciences, Szeged, and engineers in hydrogeology at the University for Heavy Industry, Miskolc (Mining Branch). Also civil engineers are trained to some extent in hydrogeology at the University of Technology, Budapest.

The public health organization has the responsibility for supervising the quality of drinking water. Through its local agencies it has the authorization of sampling and analysing the water used for drinking purposes and of closing down water supplies if quality does not meet standards.

2. Technical assistance

Ground-water exploratory works were carried out by Hungarian enterprises and experts in Democratic Yemen, Ghana, Kuwait, Libyan Arab Jamahiriya, Mongolia and the Yemen Arab Republic. An international post-graduate course sponsored by UNESCO is organized regularly in Hungary. The participants come mainly from developing countries. Various contacts and co-operative efforts have been developed with most European countries; the co-operation is realized on both the institutional and governmental level. Hungary participates in several international and scientific organizations interested in ground water.

References

Alföldi L. and A. Homonnai. Groundwater Pollution in Hungary with special regard to Nitrates. Manuscript, 1984.

Collection of results of exploratory works, Reports for 1975-1984. Institute for Water Resources Management, Budapest, 1975-1984.

Deák J. and Gy. Dénes. Analysis of ground water by environmental isotopes. Research Centre for Water Resources Development (VITUKI), Budapest, 1981.

Evaluation of ground water production possibilities and estimation of needs for future regional ground water exploration. Report of Institute for Water Resources Management, Budapest, 1981.

Evaluation of natural protection and vulnerability of ground water. Investigation of groundwater occurrences being utilized and to be utilized until the year 2000. Report of Institute for Water Resources Management, Budapest, 1982.

Geothermal wells of Hungary and Inventory of thermal waters, vols. 1-3. Research Centre for Water Resources Development, Budapest, 1977.

History of river training in Hungary. Budapest, 1973.

Hydrological characterization of Hungary: Precipitation, evaporation. Vizgazdálkodási Intézet, Budapest, 1984.

Inventory of drilled wells in Hungary, vols. 1-10. Institute for Water Resources Management, Budapest, 1981.

Master plan of water resources development in Hungary. Országos Vizügyi Hivatal, Budapest, 1984.

Water Management in Hungary. Budapest, 1976.

VII. ICELAND

Area: 102,850 km^2 Population: 200,000

A. General

The Republic of Iceland is an island in the North Atlantic Ocean, just below the Arctic Circle and lies astride the Mid-Atlantic Ridge. It is wholly of volcanic origin consisting, for the most part, of basaltic rocks (figure 37). The country is nearly oval in shape forming essentially a plateau at 300-1,000 m altitude, indented by numerous fjords and valleys and capped by many late Quaternary volcanic mountains. The higher mountain clusters are crowned by glaciers.

The age of the volcanics is Miocene to Recent, the youngest rocks occurring in an active volcanic zone across the island. The zonal arrangement in the stratigraphy is the result of ocean-floor spreading along the Mid-Atlantic Ridge and its prolongation across Iceland. The Tertiary rocks (older than 3 million years) can be described as a pile of tholeiitic and olivine-tholeiitic lava sheets, interspersed by central volcanos, characterized by acidic and intermediate rocks, geothermal alteration and dyke swarms. In most regions this lava pile is, with few exceptions, tilted a few degrees towards the present volcanic zone. The same mode of volcanism is present in the younger formations. In the Pliocene-Pleistocene, most strata dip slightly less than in the Tertiary. Signs of glaciation appear in this formation, including tillite and heaps or layers of subglacial eruptives in the form of pillow lavas and hyaloclastites, most palagonitized (móberg in Icelandic). These subglacial eruptives increase in importance with time, becoming the dominant feature in the late Quaternary (less than 0.7 m years old), and forming table mountains, ridges and mountain clusters, rising some hundred metres above the plateau. The volcanism has remained active in the Holocene, resulting in recent lavas covering wide stretches and extending into the lowlands. A comprehensive review of Icelandic geology is given by Saemundsson (1978, 1979).

The climate in Iceland is oceanic and sub-polar. The south and west coasts, and frequently the north coast, are warmed by a branch of the Gulf Stream, while colder sea currents flow along the east coast, and affecting at certain times the north coast also. On the other hand, Iceland lies in the path of moisture-laden depressions crossing the Atlantic from west to east giving rise to changeable weather and bringing in warm, wet, south-east winds, together with cooler but still wet south-west winds. North and north-east winds from depressions passing east or north-east of Iceland are cold but not very wet. The result is that the southern and eastern highlands experience annual precipitation in excess of 2,000 mm, while it is generally below 1,000 mm in the northern half of the country (figure 38). The average July temperature at the west and south coasts is 10°-11° C, but 9° C at the north and east coast. In January the average temperature is 0° C at the south-west, south and east coasts, but -1° C at the north-west and north coast. The temperature falls farther inland.

This wet and cool climate leads to a high runoff in Iceland, exceeding 100 l/s/km^2 on the south-east highlands and glaciers, although the greater part of the northern half has a runoff below 40 l/s/km^2 (Rist, 1956). The biggest rivers are all glacier-fed. At least 10 have a mean discharge of more than 100 m^3/s (figure 38). As many of them have a steep descent from the plateau they are great potential sources of hydropower. Because of that the hydrographic network is relatively dense. At present approximately 150 hydrographs are operated by the National Energy Authority.

Figure 37

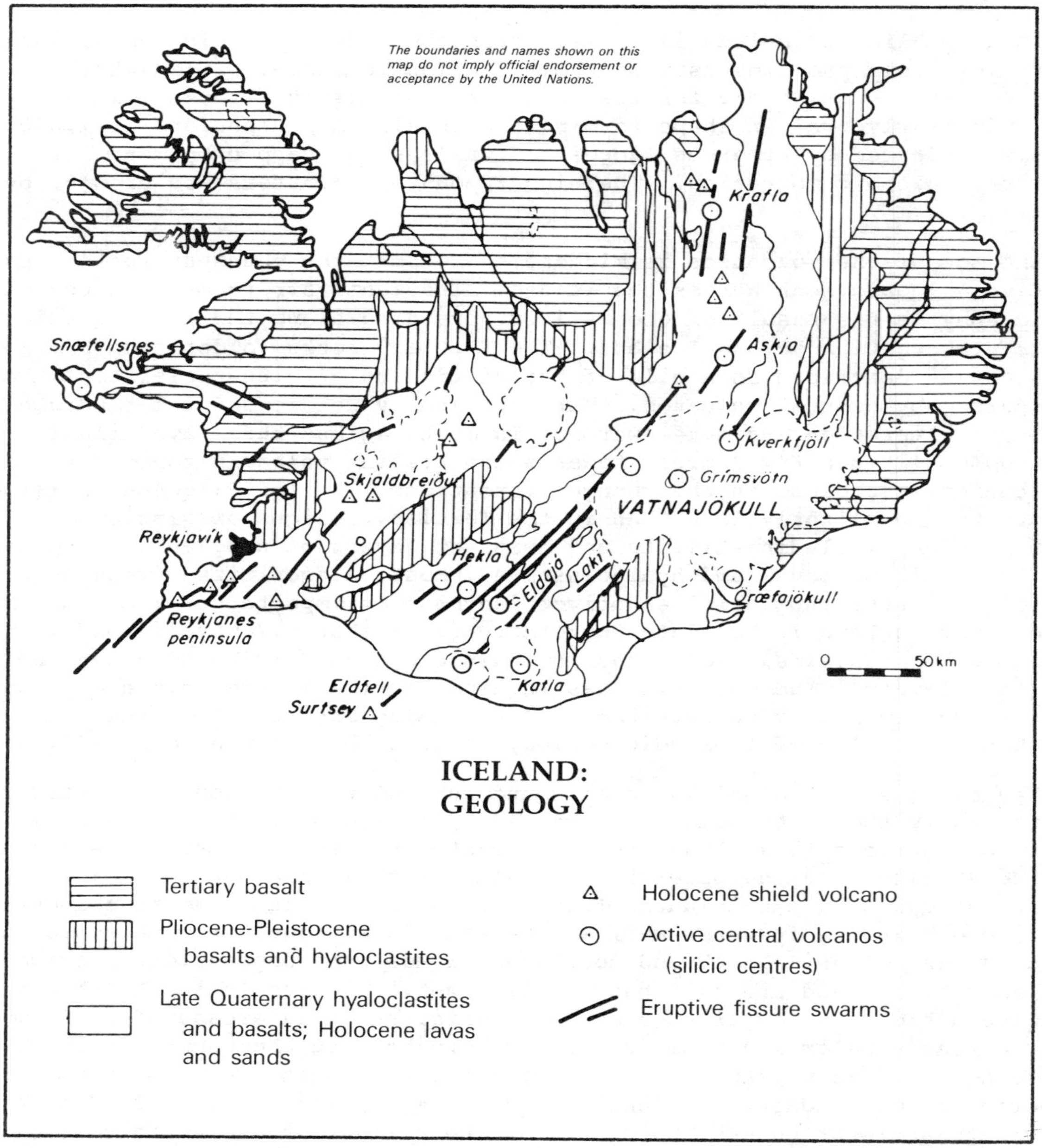

3546.27x

Figure 38

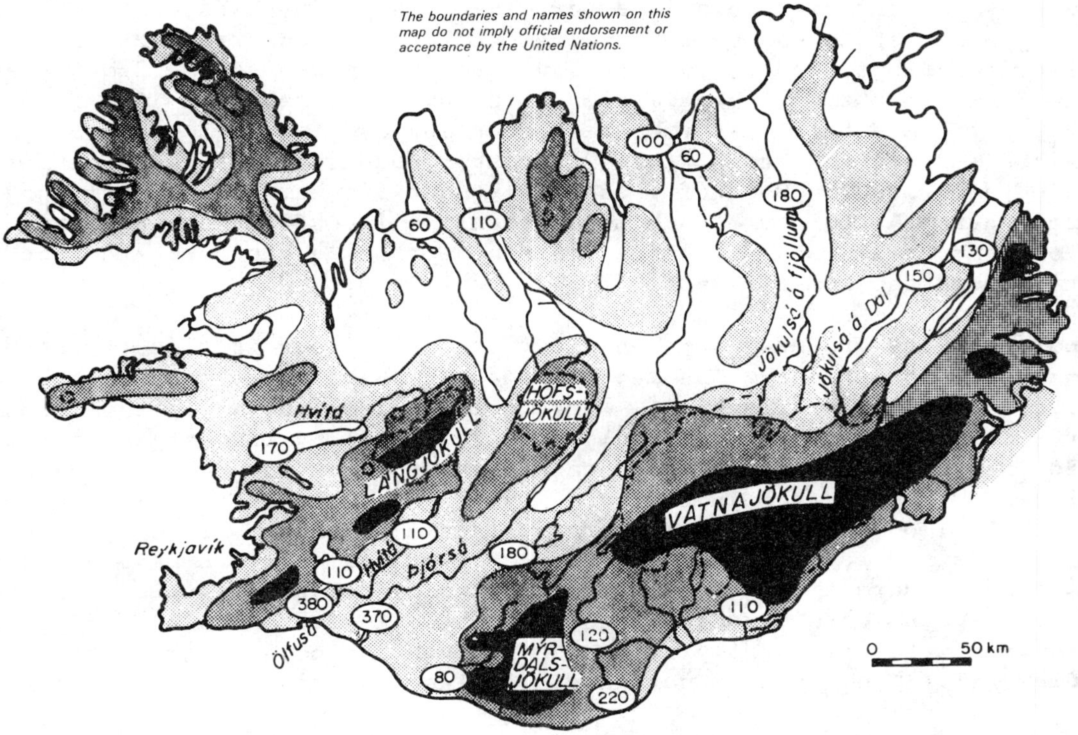

ICELAND: RAINFALL AND RIVER DISCHARGE

(110) Mean discharge in cubic metres per second

Precipitation greater than 3,200 mm per year

Precipitation between 1,600 and 3,200 mm per year

Precipitation between 800 and 1,600 mm per year

Precipitation less than 800 mm per year

3546.28x

B. Hydrogeology and water quality

1. Historical

Hydrogeologic research in Iceland has emphasized practical purposes, that is, to furnish the public with waterworks and industries with clean fresh water and to evaluate the ground-water component of rivers which are intended for hydropower. Pure academic research has been minimal. The sporadic beginnings are indeed as old as the geologic investigations in Iceland as such. Thorough work on the hydrogeology of Iceland started only a few decades ago (Kjartansson, 1945; Rist, 1956). Its present state is due to developments of the last two decades. The first real hydrogeologic map of an extensive area was presented by Sigbjarnarson (1972), the first intensive investigation of an area was begun in 1975-1976 (Sigurdsson, 1976) and the first reconnaissance survey of a region was started in 1978. The preliminary results of these and other research works were summarized by Hjartarson et al. (1980). Most of these investigations have been carried out by the National Energy Authority. Much more extensive investigations have been conducted on geothermal waters, the ultimate object being the energy content of the water, not so much the water itself.

The permeability ground-water behaviour of the bedrock is closely connected to the main geologic formations. The Tertiary basalts have low permeability, chiefly because of secondary mineralization. Where this is lacking, as is the case in the higher parts of the north-west peninsula, the permeability is markedly higher. Rejuvenated fracture zones can also be water-bearing, but they are uncommon. The chief aquifers at the surface in the Tertiary are of sedimentary origin as river gravel and rock slides. From the latter, springs may issue, virtually none exceeding 100 l/s and few more than 10 l/s. Few of the aquifers have a great areal extension (a few square kilometres) and are normally relatively thin (less than 10-30 m). The springs are usually caught in dug wells for exploitation. From river gravel the water is usually extracted through boreholes ranging from 10 to 50 cm in diameter. The yield is normally 5-20 l/s, but higher yields occur.

The rocks of the Pliocene-Pleistocene resemble the Tertiary ones in many aspects but are more permeable, although secondary mineralization can also locally reduce permeability. Many sedimentary intercalations (especially tillites) and layers of palagonitized hyoloclastites have a low permeability. The highest permeability is at the contact between lava sheets or between lava sheets and the intercalations or intrusions (Fridleifsson, 1979). This causes an anisotropy, the flow along the bedding planes being much easier than perpendicular to them. Tectonic fracture zones also have an anisotropic effect, many being highly permeable (figure 39).

Figure 39

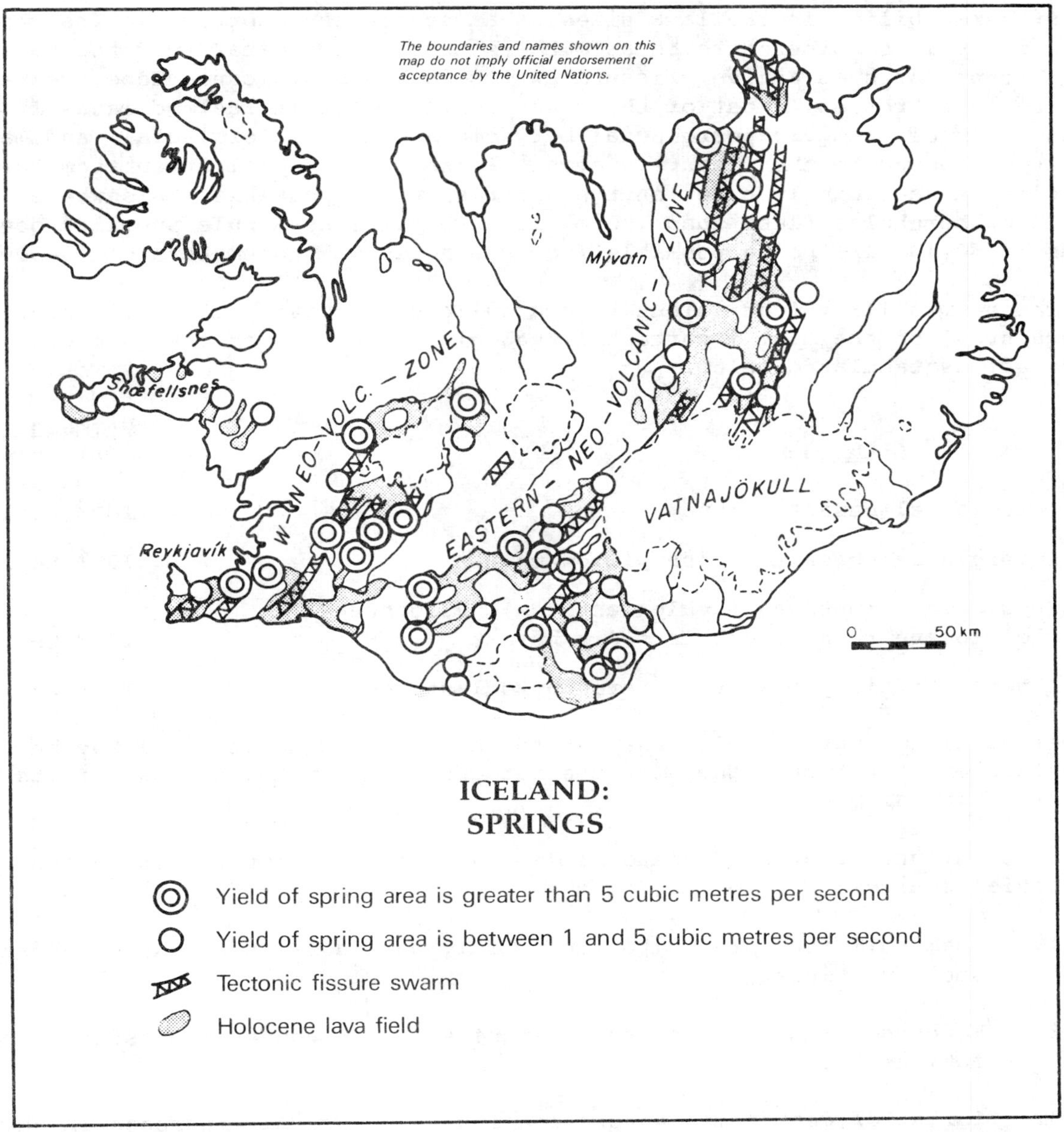

3546.29x

These anisotropy effects are also pronounced in the late Pleistocene to Recent rocks. These rocks can be divided in two main classes, subglacial rocks (Icelandic móberg) and lava piles. The subglacial eruptives are on a small scale and heterogeneous in lithology, structure and permeability. The pillow lava can be quite permeable, whereas the hyaloclastites tend to be much less permeable. On a regional scale (a whole mountain or so) the subglacial eruptives generally have a low permeability compared with the interglacial and postglacial lava piles. The highest permeability in the lava piles is at scoriaceous contacts. The horizontal permeability is considered to be as much as 100 times greater than the vertical one. Recent, or even active, fissure swarms in the Neovolcanic Zone create strong, regional anisotropies. Most of the great spring areas in Iceland issue from lava fields, most of them are postglacial but some are also interglacial, and many are closely connected to the fissure swarms. Fresh-water exploitation from this formation is accordingly based on the springs, where possible. Elsewhere, drilling of shallow boreholes (less than 100 m) in lava piles as a rule supplies good water, yields of 50-100 l/s being possible from boreholes with diameters of 25-30 cm.

The permeability (equivalent to hydraulic conductivity at normal ground-water temperatures) of the most important lithologic types (formations) has been roughly estimated, as tabulated below.

Lithology	Permeability (in m/s)
Postglacial basaltic lava	10^{-3} to 10^{-0}
Interglacial basalt, fresh pillow-lava	10^{-4} to 10^{-1}
Subglacial eruptives (hyaloclastites), Pliocene-Pleistocene rock, fresh Tertiary basalt	10^{-6} to 10^{-2}
Altered Tertiary basalt, altered pyroclastite	10^{-7} to 10^{-3}

The permeability (but in this case not the hydraulic conductivity) may be one or two orders of magnitude higher for the hot geothermal water because of its lower kinematic viscosity.

Fresh ground-water in Iceland is low in chemical content. Those contents have three chief sources:

1. Chemicals from precipitation, mainly marine salts and some industrial smoke particles,

2. Reaction with soil and rock, increasing the contents of silica and cations, and

3. Influx of geothermal water, leading to an overall increase in chemical content.

The marine component in the precipitation (NaCl, SO_4) is strongest at the south and west coast (figure 40). The influence from geothermal fields is confined to the Neovolcanic Zones, where the increase in silica and cations is, overall, somewhat stronger, probably because of the freshness and high glass content of the rocks (e.g., hyaloclastites). The ranges for common values of concentrations in fresh ground-water in the inhabited parts of the country are tabulated below.

Figure 40

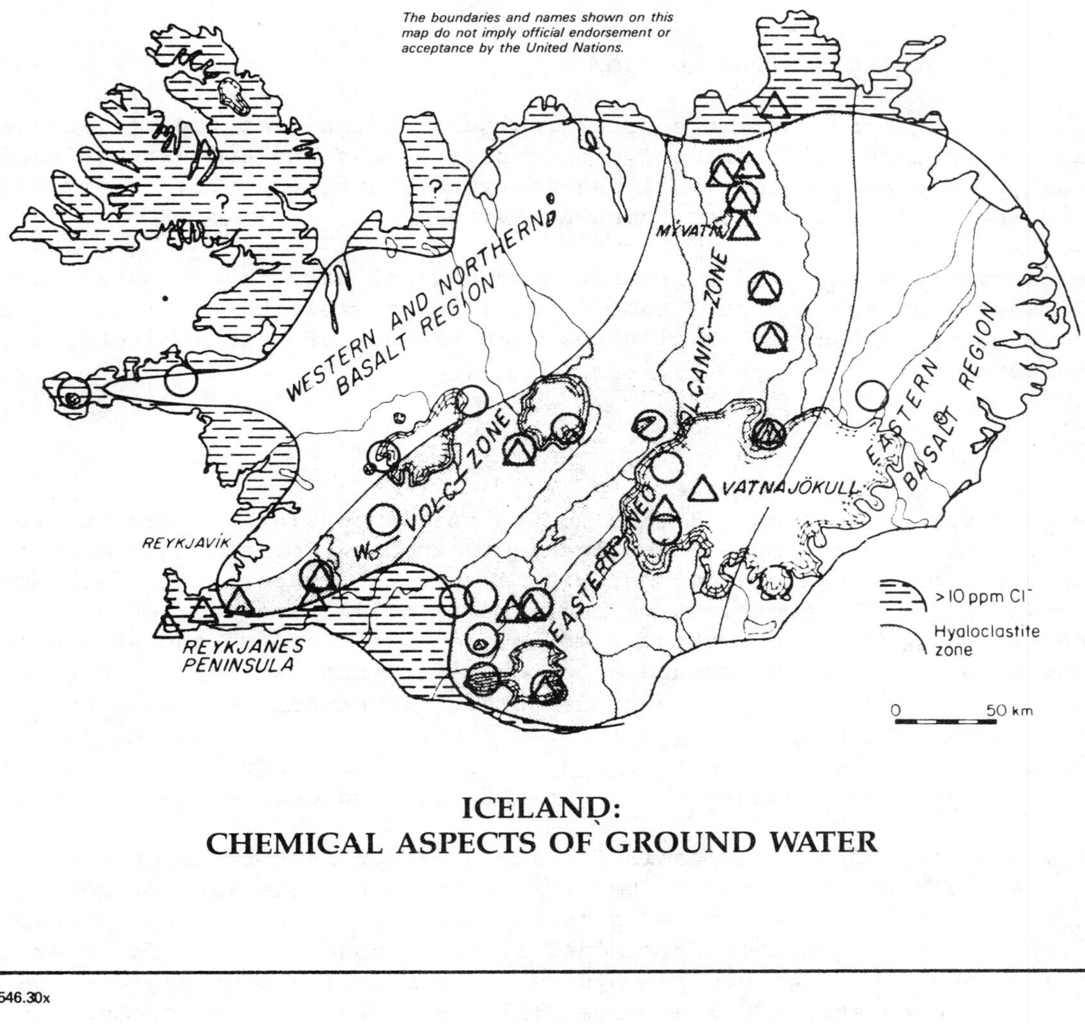

ICELAND:
CHEMICAL ASPECTS OF GROUND WATER

3546.30x

Component	Range (in mg/l)
Cl	5-15
SO_4	2-6
Na	3-15
K	0.2-1
Ca	2-7
Mg	1-5
SiO_2	10-20
CO_2	15-30
Total dissolved solids	30-100

Higher values are found in coastal regions, in the vicinity of geothermal fields and in lowlands with extensive peat bogs (carbonate and earth-alkalies). Biologically the ground water is almost everywhere pure and as yet only very small areas have been polluted through human activity.

The chemistry of Icelandic ground water has not been investigated thoroughly, but preliminary review indicates that it will be possible to distinguish between various ground-water basins. Similar use can be made of natural isotopes (Arnason, 1976; Sigbjarnarson et al., 1976).

2. Exploration

The vastness of Iceland relative to its sparse population does in many cases not allow for more than simple investigation methods. For a long time a rough hydrogeologic classification together with a surficial assessment of the ground-water hydrology was the prevalent method, and is still resorted to in many cases. In the course of time those methods have become more thorough and of a more quantitative nature. Later, ground-water chemistry and isotopes were added (Arnason, 1976). Combination of various surficial methods (hydrogeology, hydrology, chemistry, geophysics, hydraulics) have proved successful in the last 10 years (Sigurdsson, 1976; Sigurdsson et al., 1978). Intensive surveys and modelling are still reserved for ground-water basins of high economic value.

The evaluation of a water basin for exploitation is carried on stepwise, when possible. A field reconnaissance, together with interpretations of geologic, hydrologic and meteorological data is usually the first step. When necessary more sophisticated surficial methods are added like thorough sampling for chemical analysis and borehole-loggings, geophysics and eventually some exploration drilling. The third step involves more drillings, drawdown tests, other pumping tests and eventually geohydrologic modelling.

The requirements for evaluation of the ground-water component of a river for hydropower are of a more regional nature. They include the measurement of springs (discharge, temperature, chemical sampling) and other surface manifestations of the ground water. Comprehensive hydrogeologic mapping is also needed (scale 1:50,000) and when necessary exploration drilling. The data thus obtained is used for hydrologic modelling and monitoring of climatic changes for the operation of the hydropower system.

No single agency is responsible for the exploitation of fresh water in Iceland, to a noticeable harm for the users. Accordingly, no single authority keeps records of the ground-water resources and their use. The most thorough information on water works is presumably the one compiled by the scientists of the National Energy Authority, Hydro Power Division. No single ministry is responsible for ground water in Iceland as a resource, and this has caused some confusion.

3. Development

Water has always been an inexpensive and easily available commodity in Iceland, with the result that consumption per capita is probably one of the highest in the world. In the urban area of Reykjavík and its suburbs (population 130,000) consumption is estimated to be 500-700 litres per day per capita. In some settlements with intensive fish industries it is thought to be as high as 1,500 litres per day per capita (figure 41). This high use of water is not necessary because of any natural conditions, it rather has its causes in the general and common attitude of Icelanders towards the cost and price of water: water has been inexpensive and abundant in Iceland, and it is a common view that it will remain so indefinitely; the water is not paid for proportionate to consumption but through a property tax; necessary costs are not always treated as such, but this may play a disturbing role in communal politics. Thus the present mode, to leave the water supply completely in the hands of communal authorities has not proven satisfactory, from the point of view of national interest.

A second cause at these unsatisfactory conditions is to be seen in the historical development. With the rapid increase in the urban population in the twentieth century, the quantitative demands for water also increased. These demands were often met in the cheapest way possible and then equally often with the result that the quality of the water deteriorated as the quantity increased. In later years amendments of those conditions intensified with good results.

The type of wells for waterworks is chiefly dictated by the hydrogeology (figure 42). In the Tertiary regions, springs in rock slides and boreholes in river gravel are the most common types. The springs are usually caught in dug, shallow wells, where great precautions are taken not to disturb the natural conditions. The boreholes in the loose sediments are usually drilled with a cable-tool, and have a diameter of 15-60 cm. The boreholes have a whole casing down to below the supposed water-level at maximal drawdown.

In some settlements with old waterworks, and where it is difficult and expensive to obtain fresh water by other means, surface water has been used. It has nowhere been possible to obtain in that way clean and pure water all year, a fact that sometimes has been concealed by communal authorities. The chief reason is the tremendous amount of mud the surface waters carry with them after thaws and heavy rains.

Special treatment of public water is not generally practised in Iceland. In most cases it is also unnecessary, i.e., where ground water is used. As yet no installed filter works are known to have been adequate where surface water is used, i.e., to clean the muddy flood waters successfully.

Figure 41

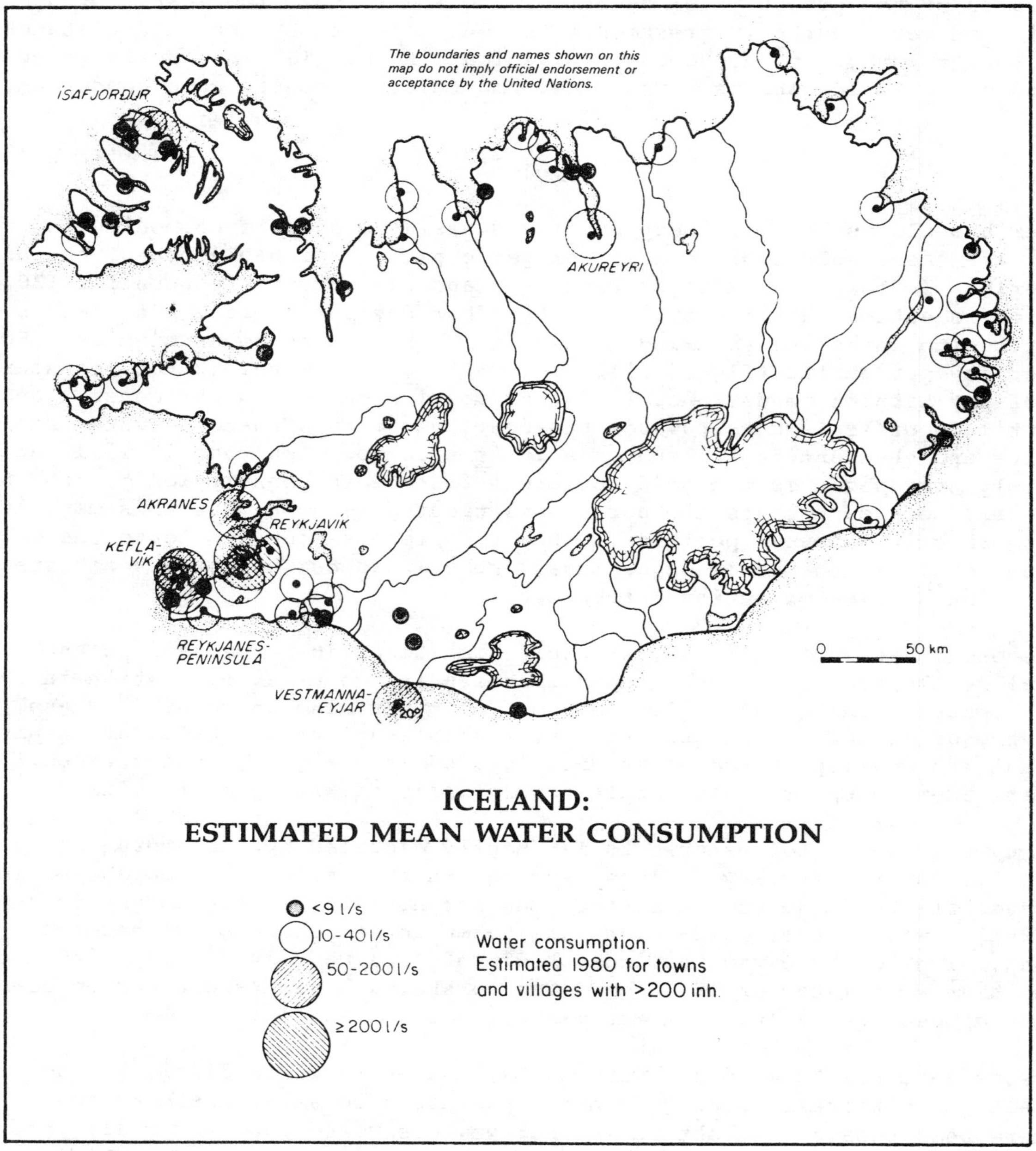

ICELAND: ESTIMATED MEAN WATER CONSUMPTION

3546.31x

Figure 42

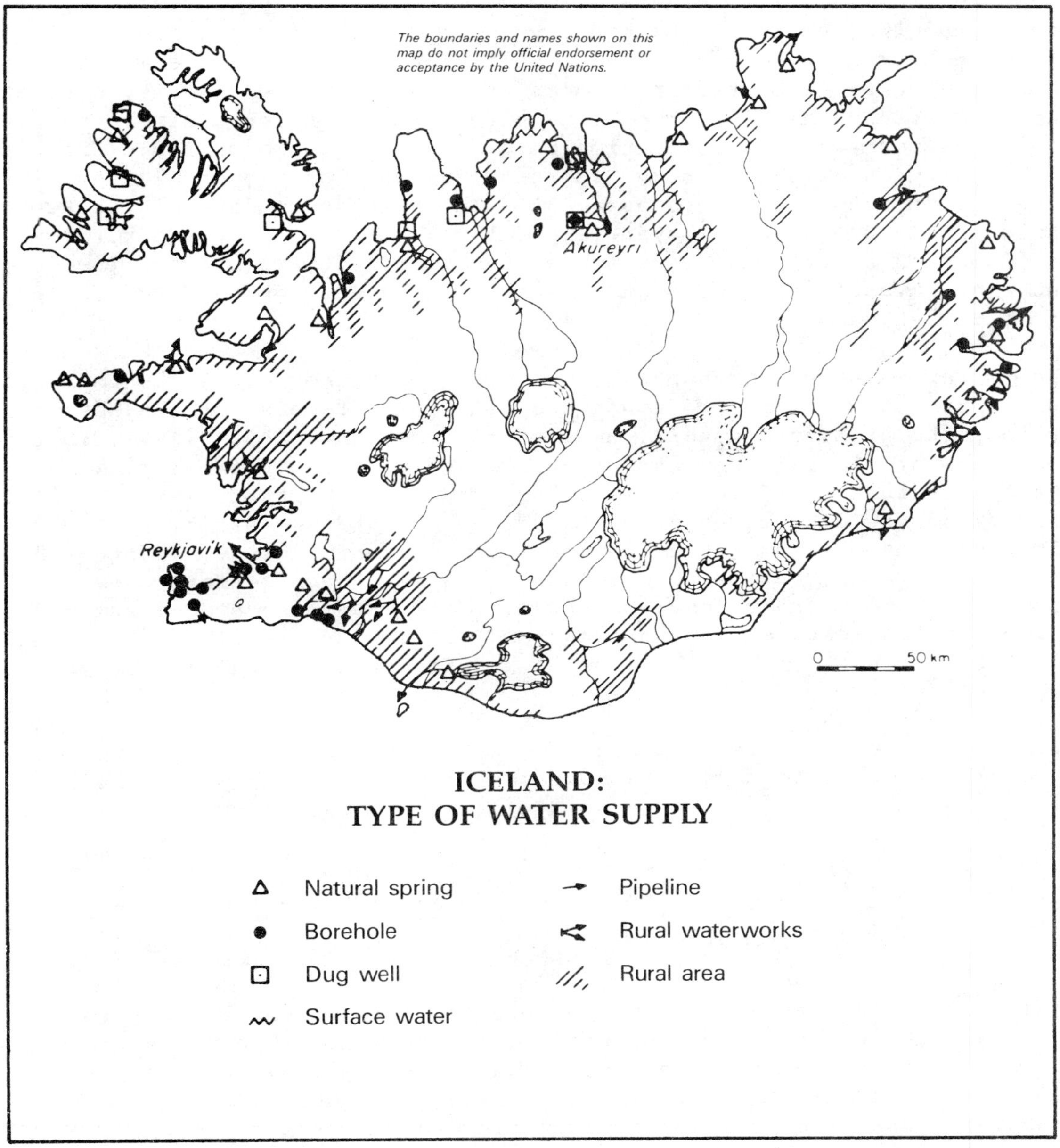

3546.32x

In the Quaternary regions, springs which issue directly from the bedrock (e.g., lava fields, fissures), or after flow for a short distance through sediments, are the chief source of supply, together with boreholes in the permeable bedrock (postglacial and interglacial lava). Various drill equipment is used depending on the lithology.

Most wells lie outside the settlements, in the thinly populated countryside or even in uninhabited parts of the country. The danger of contamination is, therefore, not great. Fencing off the well area and closing the wells is usually regarded as a sufficient preventive measure. In view of the scanty vegetation, thin soil covers and highly permeable aquifers, this is certainly in some cases inadequate. Fresh-water preservation is a problem on a major scale only in the south-west portion of the country, where 60 per cent of the population lives on 2 per cent of the total area (figure 43). In the western part of the Reykjanes Peninsula a thin fresh-water layer (appr. 50 m) floats on sea water (Sigurdsson, 1985). Cases of sea-water intrusion through overdraft are known. Investigations in that area have contributed significantly to the development of more adequate exploitation methods.

The water is seldom transported over long distances, 20-30 km being the maximum length of pipelines to towns and villages. Because of the severe climatic conditions, the pipes are usually dug and ploughed into the soils. Great use is made of pipes of plastic or other non-corrosive and flexible materials. Storage tanks are as a rule insufficient thus requiring higher capacities of well pumps and pipelines than would be necessary with adequate storage capacities. Pumping costs are high, one cause being the above-mentioned lack of storage. Another cause is a poor choice of pumps because no drawdown tests have been performed or in the misunderstood intention to lower costs. A third cause is damaged pumps due to rapid pressure changes, where the pumps are connected directly to the pipelines and the distribution net. A fourth cause is a deficient construction of wells because of a lack of proper advice.

Some rural regions are supplied with water from central water works, despite the long distances to be covered. Much use is made of slim plastic pipes, ploughed down in the soft soils.

C. Training, research and international co-operation

1. National agencies

There is no central agency responsible for fresh-water matters in Iceland. The water supply is a communal concern in the urban areas, or else a private business. Scientific advice has been given by the National Energy Authority, Hydro Power Division, and private firms, which also contribute technical advice and construction plans. No official approval is necessary for the instalment and operation of such waterworks.

Possibilities for scientific training in ground-water sciences are limited in Iceland. The University of Iceland in Reykjavík gives undergraduate courses in geology and hydraulics, but geohydrology, hydrogeology and fresh-water exploitation are not dealt with extensively. However, the United Nations Geothermal Training Programme in Reykjavík, a branch of the United Nations University, provides useful lectures and training courses for students from all continents.

Figure 43

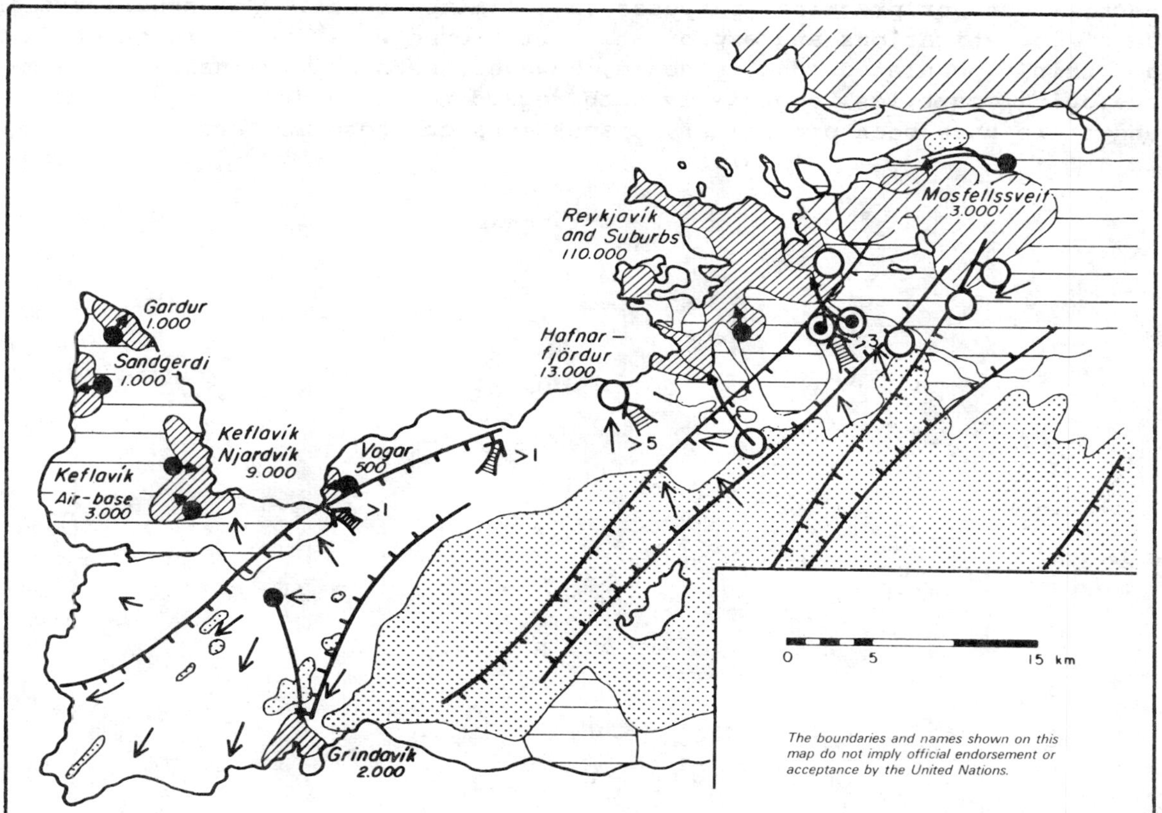

ICELAND:
GROUND WATER USE IN SOUTH-WEST

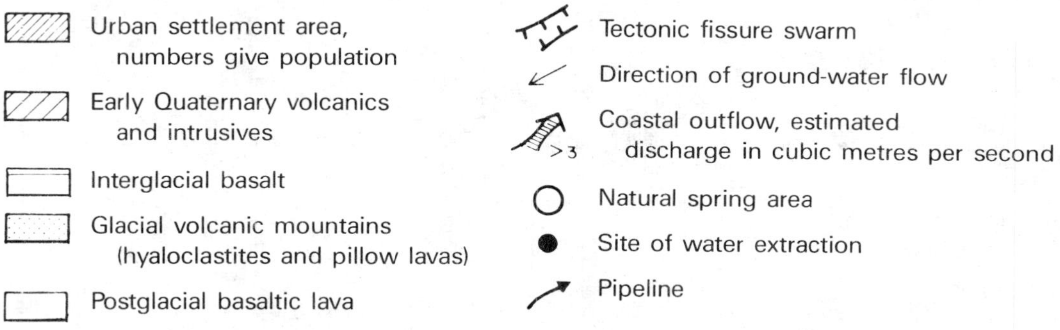

3546.33x

2. Technical assistance

Icelandic agencies participate in various international organizations, dealing with ground water and related matters. Co-operation with the Nordic countries has been carried out for practical purposes on a limited scale. The experience in Iceland of investigations and exploitation of ground water under rough conditions and in a thinly populated country could, however, be highly valuable for some of the developing countries, especially with regard to the hydrogeology of volcanic rock and there have been preliminary discussions on these matters.

References

Arnason, Bragi. Groundwater systems in Iceland. Soc. Sci. Island XLII. Reykjavík, 1976, 236 pp.

Fridleifsson, Ingvar Birgir. Geothermal activity in Iceland. Jökull 29:47-56. Reykjavík, 1979.

Hjartarson, Arni; Guttormur Sigbjarnarson and H. Karrenberg. International Hydrogeological Map of Europe. Sheet B2: Island. With Explanatory note by Hjartarson, A. et al. Bundesanstalt für Geowissenshaften und Rohstoffe, Hannover; and UNESCO, Paris, 1980.

Kjartansson, Gudmundur. Vatnsfallategundir. Náttúrufraöingurinn, vol. 15: 113-126. Reykjavík, 1945.

Rist, Sigurjón. Islensk vötn (Icelandic fresh waters), Raforkumálastjóri, Vatnamalingar. Reykjavík, 127 pp.

Saemundsson, Kristján. Fissure swarms and central volcanoes of the neovolcanic zones of Iceland. Geological Journal, Special Issue No. 10 (1978), Liverpool UK, 1978.

__________. Outline of the geology of Iceland. Jökull 29: 7-28. Reykjavík, 1979.

Sigbjarnarson, Guttormur. Vatnafraöi Pórisvatnssvaöis. NEA. Reykjavík, 1972. 64 pp. and map.

Sigbjarnarson, Guttormur; Páll Theodórsson and Bragi Arnason. Use of environmental isotope techniques together with conventional methods in regional groundwater studies. Nordic Hydrology, 7: 81-94. Copenhagen, 1976.

Sigurdsson, Freysteinn. Vatnajaröfraöi og jarövatn á utanveröum Reykjanesskaga NEA, Reykjavík, in press 1985.

Sigurdsson, Freysteinn; Freyr Thórarinsson, Snorri P. Snorrason, Kristjan Agústsson and Guttormur Sigbjarnarson. Integrated survey of a Freshwater Lens. Nordic Hydrological Conference, Preprint of Papers: I: 72-75. Helsinki, 1978.

VIII. NORWAY

Area: 323,900 km^2 | Population: 4.2 million

A. General

The Kingdom of Norway extends 1,770 km along the western part of the Scandinavian Peninsula and has the northernmost point of any European country. About 80 per cent of its population depend on water from works, each of which serve more than 100 inhabitants. Surface water is the main source for most, but many smaller units use ground water from drilled or dug wells. Only 15-20 per cent of the population depends on ground water, and only 30 or 40 ground-water works individually supply more than 1,000 inhabitants.

There is a huge loss of water through poor or old pipelines. Breakdown of the pipelines is partly caused by acidic surface water rich in humus, and is one of the reasons for a trend towards greater use of ground water.

The first municipal waterworks in Norway were constructed in the middle of the nineteenth century. However, because access to high-quality surface water was easy, ground water was not used until 100 years later, about 1955.

Increasing pollution of the surface water has led the public to demand a higher quality of water. As a consequence, the use of ground water has increased.

Tertiary uplift of the old eroded Caledonian mountain belt formed the present Scandinavian mountain range which stretches through the country from south-west to north-east. The highest mountains more than 2,000 m above sea level are located in southern Norway. The most extensive lowland areas are in south-eastern Norway but there are also areas well suited for agriculture purposes in south-western and central Norway.

The larger part of southern Norway drains south and south-east to the Oslo Fjord and the Skagerrak. A landscape dominated by long, narrow valleys is typical for these areas. Most of the other parts of Norway have drainage systems perpendicular to the coastline or to the deep fjords which cut into the country. The highest parts of the mountain range are generally near the coast and, therefore their drainage systems are, as a rule, smaller with short steep rivers.

Several times during the Quaternary period, the area which is now Norway was covered by ice. Large parts of the lowland areas were also inundated by the sea for a time after the last glaciation. As a consequence the morphology of the landscape is characterized by glacial erosion and accumulation and, in the lowlands, also partly by the levelling effect of young marine deposits. Generally, however, the cover of surficial deposits is sparse.

Norway is situated within the northern temperate zone in the belt of moving cyclones along the polar front. The cyclonic activity gives the region a climate with alternating warm and cold periods which can be either dry or humid. The climate along the west coast has a humid, maritime character. The mean annual precipitation ranges from about 1,000 mm in the outermost islands to more than 4,000 mm on high mountains near the coast of southern Norway. In south-eastern Norway and the inland areas of north-eastern Norway, the climate has a more continental character with a mean annual precipitation well below 1,000 mm.

The precipitation in most parts of Norway is fairly evenly distributed throughout the year. In the winter, however, most of the country has a temperature well below 0° C. The ground freezes and the precipitation is snow. In January only the outermost parts of western and south-western Norway have a mean temperature higher than 0° C.

These climatic variations are reflected in the hydrologic regions of the rivers as well in the variations in ground-water levels. The melting of snow in the spring causes both a flooding period in the rivers and a rise in water tables. In western Norway where winters are mild, the rise in ground-water level is insignificant.

B. Geology

Glaciation caused a well-defined division between the pre-Quaternary bedrock and the Quaternary surficial deposits, which remain unconsolidated. A weathered transition zone between soil and bedrock is generally lacking in Norway.

A simplified map of Norway's bedrock is presented in figure 44.

The Precambrian areas in the south and north-east belong to the Baltic Shield. Gneissic rocks of granitic and dioritic composition dominate, but granite plutons, metavolcanics and metasedimentary rocks also occur.

The Paleozoic sediments which once covered the Baltic Shield are now mostly eroded. Only remnants are left in areas down-faulted by Permian activity in south-eastern Norway. At the front of the mountain range, parautochthonous sediments are found underneath many of the nappes.

The Scandinavian Caledonides were formed during the mid-Paleozoic Era, and the nappes were folded and thrust towards the east and south-east over the Baltic Shield. Gneissic rocks of essentially Precambrian origin constitute most of the western part of the Caledonian belt at the present erosion level. To the east sequences of sandstone, quartzite, schist and other metasedimentary rocks of Upper Proterozoic and lower Paleozoic origin build up the thrust sheets. In the central parts of the Caledonian belt a number of rock types of sedimentary, magmatic, and metamorphic origin are represented. Most of them are from the Paleozoic Era.

The late Carboniferous and Permian activity in connection with the Oslo Rift caused faulting and jointing of the Precambrian Baltic Shield and its Paleozoic cover. The main fracture direction was N-S, but reactivation of older joints in other directions also took place. The Permian rocks of the rift zone comprise a suite of varied alkalic intrusive and extrusive rocks.

Tills from Quaternary glaciation are the most common surficial deposits in Norway, even though most of the till cover is rather sparse. In the valleys there are also glaciofluvial deposits of various form and origin. After the glaciation recent rivers have formed alluvial plains which now make up large parts of the main valley floors.

Marine clays cover most of the areas once inundated by the sea. In the same areas many of Norway's most prominent glaciofluvial and fluvial deposits occur, and in some places clay hides coarser sand and gravel.

A sketch of the surficial deposits in a representative Norwegian valley is presented in figure 45.

Figure 44

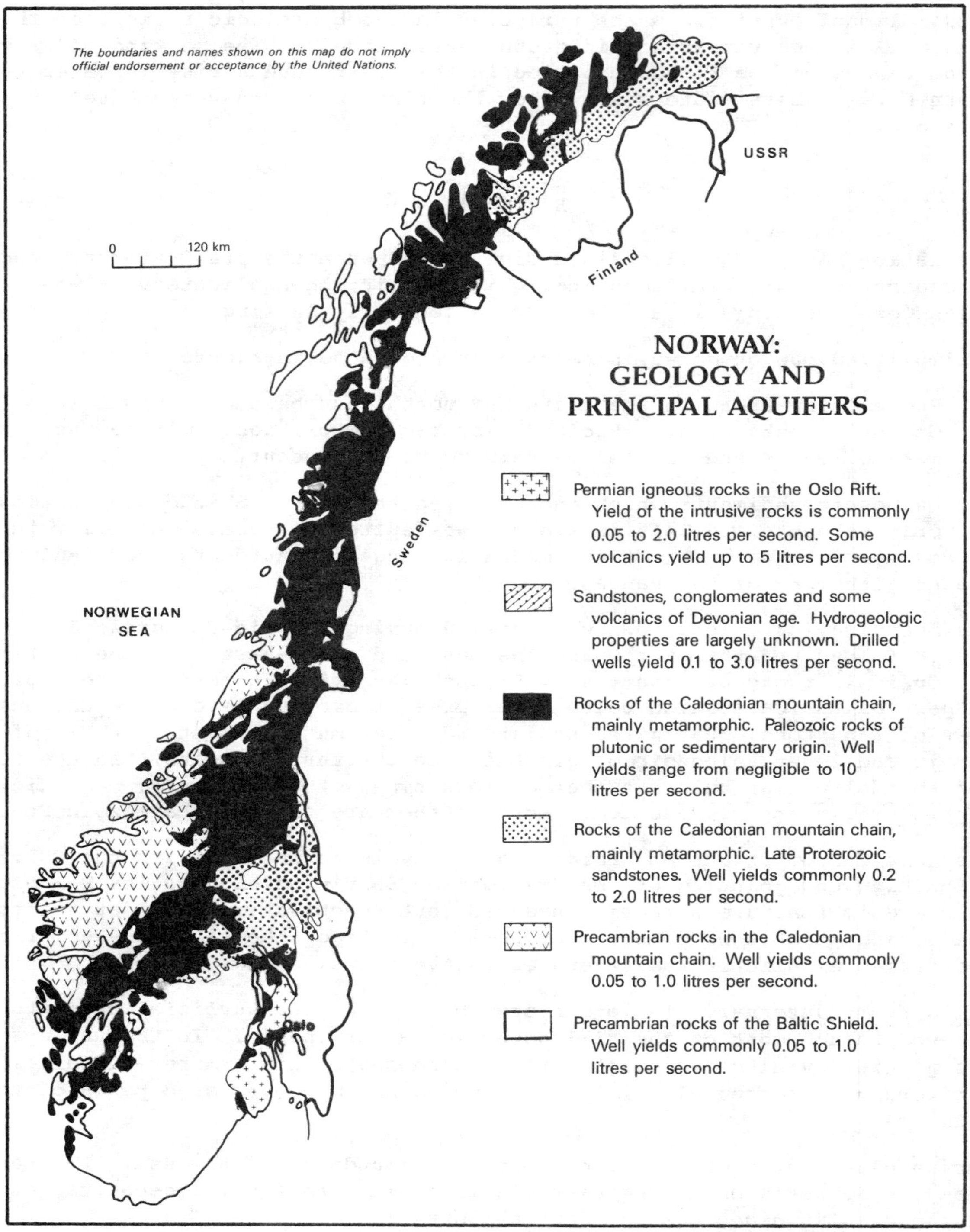

3546.34x

Source: Geological Survey of Norway

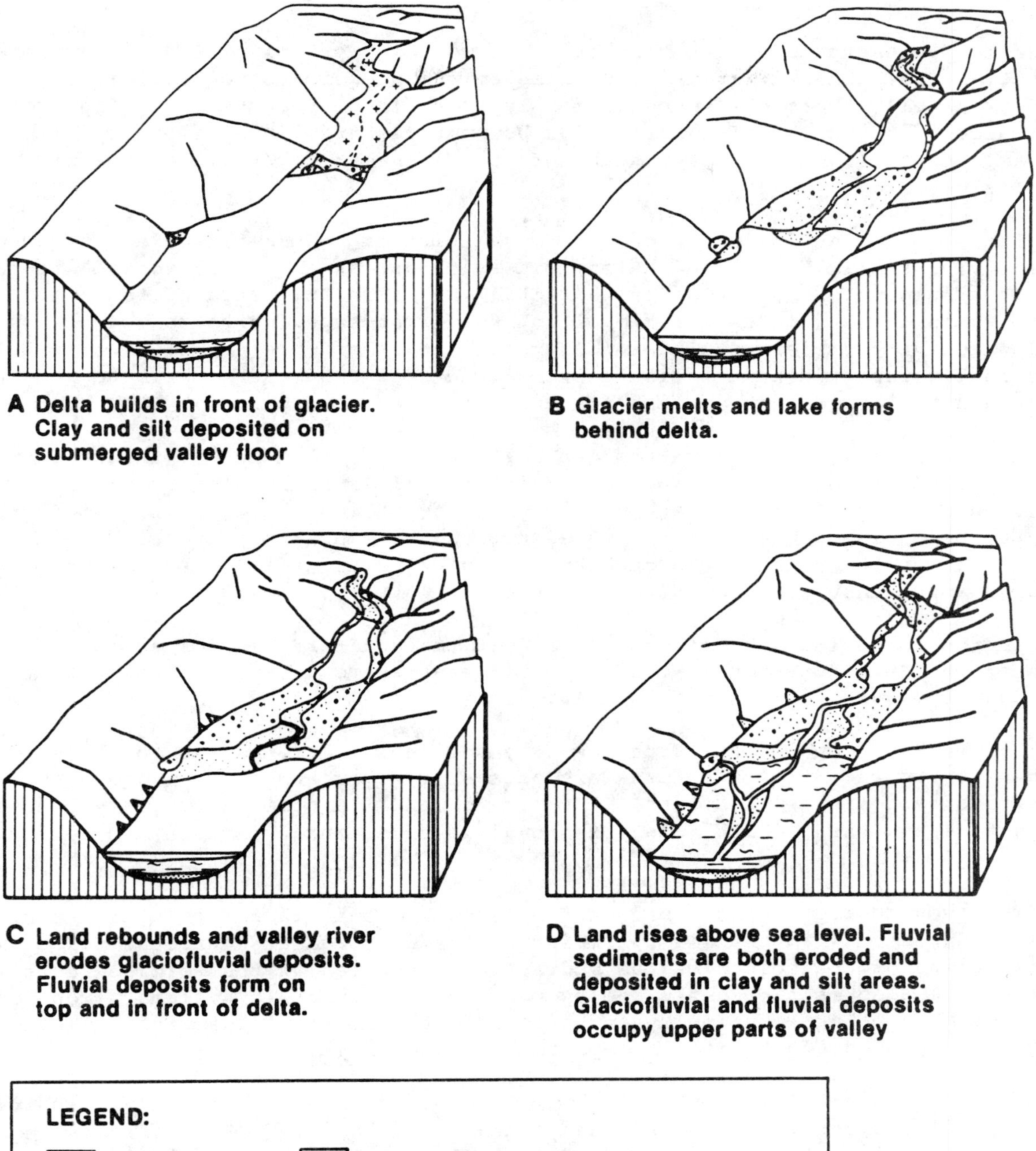

A Delta builds in front of glacier. Clay and silt deposited on submerged valley floor

B Glacier melts and lake forms behind delta.

C Land rebounds and valley river erodes glaciofluvial deposits. Fluvial deposits form on top and in front of delta.

D Land rises above sea level. Fluvial sediments are both eroded and deposited in clay and silt areas. Glaciofluvial and fluvial deposits occupy upper parts of valley

LEGEND:

- Water
- Glacier
- Basal till
- Marine clay and silt
- Glaciofluvial deposits (sand and gravel)
- Fluvial deposits (sand and gravel)

Source: Geological Survey of Norway

Figure 45. Norway, alluvial deposition in glaciated valley

C. Hydrogeology

1. Principal aquifers

With few exceptions, the bedrock in Norway is crystalline and lacks primary porosity. The permeability is controlled by the presence of open fissures. The yield of a well, therefore, depends mainly on the number and size of fissures penetrated by the well. The connection between fissures and fissure systems is also important. Some characteristics of the different crystalline aquifers are given below.

The yield of wells in the gneiss and granite of the Baltic Shield are commonly between 0.05 and 1 l/s. Mica schist, mica-rich gneiss and basic rocks like gabbro are, as a rule, less suitable for well drilling. The existence of regional fractures is in many cases more important than rock type. Fissures caused by Permian activity are of special importance. Yields of more than 10 l/s are occasionally reported from wells in such structures. On the other hand some wells drilled in the ridges between the fracture systems are dry.

Characteristic for the Caledonian belt are the importance of thrust planes and the low yield of soft rocks (such as shale, phyllite, and schist) compared to harder ones, many of higher metamorphic grade or of igneous origin. Some carbonate rocks occur as layers in sedimentary sequences, but karstic phenomena are not common. In spite of that, the carbonates are better aquifers than the sedimentary rocks to which they are commonly connected.

In the gneissic areas to the north-west the aquifers to some extent resemble those of the Baltic Shield. Within a small area there are many large variations in the yield from one locality to another.

In the metamorphic Late Proterozoic sandstones in the nappes to the east and south-east, imbrication structures are of special importance for the hydraulic properties, and yields of 0.2 to 2 l/s are common. In some places interbedded shales form aquitards which rule the ground-water drainage pattern and reduce the water yield.

The hydrogeologic importance of the Permian events in the Oslo Rift are, first of all, connected with tensile fissuring in the rift and its surroundings. With the exception of the volcanics the hydraulic properties of the Permian rocks within the rift are also mainly controlled by the regional fracture systems. Most of the volcanics have a better porosity and permeability; water yields of 1-5 l/s are common, and yields of 15 l/s are reported occasionally.

Aquifers in glaciofluvial and fluvial deposits, and occasionally in coarse-grained tills, occur most commonly along the river valleys. Many of these aquifers are connected with glacial or glaciofluvial ridges with some content of silt and clay. Erosion, sorting and resedimentation caused by recent river activity improve the hydraulic properties, and the fluvial deposits are usually preferred for ground-water withdrawal. Most of the aquifers in Quaternary deposits are small, and extensive alluvial plains are known only in a few of the larger river valleys, mainly in south-eastern Norway.

Along the west coast, Recent river deltas are found at the head of many fjords, and these are to some extent used for ground-water supply.

In the lowlands, some glacio-marine deltas constitute aquifers of considerable extent. With the exception of these deltas, valuable aquifers in surficial deposits are uncommon in the flat regions. Glacial or glaciofluvial deposits buried underneath the clay are seldom used.

2. Exploration and data collection

The initial steps in ground-water investigation generally follow conventional routines like the study of topographic, geologic and hydrogeologic maps and reports. Aerial photographs are commonly studied, and occasionally satellite pictures are used. This material commonly indicates whether there are aquifers of interest in an area, and makes up the basis for a field reconnaissance study. Further investigations depend on the type of aquifer.

To meet moderate ground-water demands from aquifers in crystalline bedrock, drilling is commonly carried out immediately after the field reconnaissance. The cost of further investigations would easily exceed the cost of drilling and often there is just one feasible drill site.

Detailed investigations with special emphasis on structural geology are needed to achieve larger ground-water volumes from crystalline bedrock. The poor storage properties, based solely on fissure porosity, cause large drawdown cones in these aquifers.

Applied geophysics, including seismic refraction and electromagnetic VLF (very low frequency) measurements, are in some cases used in the mapping of the water-bearing fissure systems. The Geological Survey of Norway, several universities, and some private firms have suitable geophysical equipment.

In Quaternary deposits the investigation routines include refraction seismic, geoelectrical work, and sounding by steel rods driven into the ground, either by repeated impact or by rotation. To sample ground water, a slotted test pipe 30-50 mm in diameter is commonly used. Solid material may also be sampled, through this pipe, either by pumping or by flushing. Even though the most coarse-grained material is left outside the pipe, and a part of the clay fraction is washed away with the water, an experienced geologist is usually able to get a good impression of the underground conditions.

The whole operation of sounding, pipe driving and sampling is normally carried out with portable equipment including a motorized hammer drill. Heavier drills must be used where the deposits include a large number of stones or blocks or where a larger-diameter pipe is necessary. Heavier equipment is also recommended if more detailed samples of the solid material are desired. Sampling is then carried out by using compressed air or special devices for taking undisturbed samples.

Preliminary test pumping may be done using the test pipes, but is usually done after construction of the production well. To test water quality, the Norwegian health authorities recommend a pumping period of one year.

The Geological Survey of Norway runs the main archive for hydrogeologic data in Norway. This archive is partly manual, but is being reorganized so that the data can be analysed by computer. The supply of well data depends upon voluntary contributions from the well drilling companies. As most data never reaches the data bank, legislation mandating the reporting of well data is needed.

The records include depth of drilling in bedrock and overburden, fissures in the bedrock, grain size distribution in Quaternary deposits, water yield, casing and filters. The records will also include digitalized areal data and can automatically plot out a range of hydrogeologic data or maps or in tables. The National Institute of Public Health also stores data concerning ground-water works and quality.

A ground-water monitoring network has recently been established by the Geological Survey of Norway and the Norwegian Water Resources and Electricity Board. Variations in ground-water levels, temperature and chemical composition are monitored by means of a nation-wide network of observation and sampling stations. About 50 stations are in operation, most of which include several measuring points. The main aim of the project is to collect data for reference purposes and for environmental control.

3. Development

The oldest tradition in the use of ground water is the utilization of natural springs, often combined with a dug well at the spring site. Sources of this type are still in use; the wells are now commonly lined with concrete rings or with poured concrete walls. To control sand and silt entry, gravel is back filled around the walls and at the base. Older dug wells lined with timber, bricks or cut stones are also known. To reach a deep water table, such wells may be 20 or 30 metres deep.

An easier way to tap deep ground water in unconsolidated deposits is by driven screened wells. Pipes with a diameter of 50 mm are commonly used. The small diameter and screen length (usually a metre) require a good hydraulic conductivity in the aquifer. Use of a suction pump requires a shallow water table. A maximum yield of about 5 l/s is reported from this type of well which is generally used only for individual water supplies or small waterworks.

Large screened wells in Quaternary aquifers were first constructed in Norway after 1950. The depth is usually less than 30 mm; only a few deeper wells are known. The diameter of the filters is normally between 100 and 400 mm.

A temporary casing of larger diameter is often used to make an artificial gravel wall around the well screen. Manufactured gravel packed well screens are now seldom used in the Norwegian fluvial and glaciofluvial aquifers, and continuous slot filters are gradually replacing other filter types. Casing tubes are usually made of steel, but the use of stainless steel well screens has become frequent; PVC casings and screens are not common.

Air-percussion rotary drilling is the usual well-construction technique for wells up to a diameter of 300 mm. For eccentric reamer drills the hole is slightly larger than the casing, which is continuously inserted into the borehole. For larger diameters, cable tool drilling is still used. The relatively high content of stones and blocks in Norwegian glaciofluvial and fluvial deposits make them unsuitable for mud-flushing rotary drilling without casing.

Large screened wells usually yield 20 to 100 l/s. Even if a higher capacity is possible, it is usually preferable to construct two or more wells within a short distance.

Most large wells are situated along the main rivers, and bank infiltration commonly takes place to compensate for the withdrawal of ground water. Sometimes artificial recharge is necessary because of the occurrence of layers with low permeability along the river bottoms. Infiltration trough excavated basins are then usually used.

The mining of open wells in hard rock has been replaced by the use of modern air-percussion rotary drills, and about 3,000 wells are drilled each year. Depths range from about 20 to 250 m, with a mean depth of 60-70 m. Well diameters of 115 to 150 mm are the most common. Eccentric reamer drilling with continuous casing a few metres into the bedrock is used to penetrate overburden with a thickness of more than a couple of metres. Subsequent drilling continues with a drill bit of smaller diameter. A PVC casing is used for overburden not more than 2 or 3 metres thick. Some boreholes are lined with a slotted PVC casing to avoid blockage by falling stones after the drilling.

As previously mentioned, in Norway the yields of most wells in bedrock areas are relatively small, and thus they are often used to supply water to just one household or a few houses. However, some municipal waterworks have also concentrated on the ground water in bedrock aquifers. Efforts have begun to evaluate the ground water potential of the crystalline bedrock in Norway.

Some inclined wells are drilled to hit and penetrate steep fissures or fracture zones. The drilling angle is normally not more than 30°. Hydraulic fracturing has proven successful in stimulating the capacities of low-yield boreholes. The use of explosives has usually been replaced by high-pressure pumping of water into the open hole below a packer. Sometimes the water injection is carried out between two packers.

Submersible centrifugal pumps and jet pumps have replaced piston pumps in deep wells in recent decades. The use of submersible pumps is certainly the cheapest for high capacity wells, but it is often difficult to lower them into, or remove them from, narrow boreholes. Occasionally pumps are lost in 110-mm wells and, for small supply units, jet pumps are often preferable.

Overdraft problems occur with bedrock aquifers, especially in areas with many small water supplies for single houses. In many cases such problems arise after the drilling, often in connection with dry summers or long, cold winters. Fissured bedrock aquifers commonly provide the ground-water supply to aquifers in surficial deposits, and drilled wells thus can affect shallow wells in the overburden.

Geotechnical problems due to overdraft for water supply are uncommon, but examples of buildings sinking in clay areas are known. Geotechnical ground-water problems are more common in connection with larger engineering works like mining tunnels or large building sites.

4. Water quality

The natural quality of Norwegian ground water is generally good. To attain drinking water quality, oxidation and an increase in the pH value are often the only measures necessary.

A high content of iron and manganese is otherwise the most common problem.

Such difficulties occur both in surficial deposits with low oxidation potentials (which are often combined with slow ground-water movement) and in bedrock with a high iron content. In municipal water works these metals are generally removed by oxidation and filtration.

High fluorine content is a problem in connection with some of the magmatic Permian rocks. Fluorine levels of more than 10 mg/l have been reported.

Some Paleozoic shale and limestone contain sulphides which render the ground water unfit for domestic use. This problem is also known from some other rocks with a high pyrite content.

Sea water intrusion into bedrock wells is frequently a serious problem in the coastal areas of Norway. Experience has shown that it is difficult to predict whether such problems will arise. Special care should be shown in pumping wells on small peninsulas and islands.

Relict (fossil) salt water exists both in the Quaternary sediments and in fissures in the bedrock in those parts of Norway inundated by the sea after the last glaciation. This sometimes causes water quality problems far from the present coast. Deep wells and the pumping of large quantities of water increase the risk for salt water problems in these regions.

The nitrate content of Norwegian ground water is, as a rule, low but some problems exist in shallow ground water in agricultural areas.

Local contamination problems are known in connection with waste disposal sites, animal farms, and poor waste-water handling. Some industries discharge poisonous chemicals into the ground along with sewage or waste water, without ensuring that decomposition or retention is sufficient.

Some acidification of ground water has occurred in the upper parts of the ground-water zone in areas where the surface water is subject to acid rain. However, most of the slightly acidic ground water is probably the result of natural geologic and climatic factors.

D. Training, research and international co-operation

A complete education in hydrogeology is not taught as a separate subject at any of the Norwegian universities. Education and research in some areas of the subject is, however, provided at several colleges and universities. Basic training in hydrogeology is given at the Agricultural University of Norway, and there is co-operation on this subject with the University of Oslo. Technical fields like water well technology and water treatment are more thoroughly dealt with at the Norwegian Institute of Technology of the University of Trondheim.

Government agencies involved in ground-water research are the Geological Survey of Norway (NGU), and the Norwegian Water Resources and Electricity Board (NVE). NGU is responsible for general ground-water mapping and registration, and is engaged in the exploitation of the ground-water resources for a variety of purposes. NVE works on ground water in connection with the regulation of rivers for hydropower development. Their main objective is to investigate the effects of river regulation on the ground-water régimes of bordering river plains.

For several years Norway has co-operated with developing countries. Development assistance is offered partly as economic support through such international organizations as the United Nations and IBRD and their special programmes and funds (e.g., UNDP, WFP, WHO, UNICEF, IDA), and partly as programmes extended directly to countries.

The bilateral co-operation is administered through the Norwegian Agency of International Development (NORAD), which is under the auspices of the Royal Norwegian Ministry of Development Co-operation.

Bilateral projects concerning ground-water supply have concentrated on Africa, primarily rural, in Kenya, the United Republic of Tanzania, Zambia and Zimbabwe. Through non-governmental Norwegian organizations, ground-water projects are also run in some West African countries, e.g., Cameroon, Mali, Senegal and Sierra Leone.

Some ground-water projects include all the work from hydrogeologic investigation to construction of the water works. The projects are usually organized through Norwegian consultant firms.

The Nordic countries (Denmark, Finland, Norway, Sweden) work in close collaboration in several aspects of hydrogeology. This co-operation is partly carried out within the Nordic Hydrological Programme (NHP) and the Nordic Association of Hydrology. The activities of the Nordic ground-water monitoring programmes have, for instance, been co-ordinated by an NHP working group.

References

Aars, Ø. Groundwater and the regulation of water courses. Striae 4, 95-98, 1978.

Aune, B. Map: Mean annual observed precipitation in Norway. Norwegian Meteorological Institute, 1981.

Englund, J.-O. Chemistry and flow patterns in some groundwaters of south-eastern Norway. Norges geol. Unders. 380, 221-234, 1983.

Englund, J.-O. and J. A. Myrstad. Groundwater chemistry of some selected areas in Southeastern Norway. Nordic Hydrology 11, 33-54, 1980.

Henriksen, A. and L. A. Kirkhusmo. Acidification of Groundwater in Norway. Nordic Hydrology 13, 183-192, 1982.

Kirkhusmo, L. A. (Ed.). International Hydrogeological Map of Europe, scale 1:1,500,000. Sheet C2, Trondheim. Bundesanstalt für Geowissenschaften und Rohstoffe, Hannover; and UNESCO, Paris, n.d. (in press).

Nordberg, L. (Ed.). National groundwater networks of the Nordic countries. International Hydrological Programme, Report 3, 1980.

Persson, G. (Ed.). International Hydrogeological Map of Europe, scale 1:1,500,000. Sheet C3, Oslo, Bundesanstalt für Geowissenschaften und Rohstoffe, Hannover; and UNESCO, Paris, 1979.

Persson, G. (Ed.). International Hydrogeological Map of Europe, scale 1:1,500,000. Sheet D2, Haparanda. Bundesanstalt für Geowissenschaften und Rohstoffe, Hannover; and UNESCO, Paris, n.d. (in press).

Sigmond, E. M. O., M. Gustavson and D. Roberts. Geological map of Norway at scale 1:1,000,000. Geological Survey of Norway, 1984.

Soveri, J. (Ed.), Acid Groundwater in the Nordic Countries. Nordic Hydrological Programme, Report 3, 1982.

IX. POLAND

Area: 312,680 km^2 Population: 37.8 million

A. General

The Polish People's Republic is mainly a flat country. The Polish Plain is the main morphologic feature (about 72 per cent) and is mostly covered with a thick cover of Tertiary and Quaternary deposits (figure 46). Mean elevation of the plain, which is of glacial origin, is 173 m above sea level. There is an assemblage of glacial topography developed in this area. The southern boundary of the Polish Plain coincides with the northern limits of the Sudetes Mountains and Malopolska Upland.

The Middle Polish Uplands consist of Paleozoic and Mesozoic formations with elevations less than 600 m. High mountains occupy the southern part of the country, the Alpine folded Carpathian Mountains with highest elevations reaching 2,500 m in the Tatra Mountains. The south-western part is occupied by the Sudetes Mountains of Variscan and Caledonian origin; they are part of the Bohemian Massif reaching 1,600 m.

These high mountains constitute the upper reaches of watersheds which drain to the Baltic Sea (99 per cent of the country), the North Sea and the Black Sea (figure 47).

The Vistula (Wisla) and the Odra (Oder) are the two rivers which, with a few small rivers falling directly into the Baltic Sea, drain the country. The Vistula River originates in the Carpathian Mountains. It is 1,047 km long and the drainage basin in Poland is 168,700 km^2. The mean flow of the river is 1,090 m^3/s. The Odra River originates in Czechoslovakia. Its total length in Poland is 742 km, its drainage basin is 106,100 km^2 and its mean flow is 580 m^3/s.

Runoff values range from 1.8 to 5 $l/s/km^2$ in the mountainous southern part of the country. In the course of a year, runoff ranges from less than 1,000 to as many as 2,000 $l/s/km^2$ in the mountains. Average annual runoff in Poland is 5.5 $l/s/km^2$.

The Odra River is navigable along its total length and is an important traffic route from industrial Silesia to the harbours of Szczecin and Swinoujscie. The lower Vistula River is navigable but upstream is only locally so.

There are some few dozen reservoirs in a country whose hydrographic system has a total capacity of 2.7 km^3. The main purpose of the reservoirs is the storage of excess water. Poland has many lakes in the northern and north-western parts of the country. There are over 9,000 of them and their total surface area is about 3,200 km^2. The storage reservoirs and lakes occupy 4,280 km^2 surface area or about 1.5 per cent of Poland. Together they contain about 19 km^3 of fresh water.

The climate of Poland is moderate with four distinctly differentiated seasons. It is transitional between the maritime and continental types. Mean annual air temperature ranges from below 6° C in the north-eastern part of the country and in the mountains to over 9° C in the south-western part. The SW-NE direction of the axis of temperature gradients is in evidence in all seasons of the year as shown by isotherms of July and January.

Figure 46

POLAND: GEOLOGY

3546.35x

Figure 47

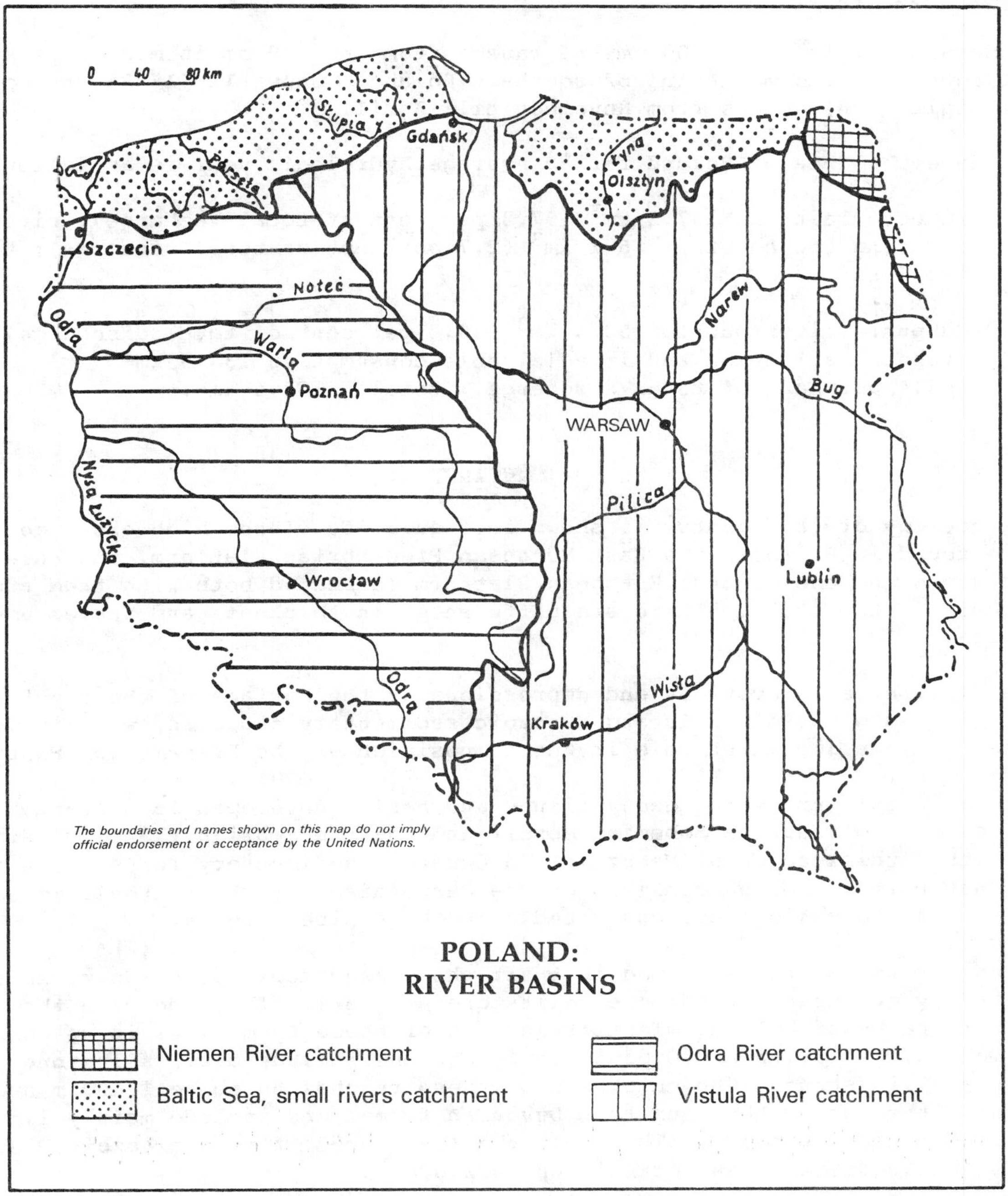

3546.36x

Because of this climatic régime the growing season ranges from about 185 days in the north-eastern part to about 225 days in the south-western part of Poland. In this period over 75 per cent of the annual potential evaporation from free-water surface (average annual value 744 mm) and 90 per cent of effective evaporation (430 to 475 mm) take place.

Mean rainfall is about 600 mm and ranges from 460-600 mm in mid-western Poland to 800-1,800 mm in the mountains of southern Poland. Snowfall (15-25 per cent of total precipitation) occurs from November till April.

The water balance of Poland for an average hydrologic year is as follows:

1. Input: rainfall 187.2 km^3 (97.3 per cent of total recharge); river water entering the country: 5.2 km^3 (2.7 per cent of total recharge); total: 192.4 km^3.

2. Output: river outflow 58.6 km^3 (30.6 per cent of total discharge); evapotranspiration and irreclaimable consumption 133.8 km^3 (69.4 per cent of total discharge); total: 192.4 km^3.

B. Geology

The geology of the country is shown in figure 46. Three high-order geologic units are found in Poland: the East European Precambrian Platform, the Paleozoic orogenic zones and the Central European Platform connected both with them and the Alpine chain in the south. There are differences in thickness and facies among these units.

The flat, broad elevations and depressions in the surface of the crystalline basement, filled with Paleozoic and Mesozoic sedimentary deposits, with reduced thickness, play an important role in the composition of the Precambrian Platform.

Paleozoic epicontinental geosynclines and basins developed in western and central Poland, between the Bohemian Massif and the Precambrian Platform, are covered with a thick cover of Mesozoic and Cenozoic sedimentary rocks. The southernmost part of Poland consists of the Carpathian flysch and their immediate foreland, whose formations are connected with the Alpine orogeny.

Paleozoic rocks are developed in their whole sequence. Cambrian rocks are represented by quartzite, sandstone, siltstone and marl. They occur in the central and northern parts of Poland. Total thickness of these formations is up to a few hundred metres. Silurian and Ordovician formations include clay, siltstone, sandstone and limestone. The thickness of these reaches to above 1,000 m and they occur nearly throughout the country. Devonian formations include mainly limestone, dolomite, marl and sandstone. They occupy a large area of the Paleozoic Platform. The total thickness of these rocks is up to 2,000 m.

The occurrence of Carboniferous strata is connected with the area of the Paleozoic Platform and the southern part of the Precambrian Platform. The thickness of formations ranges from a few hundred metres to about 10,000 m. Lower Carboniferous layers include mainly limestone and shale whereas the Upper Carboniferous is represented by a complex of shale, sandstone and siltstone with intercalations of hard coal.

The Permian formations are discordant on the Carboniferous. They are represented by differentiated deposits located in central and northern Poland. The total thickness in the central part of the sedimentary basin is more than 1,000 m. Rotliegendes (Lower and Middle Permian) formations include conglomerate, sandstone and siltstone; the Zechstein (Upper Permian) formations are represented by dolomite, marl, shale and evaporites.

Triassic rocks occur in the central and northern parts of the country. The thickness of these formations reaches, in the axial zone of the sedimentary basin, more than 3,000 m. Lower Triassic rocks are mainly represented by shale and sandstone. Middle Triassic rocks consist of limestone and dolomite and Upper Triassic rocks are developed as clay, siltstone, sandstone and locally dolomite and gypsum.

Jurassic formations occur in central and northern Poland. Their total thickness exceeds 2,000 m. Lower Jurassic formations are represented by sandstone, gravel, conglomerate, claystone and marl. Middle Jurassic rocks consist of sandstone and sand, siltstone, shale and marl. Formations of Upper Jurassic age include mainly limestone, clay, marl and minor sandstone and gypsum.

Cretaceous strata are developed mainly in eastern and central Poland. The total thickness is from 100 to more than 2,000 m in the Mogilno Trough. Lower Cretaceous rocks include siltstone, sandstone and sand; the Upper Cretaceous formations consist of limestone and marl.

Tertiary formations occur almost in the whole country excluding the Middle Polish Uplands. Their total thickness ranges from a few tens of metres to 500 m in the Polish Lowland where the rocks are mainly sandstone, sand and clay of Oligocene, Miocene and Pliocene ages. In the Carpathian Mountains and their immediate foreland the marine Tertiary sediments comprise clay, shale and sandstone. Their aggregate thickness is more than 1,000 m.

Quaternary deposits reach a maximum thickness of more than 300 m and their most complete lithofacies development, with a predominance of glacial till, glaciofluvial sand and silts as well as fluvial and ice-dammed lake deposits, in the Polish Plain.

In the Middle Polish Uplands the Quaternary formations are generally less than 50 m thick. Only in buried valleys, filled with sand mostly, are these formations 100 or more metres thick.

In the Sudetes and Carpathian Mountains the Quaternary deposits are generally not thick. Only in the intermontane basins are some 50 m thick and a few areas, due to subsidence, even more than 100 m.

C. Hydrogeology

1. Prospection and exploration

Information about ground water in the late nineteenth century was limited to springs and shallow aquifers tapped by dug wells. Intensive industrial growth at the end of the last century caused urgent demand for water. A few thousand wells were drilled in the period 1890-1914, about 500 of them to depths of 200-300 m and, in the Łódz area, even 600-1,000 m. They recovered ground water mostly from Miocene fine sands, Oligocene greensand, Lower Cretaceous sand and sandstone and from Cretaceous, Jurassic and Triassic limestone, dolomite and marl. These were mainly artesian aquifers. No systematic results of hydrogeologic investigations were, however, made then.

Hydrogeologic activities in the field of mapping, regional prospecting and exploration for potable water expanded in the years 1960-1976. A dozen or so monographs of the ground-water basins with maps at the scale of 1:100,000 and 1:500,000, a set of the hydrogeologic maps of Poland at scale 1:300,000, the ground-water resources atlas of Poland, and many studies and calculations of ground-water reserves of large communal and industrial water intakes were made. Hydrogeologic exploration for appraisal of water inflow into mining works and drainage has been carried out on a large scale.

In the last decade, further progress of hydrogeologic activity was achieved. The hydrogeologic map of Poland at scale 1:200,000 was published and detailed studies of 25 areas of importance as regards water supply were carried out. Detailed studies of ground-water conditions and drainage works for coal mines, ore deposits and entire coal basins were done; for the lignite mine in Bełchatów, among others, where the cone of depression exceeds 200 km^2 and inflow rate is 0.15 km^3 per year.

A monitoring network for potable ground-water zones in Poland has been organized which will comprise about 40 hydrogeologic observation stations of the first order with data recording instruments and almost 1,000 second- and third-order measurement sites (figure 48). By 1988 about half of the needed work had been done. A computerized data bank system for storing ground-water data called Hydro-Bank has been working for a few years. It includes data from 80,000 wells in six regional banks.

A new ground-water resources atlas of Poland was planned for 1990 to include the latest information on quantity and quality of potable water in the country. Hydrogeologic maps of Poland at scale 1:50,000 were also to be prepared.

The state of exploration of potable resources of ground water in Poland is not uniform in spite of a quite dense network of drilled wells (0.4 per square kilometre) and dug wells (4.5 per square kilometre) as well as regional investigation and mapping. Usually aquifers to depths of 50-60 m, in some cases 100 m, are well investigated, as are large urban and industrial centres where ground water is mainly recovered from deeper aquifers. Rural areas are explored very poorly.

2. Ground-water provinces and resources

Ground water with total dissolved solids less than 1,000 mg/l occur in various geologic structures down to 300 m and occasionally to 600-1,000 m. Compared with surface reservoirs, aquifers offer less possibility for large-scale projects; nevertheless, they are growing in significance for urban and agricultural purposes due to their commonness, stable quality and, above all, resistance to degradation by industrial and urban pollution.

Figure 48

The boundaries and names shown on this map do not imply official endorsement or acceptance by the United Nations.

POLAND: HYDROGEOLOGIC OBSERVATON STATIONS

[1] Hydrogeologic observation station of the first order with data recording instruments

• Hydrogeologic measurement site of the second and third orders

ŁÓDŹ Regional Hydro-Bank seat

⊢⊣⊢⊣ Boundary of regional bank area

3546.37x

Resources in aquifers suitable for development by means of drilled wells are estimated to be 1,060 km^3. This amount, the equivalent of 20 years of discharge, occurs mainly (67 per cent) in Cenozoic structures, renewable at the rate of 20 km^3 per year in the main aquifers. The safe yield of aquifers does not exceed 70 per cent and is thus relatively limited. In addition, their distribution is uneven and they concentrate chiefly in north and east central Poland. In the western, west central and southern parts of the country resources suitable for an effective concentrated exploitation are lacking.

Figure 49 shows the distribution of fresh ground water against six typologic units and thirty (I to XXX) hydrogeologic regions.

Of largest extent, particularly in north and west Poland, are Cenozoic (including marine Tertiary) aquifers which form seven relatively shallow (up to 300 m) aquifers classified as macro- and meso-regions (I-VII). These are structurally and hydrodynamically differentiated Tertiary aquifers with overlapping and locally connected Quaternary aquifers and with a slight contribution from Cretaceous or Jurassic fissure waters at the base. This group of regions is characterized by a usually high ground-water resources potential except for the western area (Wielkopolska and Wrocław regions) where the index of ground-water potential is lower.

A similar typologic unit consists of an assemblage of Cenozoic aquifers, including continental Tertiary, and of more numerous Mesozoic (Gdansk, Kujawy-Mazowsze and Kalisz regions) or Paleozoic aquifers mainly of a fissure system (XV).

In the successive typologic unit of the northern part of the country, the continental Tertiary lacks aquifers, but some Quaternary multilayered aquifers occur along with an equal number of Mesozoic fissure water systems (Podlasic-Mazury and West Pomorze regions). Generally these are small units of the size of micro- and meso-regions. In north-east Poland (Mazowsze-Poldasie region) where the Quaternary sediments are thick the ground-water resources potential is high, but where thin and fissure water are the rule, the potential is considerably less.

A particular structural position is held by the Carpathian foredeep (XXVIII, XXVII) filled with marine Tertiary of low storage capacity, and with fissure waters occurring locally in the Mesozoic or Paleozoic basement. This deep basin contains mineral waters with hydrogen sulphide. Fresh water with total dissolved solids less than 1,000 mg/l occurs almost exclusively in shallow (20-30 m) Quaternary aquifers. In general, this type of aquifer counts among the least productive.

There is a genetic and structural connection between the Carpathian foredeep and the next typologic unit, the Carpathian flysch (XXIX, XXX). They form an elevated fold-nappe structure with discontinuous fissured aquifer systems occurring at depths of 100-200 m in flysch sandstone and fissured shale. The Mesozoic formations of the Tatras overlain by the Eocene of the Podhale Valley are characterized by better storage capacity and higher temperatures. In bigger river valleys as well as in the Nowy Targ and Nowy Saçz valleys the Quaternary sandy gravel sediments play a major role. But, in general, these regions are poorest in ground waters with the index of potential below 50 or even below 10 $m^3/d/km^2$.

The last type of Mesozoic of Paleozoic fissured aquifers includes a large and hydrogeologically diverse group of regions in south-west and east central Poland (XVI, XVII, XVIII, XIX, XXVI). Here two subtypes can be distinguished:

Figure 49

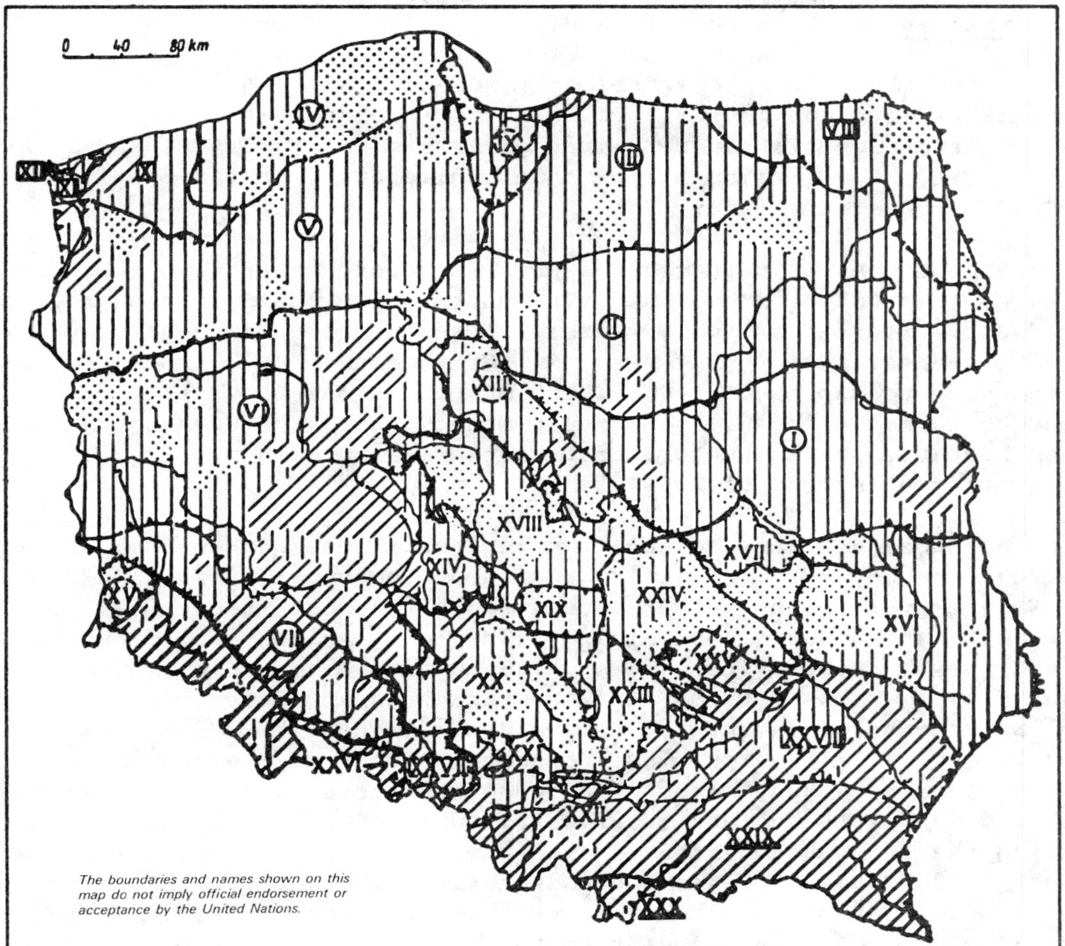

POLAND: HYDROGEOLOGIC REGIONS

HYDROGEOLOGIC REGIONS AND THEIR GROUND:

- Number of hydrogeologic region
- XVI Boundary of closed hydrogeologic region
- Boundary of open hydrogeologic region
- Quaternary and Tertiary continental aquifer
- XXVII Cenozoic and Mesozoic aquifer
- Quaternary and Mesozoic aquifer
- IX Quaternary aquifer with negligible number of marine Tertiary, Mesozoic and Paleozoic multilayered aquifers
- XVI Carpathian flysch and Quaternary aquifer with negligible number of Mesozoic multilayered aquifers
- ||||| Mesozoic and Paleozoic aquifer with Quaternary multilayered aquifers

GROUND WATER RESOURCES POTENTIAL:

- VIII High
- XXIX Moderate
- //// Low

3546.38x

1. Fold-block structures, generally elevated, Paleozoic or older, occurring in the Sudetes (XXVI) and in the Swietokrzyski (XXV) regions, i.e., in old mountain massifs, and

2. Fold structures of a highland or plateau embracing the other regions of central and east central Poland.

The first subtype is characterized by a scarce ground water found in relatively shallow (200-300 m) early Paleozoic or Proterozoic structures. More favourable conditions persist in synclines and troughs filled with Middle Devonian rocks in the Swiety Krzyz region and with Carboniferous, Permian and locally Cretaceous in the Sudeten region.

The second subtype includes regions where Mesozoic aquifers prevail and which are known to be the richest in Poland. The units (Lodz and Nida regions) form multilayered aquifers in the Upper Jurassic, Lower Cretaceous, Upper Cretaceous and Cenozoic. The deepest ones (Upper Jurassic, Cretaceous), still bearing fresh water, extend down to depths of 1,000-1,200 m. The elevated units (e.g., the Triassic and Jurassic of the Silesia-Cracow monocline or the Cretaceous of the Lublin-Radom Plateau) are characterized by the presence of one, rarely two, aquifers occurring at depths of 100-200 m.

In central and south-west Poland, two regions where the human impact is the most pronounced, occupy a quite special position, the Upper Silesian (XXI) and the Bełchatów (XIX) regions. The former counts among the most industrialized Polish areas due to extensive underground mining. The long-lasting mining operations have generated a 1,000-km^2 dewatering zone several hundred metres deep in addition to surface degradation. This process still continues as a result of mining deeper and deeper coal seams and causes the regional depression cone to shift downwards at the rate of one to several metres per year. Another effect of the environmental degradation is the increasing mineralization of waters drained due to mining activity then dumped into surface reservoirs. The Bełchatów region almost entirely coincides with the forecasted extent of the dewatering zone of the Bełchatów lignite mine. After only five years of dewatering, the depression cone whose total surface exceeds 200 km^2 and depth 100 m embraces the entire Bełchatów graben and part of the Mesozoic aquifers adjacent to the north and south, the water production being 0.16 km^3 per year and the total amount of water pumped out being 1 km^3 approximately. The extent of the depression cone in 1985-1990 at the time of maximum depression (up to 300 m in the centre of the dewatering system) is expected to cover an area more than 1,000 km^2, which points to the scale of changes in the natural environment and hydrogeologic conditions.

Of special economic significance are the Quaternary aquifers which make up to 70 per cent of Poland's ground-water resources. In north and east Poland the aquifers of valleys, and buried structures are the most important source of urban and agricultural water supply. Besides they are the subject of detailed division into lower-order hydrogeologic units - subregions and regions:

1. Valley delta: Lower Narew, Vistula, Warsaw Trough, Lower Vistula, Zuławy, and Odra,

2. Urstromtal: Biebrza, Torún-Eberswald, Kaszuby, Warsaw-Berlin, Barycz-Głógów, Wrocław-Bremen, and

3. Buried structures: Wielkopolska buried valley, Rega buried structures, Mała Panew buried valley.

Recent studies have pointed out the importance of Quaternary buried structures. The Wielkopolska buried valley, for example, recognized for 200 km and over an area of more than 3,000 km^2 contains close to 2 per cent of the total Quaternary water resources of Poland. Considering the general ground-water scarcity in the Wielkopolska area, this can be regarded as rather high.

3. Water quality

Ground water with total dissolved solids below 1,000 mg/l occurs in Poland usually at depths of 200-300 m. High carbonate content is a characteristic feature of this water. Locally, for example, in the Vistula delta at seaside zone or near the Zechstein salt domes, higher mineralized water is close to the surface. There are, however, some geologic structures where fresh water extends to a depth of 1,000 or even 1,200 m as for example in the Lódz artesian basin in central Poland. Because of the typical hydrochemical zonation in Polish artesian basins, below a fresh-water zone, saline (NaCl) water is usually found.

Ground water recovered from shallow aquifers by more than over 1.5 million dug wells and so-called Abyssinian wells in rural areas and small towns is usually of poor quality. The water is, in many cases, polluted and contains nitrates dangerous to health. Most waters from Quaternary formations of river valleys are also of poor quality, for example in the area of the Vistula delta (Żuławy) the water contains high amount of iron ions, locally over 100 mg/l, and manganese, sulphur and organic compounds which discolour the water and increase its hardness. Most deeper aquifers from the fresh water zone hold water of high quality. These mainly confined aquifers of Cenozoic, Mesozoic and in places Paleozoic age, provide potable water with total dissolved solids of 300-600 mg/l. In the case of water from Miocene silty sands of lignite formations, because of their high organic content, intensive discolouration precludes their use.

Limited ground-water contamination has occurred in and around such cities as Warsaw, Wrocław, Lublin, Płock, Radom and some Upper Silesian industrial centres. There are, however, several thousand cases of local pollution of shallow, mostly unconfined, aquifers. The principal causes of such pollution are oil products, fertilizers and communal and industrial wastes.

4. Development and management

There are more than 52,000 wells registered in Poland, half of them in rural settlements. About 5,000 new wells are built and almost 1,000 wells are renovated yearly. Wells are drilled mainly by government enterprises subordinate to the Department of Environmental Protection and Natural Resources (16 enterprises) and to Department of Agriculture, Forestry and Food Control (18 enterprises). These have well trained hydrogeologists, engineers, technicians and drillers. Only about 30 per cent of shallow wells are drilled by co-operatives and local private firms.

Hydrogeologic prospection and exploration for water supply are performed mainly by geologic and geophysical enterprises of both Departments. They also carry our regional hydrogeologic investigations and calculations of ground-water reserves at a scale of 1:25,000 and 1:50,000. The Geological Enterprise in Poznan is the centre for analog and digital computer studies.

Hydrogeologic mapping of Poland, monitoring of the ground-water régime and evaluation and inventory of the nation's ground-water resources are carried out by the Geological Institute. The results of these studies provide a basis for the development, conservation and management of water resources. The Institute is also responsible for research activities of special problems in the field of geohydrology within the country.

Research into the chemical and physical quality of water is conducted by the Environmental Protection Institute. It also deals with water protection problems.

Basic research in selected hydrogeologic problems are carried out by the universities in Warsaw, Cracow, Vrocław and Gdansk. They are good centres for the study of environmental isotopes, ground-water flow, hydrochemistry, mine water and other problems. Hydrogeologic problems dealing with mining are studied mainly by the Mine Institute in Katowice. Selected research studies in the field of hydrogeochemistry and isotope hydrology are carried out by the Academy of Sciences in Warsaw.

Shallow ground water and surface water are investigated and observed by the Institute of Meteorology and Water Management. Hydrologic data networks, measure of streamflow and the water table are handled by this Institute.

The Department of Environmental Protection and Natural Resources is responsible for water management and conservation within the country. Research activities in the field of hydrogeology are also conducted by the Department of Sciences and Education and the Academy of Sciences. Other departments carry out ground-water investigations for their own purposes but not for ground-water development. The Department of Mines and Energy is responsible for hydrogeology as related to mining and the Department of Health and Social Protection deals with prospection and captation of mineral waters.

5. Ground-water use

The total surface- and ground-water consumption in Poland in 1984 amounted to about 15.9 km^3. Of this industry used 11.2 km^3; agriculture and forestry used 1.8 km^3; and urban and rural settlements used 2.9 km^3.

Effective ground-water withdrawal in Poland, including construction and land drainage, leaks in the municipal water systems and mine drainage, amounts to 7.5 km^3 per year. Very little of the water from mine and construction drainage is used, however. Most is diverted into rivers regardless of its good quality. Greater use is sought of the approximately 1.2 km^3 of water drained from mines.

The ground-water share in supplying population and industry is about a third of the total consumption of water, i.e., 5.2 km^3 per year. From this amount 40 per cent is used for municipal needs, 25 per cent by industry and mining and 35 per cent by agriculture.

Municipalities extract ground water through 8,660 drilled wells. These intakes supply small and medium towns of central and north Poland. Cities larger than 300,000 population and smaller in south Poland mainly rely on surface water. According to data from 1981, mean water consumption by the urban population for domestic use was 208 litres per inhabitant. About 15 per cent of Poland's towns and settlements periodically fail to meet their needs for water.

The rural populace relies almost completely on ground water taken from 1,500,000 shallow dug and drilled wells. Of a total of more than 2.5 million farms in Poland only about a million are provided with a piped water supply. Most others take water directly from wells, a few from springs. The supply does not meet the needs during some period of the year for 19 per cent of the farms (about 500,000).

It is estimated that ground-water extraction for community and industrial supplies will increase from 5.2 km^3 in 1984 to 6.7 km^3 in 1990 and 8.1 km^3 in 2000. The supply will be piped almost exclusively.

D. Conclusions

1. Among all countries of the temperate zone, Poland, with ground-water resources amounting to about 14 km^3 per year, appears to be relatively disadvantaged. The distribution of ground-water resources over the country is uneven and insufficient to meet all needs everywhere. Ground-water resources reserves greater than 8 km^3 per year occur mainly in the northern rural part of the country.

2. Large ground-water reservoirs of special economic significance, amenable to easy recovery, occur in Recent and buried valleys. Unfortunately such water usually needs expensive treatment before it can be used owing to the high vulnerability of such aquifers to pollution from the surface.

3. A favourable factor for water management in Poland is the pattern of occurrence of surface and ground-water resources over the country. Good ground-water resources are in central and north Poland and good surface sources in the mountains of southern Poland (Carpathian, Sudetes). Unfortunately, in many regions of central Poland surface water cannot be used because of contamination. Water from mountain rivers must often be piped over long distances to regions whose sometimes considerable local resources cannot be used.

4. The problem of ground-water contamination is fast growing in Poland. It especially applies to shallow unconfined aquifers and karst-fissure aquifers. The sources of contamination are mainly fertilizers, leakage of oil products from pipe systems and installation of piped water supply systems without providing adequate drainage systems for waste waters.

5. Highly mineralized waste water is still a difficult problem. The amount increases as mines go deeper. Mine waste water is often diverted to rivers upstream of intakes for community water supply, irrigation and industrial uses.

References

Aktualne problemy hydrogeologii. at. Symp. Kraków-Karniowice, 1985.

Atlas zasobów wód podziemnych i surowców skalnych dorzecza Wisly 1:750,000. Institut Geologiczny, Warszawa, 1980.

Biedrzycki, W., M. Malaga, D. Poprawa and J. Sokolowski, Geological conditions and problems of thermal waters exploitation in Podhale region. Kwart. geol. t.29, No. 1, p. 179-192. Warszawa, 1985.

Hydrogeological research in Poland. Instytut Geologiczny. Bull. 227. Warszawa, 1973.

Kaczmarek, Z. Water resources in Poland today and in the future. Journal of Hydrological Sc. Pan. 1, Nos. 1-2. 1974.

Kleczkowski, A., J. Jureczko and J. Myszka. Some hydrodynamic phenomena in fissured Triassic and Jurassic rocks. Bull. de L'Ac. Pol. des Sc. Ser. des sci. geol. et geogr. vol. 17, Nos. 3-4, 1969.

Niedzielski, H. Characteristics of water capacity of the Carpathian flysch on the basis of well discharge. Ann. de la Société Géologique de Pologne. vol. L-1. Kraków, 1980.

Paczynski, B. Podzial regionalny i glebokosc wystepowania zwyklych wód podziemnych Polski. Mat. na Konferencje nauk techn. Warszawa, 1978.

__________. Woprosy metodiki issledowanji pri ocenkie zapasow podzimnych wod czetwierticznych pogrebjonnych pateowrezow. AIH Memoire. vol. XV, 1. Methods for evaluation of ground-water resources. Congress of Vilnius, USSR, 1979.

Rozkowski, A. and K. Przewłocki. Application of stable environmental isotopes in mine hydrogeology taking Polish Coal Basins as an example. Isotope Techniques in groundwater hydrology, vol. 1. International Atomic Energy Agency, Vienna, 1974.

X. ROMANIA

Area: 237,500 km^2 Population: 22.9 million

A. General

A hydrogeologic map of the Socialist Republic of Romania at the scale of 1:1,000,000 was published in 1969. It incorporated data from several maps at the scale of 1:100,000 and more specifically related to Quaternary formations which contain important reserves of ground water.

The mapping of various types of Quaternary formations received special attention and no less than 28 different combinations and colour lithology symbols are used for their identification. The map incorporates practically all the hydrogeologic data known to the date of its publication from open file reports and notes on drilling operations, to hydrogeologic surveys and related operations, as well as surface-water data including values of surface-water flow, and the location of dams and irrigation works.

Five types of karstic aquifers, which are important but of limited extent, are distinguished according to relative composition of limestone and dolomite.

Other types of aquifers represented are sandstone, marl-limestone and conglomerate (16 different types). Impervious rocks are represented as either aquitards or aquicludes.

B. Hydrogeology

The country has been divided into eight hydrogeologic structural regions (figure 50).

1. Moldavian Plateau

The Precambrian basement rocks which were eroded to a peneplain are overlain by a sedimentary cover with gaps in the stratigraphic sequence and which include several aquifers. Deep boreholes near Iasi have encountered artesian waters in Silurian sandstone and limestone. These waters hold high concentrations of minerals (9,000 to 12,000 mg/l) and hydrogen sulphide gas. Highly saline deep waters are also encountered in Miocene formations. Such occurrences have allowed delineation on the maps of the northern part of the Moldavian Plateau, a zone which is likely to hold deep ground water within the "zone two"; water-bearing subzones with a limited areal extent have also been identified, one corresponding to Tortonian limestone and sandstone outcroppings near the Prut River, the other to Sarmatian limy-sandstone with outcropping south of Iasi.

Almost all the southern part of the Moldavian Plateau is characterized by a sequence of sand and clay of upper Sarmatian-Meotian and Pontian age. It includes a number of sandy layers with infiltrations of meteoric and river water. Ground-water movement is controlled by the dipping of sediments towards the south, ground-water heads increasing accordingly. The existence of artesian ground water was confirmed by the results of drilling in the Birlad region.

Figure 50

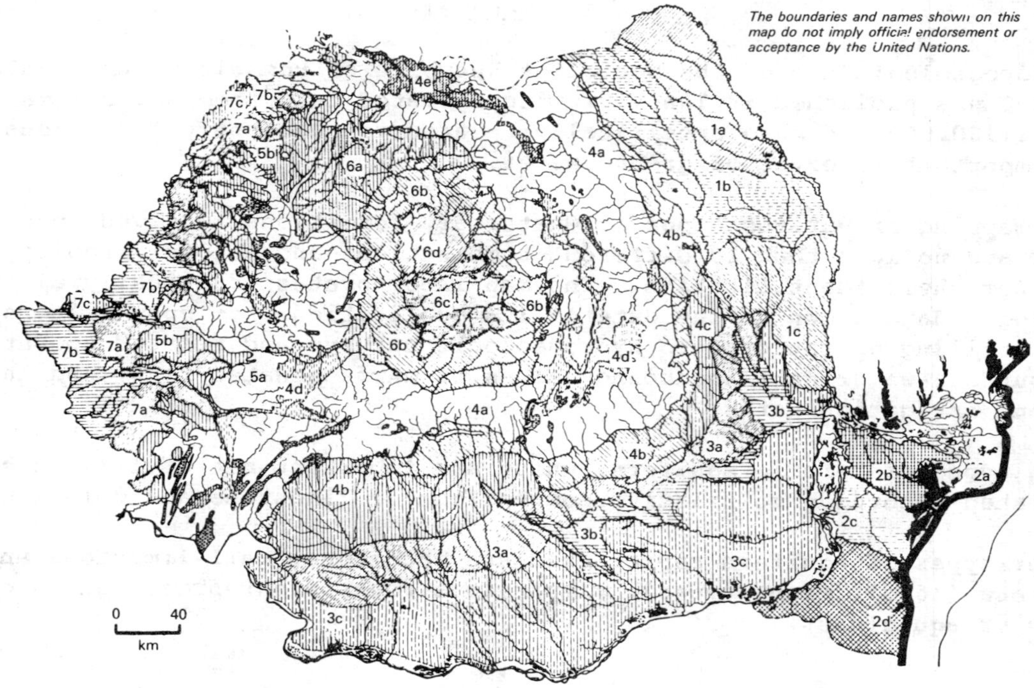

ROMANIA: PRINCIPAL PHREATIC AQUIFERS

Moldavian Plateau
- 1a Neogene marl and clay
- 1b Neogene clay and sand
- 1c Sand

Dobrogea
- 2a Danube delta alluvium
- 2b Northern Dobrogea hard rocks
- 2c Middle Dobrogea metamorphics
- 2d Lower Dobrogea Mesozoic carbonate rocks

Romanian Plain (mainly alluvium)
- 3a Piedmont area
- 3b Former Danube valley
- 3c Danube Plain

Eastern Carpathian Mountains
- 4a Crystalline rocks, schists volcanic rocks, flysch
- 4b Outer zone Neogene hills
- 4c Outer zone Neogene (monoclinal area)
- 4d Valley alluvium
- 4e Complex sedimentary deposits

Apuseni Mountains
- 5a Mountains (complex formations, mostly hard-rocks)
- 5b Piedmont

Transylvanian Depression
- 6a Paleogene-Neogene complex deposits: plains
- 6b Paleogene-Neogene complex deposits: rolling hills
- 6c Tirnave Plateau (mainly marl, clay)
- 6d Transylvanian plain (Neogene, quaternary, clay, sand)

Pannonian depression (mainly Quaternary unconsolidated)
- 7a High plain
- 7b Low subsidence plain
- 7c Dune plain

3546.39x

Source: Institute for Hydrotechnical Research, Bucharest

The southernmost part of the Moldavian Plateau is characterized by sandy deposits which include upper Pleistocene gravel which in the Touci area allow the development of ground water with specific capacity rates of 5 to 10 l/s. Also, in the Moldavian Plateau is abundant phreatic water in the Siret River alluvium containing coarse sand and gravel, but less abundant water in Prut River alluvium containing finer alluvium, mainly fine sand, occasionally clayey.

2. Dobrogea

Dobrogea can be separated hydrogeologically into several zones: the Danube Delta, Northern Dobrogea, Central Dobrogea and Southern Dobrogea.

The Danube Delta overlies the pre-Dobrogea depression which includes Mesozoic and Neogene deposits. During the lower Pleistocene the Danube Delta started to develop; fluvial sedimentation alternating with deltaic and marine. Ground water is contained in the older deposits; Jurassic, Sarmatian or Pleistocene are characterized by smaller yields and a high mineralization. In deltaic deposits highly saline ground waters are limited to a lower aquifer complex which includes high yielding gravel and sand. Potable water is secured by a local phreatic aquifer contained in dune sands.

In Northern Dobrogea, complex geologic conditions led to quite different hydrogeologic formations. In the Tulcea area, Triassic detrital and carbonate formations hold water in a fissure and fracture system. Water quality is good; well yields which amount to 5 l/s are sufficient to meet the community's needs. In Upper Cretaceous formations of the Babadag basin, highly producing aquifers also occur. The Paleozoic area includes impervious formations of different origin including metamorphosed sedimentary rocks, granite, quartziferous porphyric rocks and others. Loess deposits also occur with limited water resources in Northern Dobrogea.

Central Dobrogea is mainly constituted of formations such as green schists with limited metamorphism, dating from the Late Proterozoic and which are practically impermeable. In the region ground water is found only at the bottom of loess formations and in Recent alluvium of the Casimcea valley. Such resources are only temporary with low yields of 0.5 to 0.7 l/s.

In Southern Dobrogea, Mesozoic formations include dolomite, limestone, limy sandstone and conglomerate, which were deposited from the Late Jurassic to Cretaceous, contain ground water as evidenced by springs and the successful results of a number of boreholes. Water accumulates in the Upper Jurassic sequence (Barremian-Cenomanian). In many locations it may be considered as a single aquifer complex as no impermeable horizons are encountered in the Mesozoic limestones.

In southernmost Dobrogea (Mangalia), ground water held in Mesozoic formations are artesian, abundant, and relatively hot (26° C). In Southern Dobrogea, Mesozoic deposits are covered locally by Eocene deposits, widely extended Sarmatian deposits, and Pliocene deposits of local importance.

Sarmatian limestone and sandstone hold the best aquifer, on top of clay layers. Small quantities of ground water are also found at the base of loess deposits. In a small area on the right bank of the Danube ground water can also be obtained from Pliocene deposits, mainly of sand.

3. Romanian Plain

Water-bearing geologic formations of the Romainian Plain are dated from:

1. Lower Pleistocene (Cindesti and Fratesti layers),

2. Upper Pleistocene (Colentina gravel, Mostistea sand and gravel), and

3. Holocene (alluvial gravel and sand).

All these aquifers are extremely permeable and contain large quantities of ground water. The Cindesti strata occupy in the outer Carpathian Mountains and in the Romanian Plain, a vast area from west of the Jiu to north of Trotus, with the exception of one sector (Cricov-Buzau) where they are missing. Such strata, which are of fluvio-lacustrine origin, include gravel layers several tens of metres thick alternating with marly-clay and sand. Away from the Carpathian arch and towards the exterior, grain size becomes smaller.

The external edge of the sub-Carpathian area is a recharge area for the Cindesti deep aquifer. There, ground water is encountered at depths of 50-200 m and is drained in the interior of the plain. In areas where formations are entirely saturated, excess water is discharged through springs or in Recent alluvium.

Variations in grain-size of the Cindesti layers from the interior to the exterior is reflected into their hydrogeologic characteristics. In the interior part of the sub-Carpathian area where gravel prevails, two thick aquifers occur which allow for relatively high-yield boreholes (5 to 10 l/s) while towards the exterior the thinness of the elements leads to the occurrence of several low-yielding (1 l/s per well) aquifers.

The Cindesti aquifer is artesian in such lower-lying areas as the Jiu alluvial plain west of Craiova, the alluvial plain and lower terrace of the Olt at Peska, Brezvica, the alluvial plain of Vedea at Potcoava and in any area parallel to the Carpathian chain from the alluvial plain of the Arges to the east through Gaiesti, Titu, to the south of the cities of Ploiesti, Mizil, Buzan and east and north of Rimnicu Sarat. Waters of the Cindesti aquifer are potable with total mineral content of 650 mg/l, mainly bicarbonate.

The Fratesti aquifers extend from the Jiu River to the Danube Delta; they include sand and gravel with some interbedded clay. Between Jiu and Vedea and also in areas adjoining the Danube valley such layers are directly below loess deposits at relatively shallow depths (20 to 25 m); they deepen towards the interior of the plain, down to 200 meters in Bucharest. Where they occur at shallow depths they constitute a single sand-gravel layer up to 25 metres thick. At greater depths they subdivide into three layers separated by clay beds.

East of Vedea and in the Burnas plain, the Fratesti aquifer is unconfined phreatic, at depths of 20 to 30 metres. Towards the interior, artesian conditions develop, increasing from east to west: from three to four metres in Galati to 70 m west of Bucharest. Ground-water flow is oriented from north-west to south-west between the Jiu and the Arges rivers; from west to east, in the area east of the Arges, and from south to north in the Fetesti-Galati area.

Near the Danube, well yields average 10 l/s but in the interior of the plain they are about 3 l/s. Mineral concentration ranges from 300 to 600 mg/l (mainly sodium bicarbonate). Water quality is better than that of western Romanian plain-terrace (Pleistocene) and alluvium (Holocene) deposits; sands, gravel, broken stone deposits are also coarse grained and highly productive (2 to 12 l/s; the piezometric level is at shallow or moderate depth (2 to 20 m).

In the Romanian Plain, some ground water occurs in medium-grain formations, that is, the Mostistea sands south of the Ialomita, and sands alternating with sandy clay north of Ialomita. Water levels are at 5-20 metres depth and yields are generally limited, less than a litre per second.

In the north-east part of the Plain is an aquifer in fine-grained deposits (fine sand, loess and clayey sand of Pleistocene-Holocene age). Yields are moderate and mineralization is high. In this part of the Plain, salt lakes are found in endorheic depressions. Such lakes are the outlet of aquifers and the salinity results from evaporation.

4. Western Carpathians

The Western Carpathian Mountains include numerous distinct geologic units. One is crystalline schist (Precambrian-Paleozoic) which is practically impermeable. Ground water is contained only in a thin layer of colluvium covering the mountain slopes, which give way to some springs.

In this region, crystalline limestone and dolomite and Mesozoic limestone are in synclines which locally give way to karstic springs. Important mineral springs occur in the crystalline and Mesozoic areas of the Western Carpathians. Ground water flows along fractures or fissures in the crystalline schist or on contact surfaces between crystalline and sedimentary formations. A first sector of mineral springs is in Vatra Dornei where the chemical component includes bicarbonate, iron, arsenic and metals of the calcium group. In Iacobeni, waters flow through Triassic limestone containing tar and are therefore sulphurous. A second area with mineralized water is located around Tulghes, Bilbor and Borsec. In Tulghes, the water occurs in crystalline schist; in Bilbor and Borsec, from dolomitic limestone. These waters contain carbon dioxide.

A second unit includes mainly Neogene-Quaternary volcanic rocks, especially along the inner rim of the Eastern Carpathians. Volcanic activity produced andesitic and basaltic lava flows and also pyroclastic breccia and conglomerate, alternating with sandstone and sand of andesitic nature. The Miocene pyroclastic formations are porous and permeable, but the lava flows are largely impermeable.

In the eruptive massif of Calimani-Gurghiu-Harghita, mineral springs occur mainly in pyroclastic layers. In general, they occur along dislocations which affect the western slopes of the chain, close to intra-Carpathian basins (Gheorghieni, Cive) and also along certain transverse faults along which carbon dioxide moves upwards, mineralizing the infiltrating water. In the Toplita area, hypothermal mineralized springs (26° to 27° C) occur, with bicarbonate type composition emerging from terraces in the Mures River, which are cut in andesitic agglomerate and overlay crystalline schist.

A third unit is that of flysch. An internal flysch zone is represented in Sinaia layers by sandstone and limy sandstone, which are practically impermeable. Small amounts of ground water flows through fissures and stratification joints and, thus, is highly mineralized (Sinaia, Moroieni, Busteni). Another Lower Cretaceous facies is represented by schist-sandstone and flysch which include rocks abundantly fissured in which ground water easily flows. In Cretaceous flysch, coarse-grained conglomeratic complexes (Ceahlau, Ciucas, Bucegi) have been identified, with high storativity and high-yield springs used for water supply.

In the Carpathian flysch area, springs with mineralized waters occur along two major faulted areas, one located on the western flank of the Ciuc and Bodoc mountains (around the population centres of Ciuc-Singeorz, Sinmartin, Lazaresti, Toria, Malnas and Bodoc) and the other between the Audia and Tarcau mountains, (Iacobeni, Casin, Poiana, Peteni and Covasna).

Paleogene flysch includes rocks with many fissures which allow for good storativity and transmissivity. These include sandstone and pelite containing tar sands. Some hydrogeologic structures within this flysch unit contain mineral water with a high content of hydrogen sulphide due to iron pyrites disseminated in schist (Slavic Moldova, Manesti, Lopatari).

The fourth unit is the outer zone of the trough. Two hydrogeologic units have been identified with the Outer Carpathians: First, Miocene deposits, thick sandstone with lagoonal beds containing gypsum and salt, largely impermeable, overlying sand and conglomerate with ground-water potential. Waters with high concentrations of sodium chloride issues in Brusturi, Oglinzi, Baltatesti, Soloni, Sarata, Bacau and Slanic Prahova. Second, Pliocene deposits which in the upper parts contain sands with clayey layers.

Internal depressions within the Eastern Carpathians are extremely important from the hydrogeologic point of view as they include formations with good storativity. The waters contain carbon dioxide. Some are artesian, others not. They are contained in Pleistocene deposits, many of which are thick as in the Gheorghieni, Ciue and Birsa basins.

5. Southern Carpathians

The Southern Carpathian area is composed of crystalline rocks and sedimentary rocks of Paleozoic, Mesozoic and Cenozoic ages. The crystalline rocks, with the exception of crystalline limestone, are impermeable. The only aquifers are related to alteration layers and diluvium which give way to some small springs. Carboniferous and Permian rocks include sandstone conglomerate and limestone. Water occurs only in fractures and yields are small. Deposits of Upper Jurassic and Lower Cretaceous include karstic limestone and dolomitic limestone in which high-yield springs occur.

In the Jiu River basin, springs at Isvarna yield some 1,000 to 2,000 m^3/s, springs at Runcu 200 to 3,000 l/s and in Nera Cerna basins 70 to 250 l/s. Upper Cretaceous deposits which include alternating limestone, sandstone, marly limestone and conglomerate hold low-yielding fissure aquifers. Conversely, Burdigalian conglomerate and gravel located in the western Dimbovita valley are highly productive. From Pliocene sandy clays and fine sands only small yields of highly mineralized water are available.

6. Apuseni Mountains

Metamorphic eruptive and sedimentary rocks of various ages are found in the Apuseni Mountains which have complex geologic structure. Most of the formations are impermeable and water occurs only in the weathered and fissured upper layers.

Eruptive rocks outcrop widely in mounts Highis, Gilav, Vlades, Bihor and Metalici. They include andesitic, rhyolitic and basaltic lava and also thick pyroclastic complexes, into which rain water infiltrates and which give way to many springs.

In the Apuseni Mountains, sedimentary formations include Paleozoic clays, phyllitic schist, sandstone, tuff, mostly impermeable, and Mesozoic karstic limestone, dolomite and marly limestone. Karstic channels are in open communication with Triassic limestone and dolomite karst which occur in the Padurea Craiului, Codru and Bihor massifs, and also with Lower Jurassic limestone massifs of mounts Trascau, Bihor and Padurea Craiului. Considerable seasonal fluctuations in yields are observed on springs which have yields up to 100 l/s. Mesozoic formations include sandstone, marly limestone and conglomerate highly fissured which give way to small springs (0.05 l/s to 1 l/s).

In the massifs of Trascau, Codru, Drocea, Metalici and Parurea Craiului impermeable marls, clays, sandstone of Jurassic and Early Cretaceous age occur. Mesozoic deposits in intermontane basins such as Beius, Borod, Simleul, Silvaniei are lithologically similar.

7. Transylvanian Depression

The Transylvanian Depression includes mostly continental deposits of Paleocene, Miocene and Pliocene ages. The best aquifers are the karstified Eocene limestone as shown by drilling and the occurrence of springs (5 l/s). Less productive are sandstone conglomerate and limestone of Oligocene and Burdigalian age in which small springs occur. The southern boundary of the basins include permeable rocks such as Miocene micro-conglomerate (Tortonian and Sarmatian), sand and gravel. Several springs issuing from Miocene deposits are mineralized as a result of the occurrence of evaporites in the sequence. Lakes located above these saline deposits contain waters with therapeutic properties (Sovata, Ocna, Sibiului).

South-west and east of the Transylvanian Depression and on contact with the mountainous zone, Pannonian deposits of sand, gravel and some conglomerate have springs yielding up to 10 l/s; drilling for water in such formations gives good results.

A large part of the Depression is occupied by a thick series of marl, red and variegated clay, clayey sand, tuff and gypsum of Paleogene, Miocene and Pannonian age, mostly impermeable, especially in the centre of the Depression. However, the occurrence of deep ground water has been noted in some locations.

8. Pannonian Depression

In the Pannonian plain of the western part of the country, Quaternary, Cenozoic and Mesozoic formations overlay the crystalline basement. The best aquifers are encountered in Mesozoic formations as revealed by drill holes at Oradea, Toboliv, Baile Victoria and May Day; these are in Triassic, Jurassic and Cretaceous limestones. Deep water is artesian; yields range from 130 to 17,000 m^3/day and water temperatures can be as high as 90° C. (No specific explanation has been given for these high temperatures.)

The lower beds include impermeable clay, marl and clayey sand; the upper part includes sand and gravel with alternating clay and marl which facilitate the infiltration of surface water, and a complex aquifer at depths not normally exceeding 300 metres. They can be artesian as a result of lowering, by degrees, west of these formations (Timisoara, Arad, Beius, Salonta, Oradea). Waters have a high content of carbon dioxide and hydrogen sulphide (Tinca, Toboliu, Salard, Osorhei, Inand).

In general, yields are small (0.5 l/s). Higher yields (5 l/s to 10 l/s) are known in Timisoara, Arad, north of Oradea and in the Timisoara-Foieni sector. Pleistocene deposits including sand, and gravel with intercalations of clay and marl are of fluvio-lacustrine, alluvial and proluvial origin. These coarse-grained formations are good aquifers. Boreholes not more than 60 metres deep are highly productive (Bors, Episcopia, Bihor, Sintion, Tamaseus, Arad, Salonta). Waters occur under artesian conditions, mainly non-flowing, in some cases flowing, in alluvial plains and in terraces of main rivers (Mures, Timis, Crisuri, Somes) extended aquifers are found; borehole yields are in the range of 0.4 to 1 l/s.

C. Ground-water development

Some 50 years ago the knowledge of water resources in Romania was limited primarily to the well fields supplying such large cities as Bucharest, Ploiesti, Iasi, Craiova, Timisoara, Arad and Cluj-Napoca. Investigations since have been developed to serve community water supply and industrial needs and also agro-pastoral and agro-industrial needs. The use of ground water for irrigation is still limited in space and time (dry periods).

Ground water is extracted by means of boreholes, drains, wells, drainage galleries and springs. In the Moldavian Plateau ground water is mainly developed from phreatic aquifers and to a much lesser extent from deep aquifers (table 5 and figures 51 and 52).

Table 5. Romania, aquifer characteristics, Moldavian Plateau

Location	Aquifer description and development	Specific capacity (l/s/m of drawdown)
Siret plain	Shallow boreholes in Recent alluvium	0.5 to 15
Bistrita plain	Shallow boreholes in Recent alluvium	2 to 57
Bistrita lower terrace	Shallow boreholes in Pleistocene alluvium	0.3 to 3
Prut plain	Shallow boreholes in Recent alluvium	0.6 to 5
Ozana plain	Shallow boreholes in Recent alluvium	1.5 to 40
Moldova plain	Shallow boreholes in Recent alluvium	0.6 to 17
Billad plain	Shallow boreholes in Recent alluvium	0.5 to 2.8
Nigrilesti	Boreholes 75-100 m in Pliocene sand	3 to 33
Tecuci	Boreholes 130-140 m in Pliocene sand	1 to 11
Iasi	Boreholes 378-513 m to mineral water in Silurian limestone	0.8

Figure 51

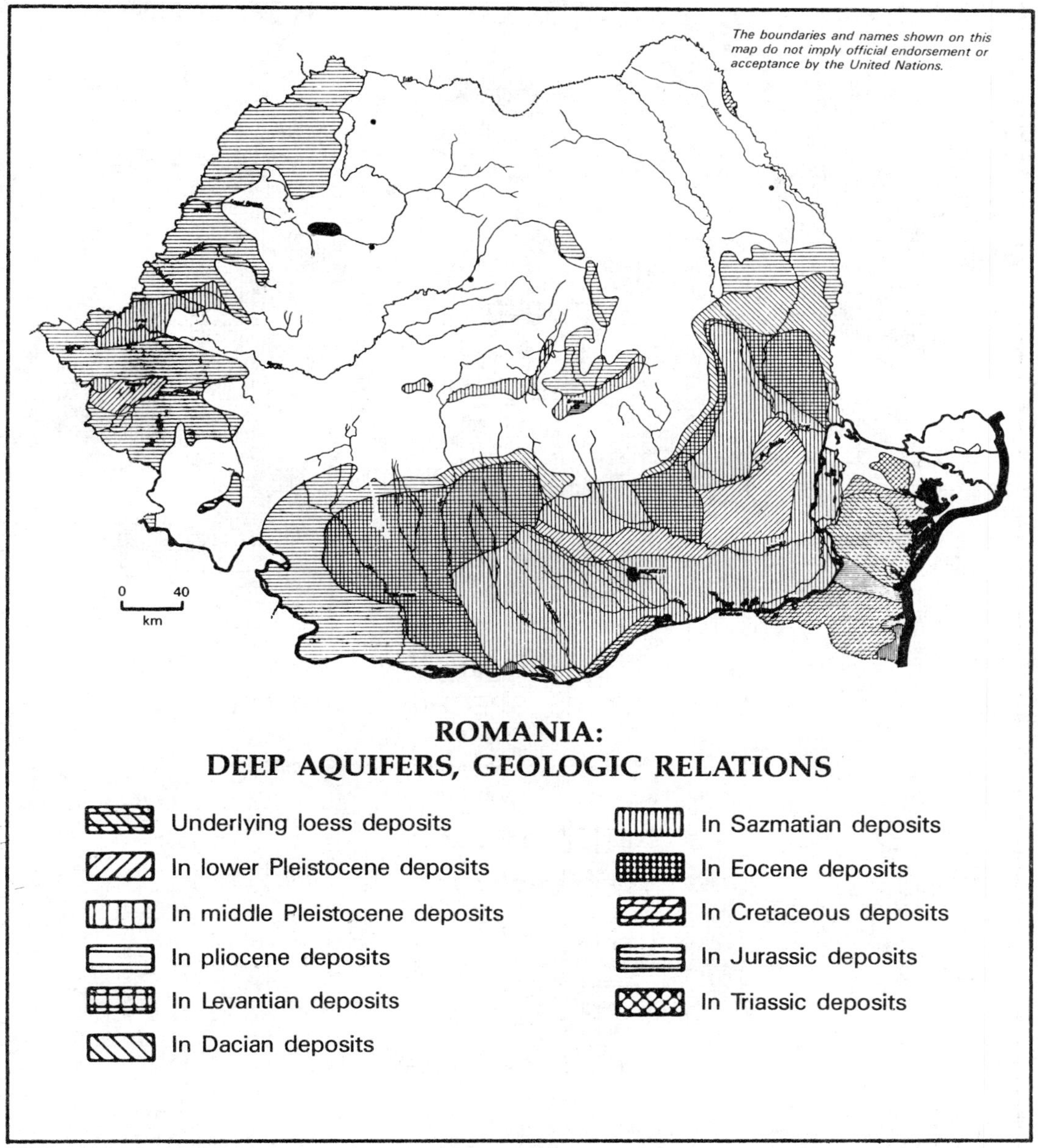

3546.40x

Figure 52

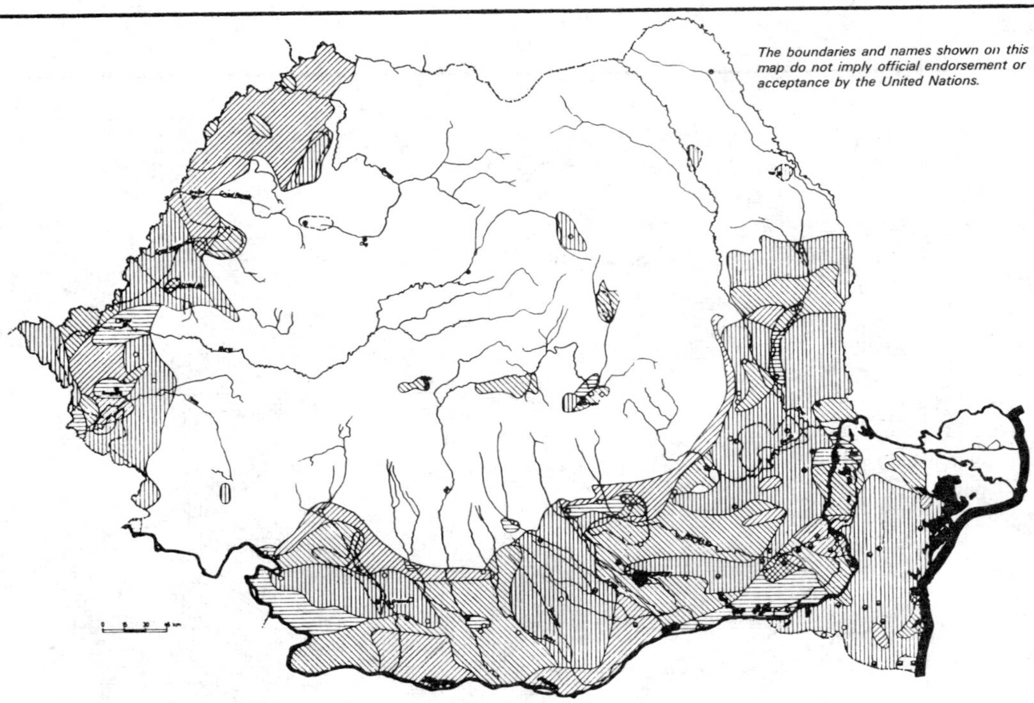

ROMANIA:
DEEP AQUIFERS, SPECIFIC CAPACITY

Drawdown expressed in
litres per second per metre

- up to 0.5
- 0.5 to 1
- I to 5
- 5 to 10
- above 10
- artesian areas

3546.41x

In the karstic Sarmatian, Cretaceous and Jurassic rocks of Dobrogea, ground water is developed by means of wells (down to 40 m) and boreholes (down to 120 m). Wells and boreholes are located both north and south of Constanza city and supply all the Black Sea coastal area. Unit yields are 10 to 60 l/s. The overall rate of extraction is 4.5 m^3/s.

In the Romanian Plains, ground water is developed mainly from alluvium and terraces of river valleys and from the Pliocene-Pleistocene beds of Fratesti and Cindesti. The highly mineralized water of Miocene, Pliocene and Mesozoic formations are tapped only for balneo-therapeutic purposes.

In table 6 are presented general data on shallow and deep water resources of the Romanian Plains developed for community, industrial and agricultural purposes.

Table 6. Romania, aquifer characteristics, Romanian Plain

Location	Aquifer description and development	Specific capacity (l/s/m of drawdown)
Danube plain	Shallow boreholes in Recent alluvium	0.5 to 30
Olt plain	Shallow boreholes in Recent alluvium	2 to 16
Jiu plain	Shallow boreholes in Recent alluvium	1.2 to 11
Arges plain	Shallow boreholes in Recent alluvium	2.3 to 16
Ialomita plain	Shallow boreholes in Recent alluvium	2 to 13.5
Prahova plains	Shallow boreholes in Recent alluvium	1 to 8
Prahova alluvial cone	Boreholes 30-70 m in Pleistocene alluvium	2 to 15
Buzan-Putna piedmont	Boreholes 20-250 m in Pleistocene alluvium	2 to 30
Fratesti beds	Boreholes 25-120 m in Pleistocene gravel	2 to 30
Cindesti beds	Boreholes 20-120 m in Pleistocene gravel	1 to 10

In the Eastern Carpathian mountains low-yielding and mineralized flysch aquifers are developed only for balneo-therapeutic purposes. Conversely, Cretaceous conglomerates (Ceahlan, Ciucas-Zaganu and Persani) and karstified Cenozoic limestone of Haghimas, Bucegi and Piatra Craiului hold important ground-water resources which are tapped for community and industrial purposes, up to 460 l/s are extracted from the karstic springs of the Piatra Craiului massif and 400 l/s from the eastern flank of the Bucegi Mountains. The sands and gravels filling in the intermontane depressions of Gheorghieni, Ciuc and Birsa are exploited by means of boreholes 50 to 250 metres deep yielding 1 to 10 l/s with 0.5 to 5 m of drawdown.

In the Southern Carpathians the Jurassic and Cretaceous massifs of Vulcan and Cerna are drained by many karstic springs yielding 20 to 50 l/s for community water supply. In Baile Herculane, thermo-mineral waters from Mesozoic limestone are developed for balneotherapy. Triassic limestone and dolomite and Jurassic-Cretaceous limestone are deeply karstified and feed springs yielding 20 to 300 l/s. In the Moneasa, Felix and May Day areas, karstic waters which have 30° to 49° C temperatures are utilized in balneotherapy. Cretaceous and Paleogene rocks and crystalline limestone give way to springs with moderate yields of 0.5 to 5 l/s.

In the Transylvanian Depression Quaternary sand and gravel of the Fagaras piedmont, pyroclastics of the volcanic chain of Calimani-Marghita, and Quaternary alluvium of low-lying plains and terraces of main rivers constitute significant aquifers (table 7).

Table 7. Romania, aquifer characteristics, Transylvanian Depression

Location	Aquifer description and development	Specific capacity (l/s/m of drawdown)
Fagaras piedmont	Shallow boreholes in Pleistocene gravel	1 to 12
Olt plains	Shallow boreholes in Recent gravel	1 to 20
Mures plains	Shallow boreholes in Recent gravel	1 to 10
Somes plains	Shallow boreholes in Recent gravel	1 to 10
Vlahita	Boreholes 250-300 m in Pliocene pyroclastics	0.5 to 1

In the western plain, that is, the eastern part of the Pannonian Basin, ground water is exploited from porous and karstic rocks. Boreholes drilled in flood plains and alluvium yield 50 to 30 l/s each, with drawdown ranging from 0.5 to 3 metres. Upper Pliocene sands, tapped at 300 metres depth yield 1 to 5 l/s per borehole, drawdown is 1 to 3 m.

Artesian thermal waters (up to 90° C) are tapped at depths of 700 to 2,000 metres in Pontian sands. Borehole yields are from 10 to 40 l/s. In the Triassic high karstic compartment of Oradea, thermal artesian water occurs at depths exceeding 2,000 metres, temperatures are 75° to 80° C, and total yield is about 750 m^3/per day. These waters are utilized mainly for heating purposes (including greenhouses). Most of the ground water occurring there is of good quality with low hardness and TDS concentration of 1,000 to 2,000 mg/l.

References

Carte hydrogéologique: échelle 1:1,000,000. Comité d'Etat pour la Géologie, Note explicative par C. Ghenea, T. Bandrabur, N. Mihaila, Ana Ghenea and P. Giurgea. Bucarest, 1969.

XI. SWEDEN

Area: 449,960 km^2 Population: 8.4 million

A. General

The Kingdom of Sweden has access to the North Sea to the south-west through the Skagerrak and the Kattegat. To the south and east is the Baltic Sea and Gulf of Bothnia. The central and northern parts of Sweden can be divided into three main units (1) the Scandinavian (Caledonian) Mountains, (2) the uplands of northern and central Sweden, and (3) the coastal plain along the Gulf of Bothnia. The Scandinavian Mountains are situated to the west along the Norwegian border. The highest peaks within Sweden reach altitudes of about 2,100 m. The uplands of northern and central Sweden east and south-east of the mountains form a dissected plateau gently tilted towards the south-east or east. The altitude is 200-500 m. The transition zone between the mountains and the plateau is characterized by a number of elongate, rather deep lake basins, formed by glacial erosion during Pleistocene time. To the east the upland plateau descends and grades into the coastal plain along the Gulf of Bothnia.

The central part of southern Sweden consists of a small upland area with altitudes above 300 m. It is separated from the northern uplands by a lowland area crossing the peninsula. In this lowland are four of Sweden's largest lakes (Vänern, Vättern, Mälaren and Hjälmaren). Other lowlands occur along the coasts of southern Sweden and in the Baltic islands (Gotland and Oland).

The northern and eastern parts of Sweden drain towards the Gulf of Bothnia and the Baltic Sea, the south-western part towards the Skagerrak and the Kattegat. Compared to the largest European rivers the largest ones in Sweden are relatively small (mean flow amounting to about 400-500 m^3/s). The landscape includes many lakes, most of whose basins were formed by a combination of glacial erosion and damming by drift.

All of what is now Sweden was covered by glacial ice during the Pleistocene. The morphology of the Swedish landscape is characterized by glacial accumulation and to a certain degree by glacial erosion. The lowland areas were extensively covered by the sea and by ice-dammed lakes during a longer or shorter period after retreat of the glaciers. This former existence of sea or lakes also influenced the morphology of the landscape.

Sweden, situated within the northern temperate zone in a belt of moving cyclones along the Polar front, has a climate characterized by alternating warm and cold periods, either dry or humid. The climate of southern and, especially, south-western Sweden has a maritime character; the northern part is more continental. Annual mean precipitation ranges from 1,500 mm in parts of the mountain range to 500 mm in central northern Sweden. Generally, less precipitation is received in the eastern part of the country. Precipitation is fairly evenly distributed throughout the year.

The mean temperature for July varies between 17° C in southern Sweden and 12° C in the northern part. In the mountain range mean July temperatures down to 5° C can be found. Mean temperature during January varies between 0° C in the south-westernmost part and -16° C in the inland areas of northern Sweden. Hence, temperatures below freezing are generally found for shorter or longer periods during the winter throughout Sweden. In the south-west these periods are generally of short duration, interrupted by mild periods. In northern Sweden temperatures below 0° C may last for several months. The climate is reflected in the hydrologic régimes of the rivers. In most parts of Sweden a flood period connected with the melting of the snow and ground frost occurs in the spring.

B. Geology

Most of Sweden is within the Baltic Shield consisting mainly of Precambrian crystalline rocks which are locally covered by sedimentary rocks of Jotnian, Eocambrian or early Paleozoic age. These rocks form parts of earlier, more extensive sedimentary covers. In the Baltic islands the crystalline basement is totally covered with sedimentary rocks of Cambrian-Silurian age.

In southern Sweden, an area with a series of fault lines trending NW-SE forms a transition zone to the North European Lowland, with relatively flat-lying sedimentary rocks having a thickness of 1,500-3,000 m. Their ages are from Cambrian to Tertiary.

In the north-west, the Baltic Shield borders the Scandinavian Mountains. This range consists of sedimentary, metamorphic and volcanic rocks of late Precambrian and early Paleozoic age which were folded and overthrust during the Caledonian orogeny.

Although all of Sweden was covered by land ice several times during the Pleistocene, most of the glacial deposits (till and stratified drift) are from the last glaciation. Owing to subsidence from the land-ice pressure large coastal areas were covered by the sea after glaciers retreated, and marine clays and silts from this time are abundant.

C. Hydrogeology

1. Principal aquifers

Three main types of aquifers can be distinguished in Sweden, namely aquifers in the crystalline Precambrian rocks, in sedimentary rocks and in unconsolidated Quaternary deposits. Most of the pre-Quaternary rocks consist of different crystalline rocks belonging to the Baltic Shield (figure 53). The permeability of these rocks is controlled by the presence of open fissures, fractures and faults, although the rock mass itself is practically impermeable. Thus the yield of a well in such rocks is mainly dependent on the number and size of open fissures and fractures which the well happens to intercept. The yield is also influenced by infiltration conditions, especially if the capacity of the water-bearing fractures is great. Common yields of drilled wells in the crystalline basement of Sweden are from 0.1 to 1 l/s, rarely up to 10 l/s and exceptionally above 10 l/s. The ten most productive wells show values of 20-30 l/s (figure 54). Drilled wells in the crystalline basement are common in Sweden and widely used for water supply to individual homes, farms and villages. The total number of such wells is not exactly known but a qualified estimate suggests some hundred thousand.

Figure 53

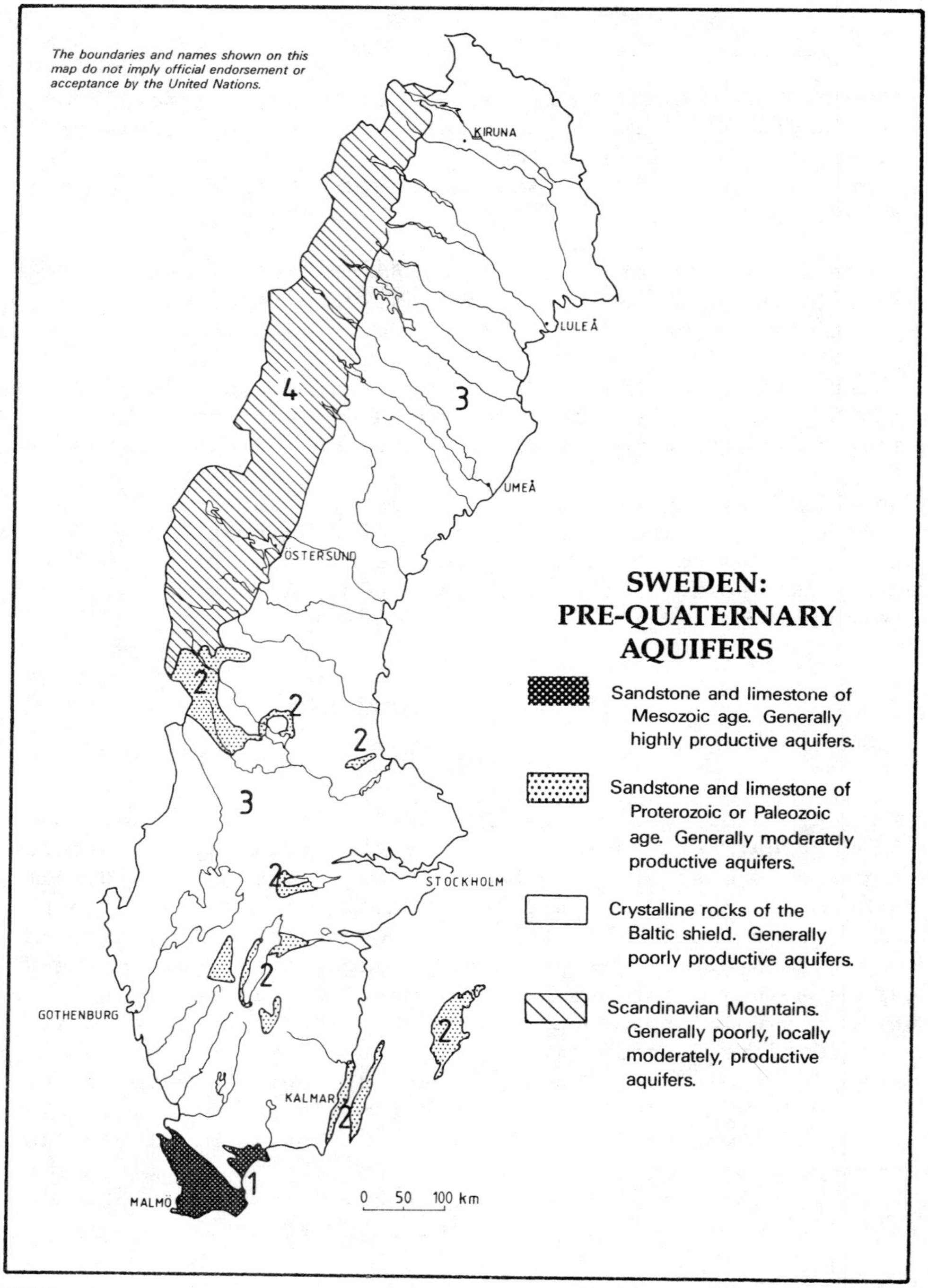

3546.42x

Source: Knutsson and Fagerlind, 1977

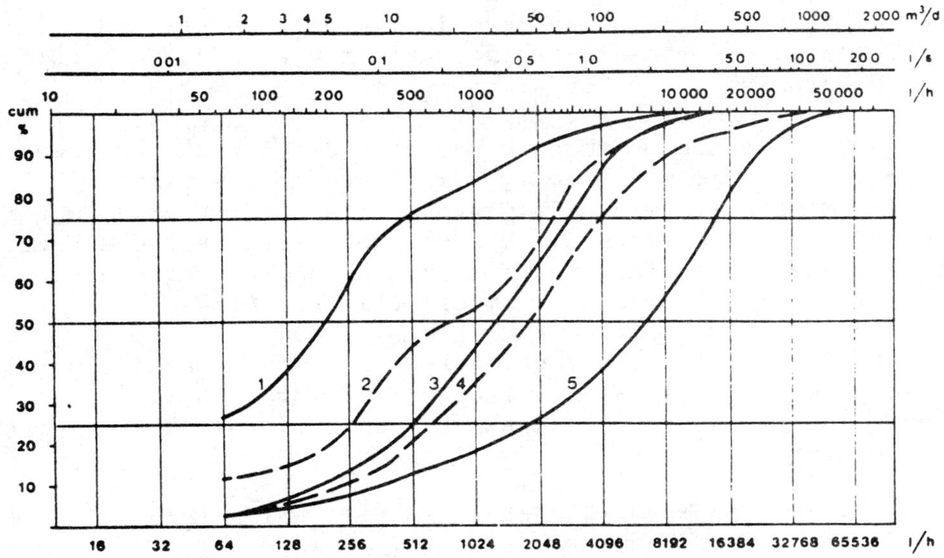

1. Gneiss west of Lake Vänern; the median yield (Q_M) of 256 (n) wells is 0.06 l/s (= 200 l/h).

2. Gneiss in south-east Sweden; Q_M = 0.20 l/s (= 700 l/h), n = 74.

3. Granite north and north-east of Lake Vänern; Q_M = 0.33 l/s (= 200 l/h), n = 190.

4. Gneiss north and south of Lake Vänern: Q_M = 0.5 l/s (= 1 800 l/h), n = 101.

5. Eocambrian sandstone (Visingsö series) at Lake Vättern; Q_M = 1.75 l/s (= 6 300 l/h), n = 64.

Source: Sheet C3, International Hydrological Map of Europe

Figure 54. Southern Sweden, well yields by rock type

Sedimentary rocks have various aquifer properties. The most important and highly productive are the Cretaceous (glauconitic sand and sandstone) and Tertiary (limestone) aquifers of southernmost Sweden. These aquifers are used for municipal and individual water supplies. Typical yields for wells in these aquifers are in the range of 5 to 30 l/s. Also in other parts of Sweden (for example, the Baltic islands) sandstone and limestone beds, mainly of Paleozoic age, form important aquifers.

On the whole, karst aquifers are rare in Sweden. In certain localities, especially in the limestone beds of the mountain range and on the island of Gotland, moderate karst phenomena occur.

The main aquifers in the Quaternary unconsolidated rocks are sand and gravel deposits of glaciofluvial origin, namely eskers and deltas (figure 55). The total withdrawal of ground water from a highly productive glaciofluvial aquifer normally amounts to tens of litres per second, in favourable sites to some hundreds of litres per second. Glaciofluvial aquifers are the dominating source for municipal ground-water supplies in most parts of Sweden (figure 56).

The yield of the glaciofluvial aquifers is, in many places, increased by artificial recharge of surface water. Induced infiltration from rivers and lakes frequently occurs and plays a substantial role in the productivity of the glaciofluvial aquifers.

The most widespread Quaternary deposit is till which, generally, is a poor aquifer but, because of its abundance, is a common source for water supply to individual homes and farms. In till, generally sand and silt dominate. Lenses and layers of sorted material are commonly found within the till, which may considerably improve its productivity in terms of water supply. In Sweden, ground water generally has low salt content, with calcium and bicarbonate as dominating ions. Mineral water of sodium bicarbonate type is found locally, for example in Mesozoic aquifers, but most ground water with high mineral content is of sodium chloride type. Such salt or brackish ground water is locally found both in pre-Quaternary and Quaternary aquifers in those parts of the country that were covered by the sea after the deglaciation (figure 57). High salt concentrations are also found in the deeper parts of the sedimentary rock series.

2. Exploration and assessment

(a) Preliminaries

In Sweden the initial steps of a ground-water investigation programme generally comprise conventional routines like the study of published geologic, hydrogeologic and topographic maps and ground-water data from technical reports and data banks. Examples of other preliminary approaches are the study of aerial photographs and satellite pictures supplemented by field reconnaissance.

These studies are then extended to include a preliminary field programme consisting of geophysical surveys and test drilling to obtain more detailed information of the hydrogeologic conditions like depth, extension, ground-water levels and gradients. Test drilling usually includes permeability tests on-site, sampling of the aquifer material and also water sampling to get a preliminary idea of quality. Laboratory tests may include sieve analyses, mineralogical composition and permeability, soil moisture and organic substance.

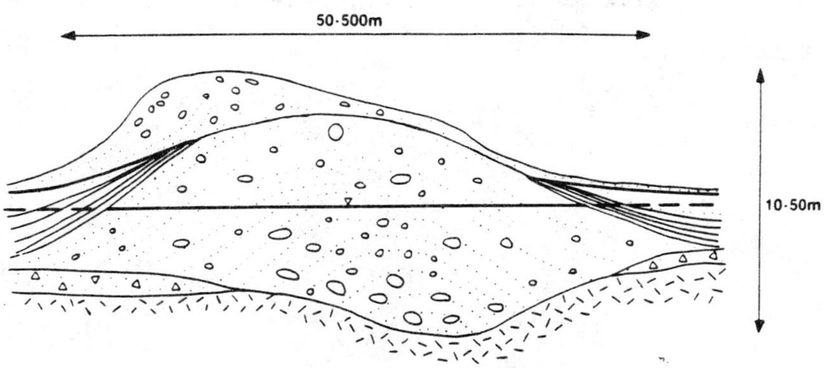

a) Esker situated in area once covered by the sea during deglaciation. Clay was deposited and the esker later eroded by wave action. Eskers of this type are large and characterized by a complex stratigraphy. May form extensive and highly productive aquifers.

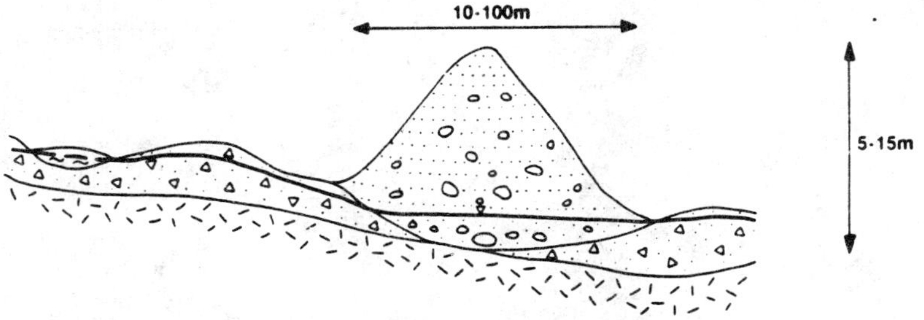

b) Esker deposited above highest shore line. Generally small and in places contains material with low permeability. Generally forms aquifers of only local importance.

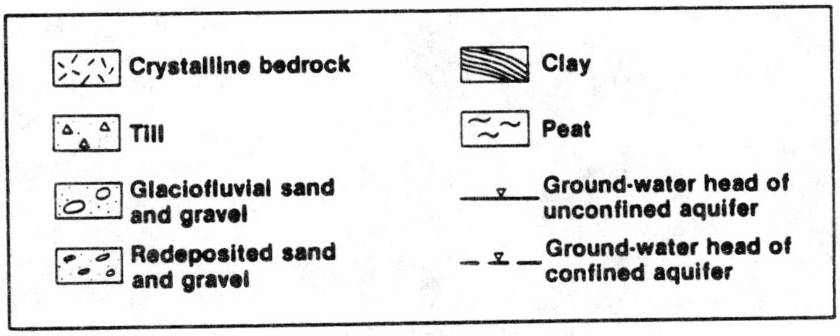

Source: International Hydrogeological Map of Europe, Sheet 0 3

Figure 55. Cross-sections of typical Swedish eskers

Figure 56

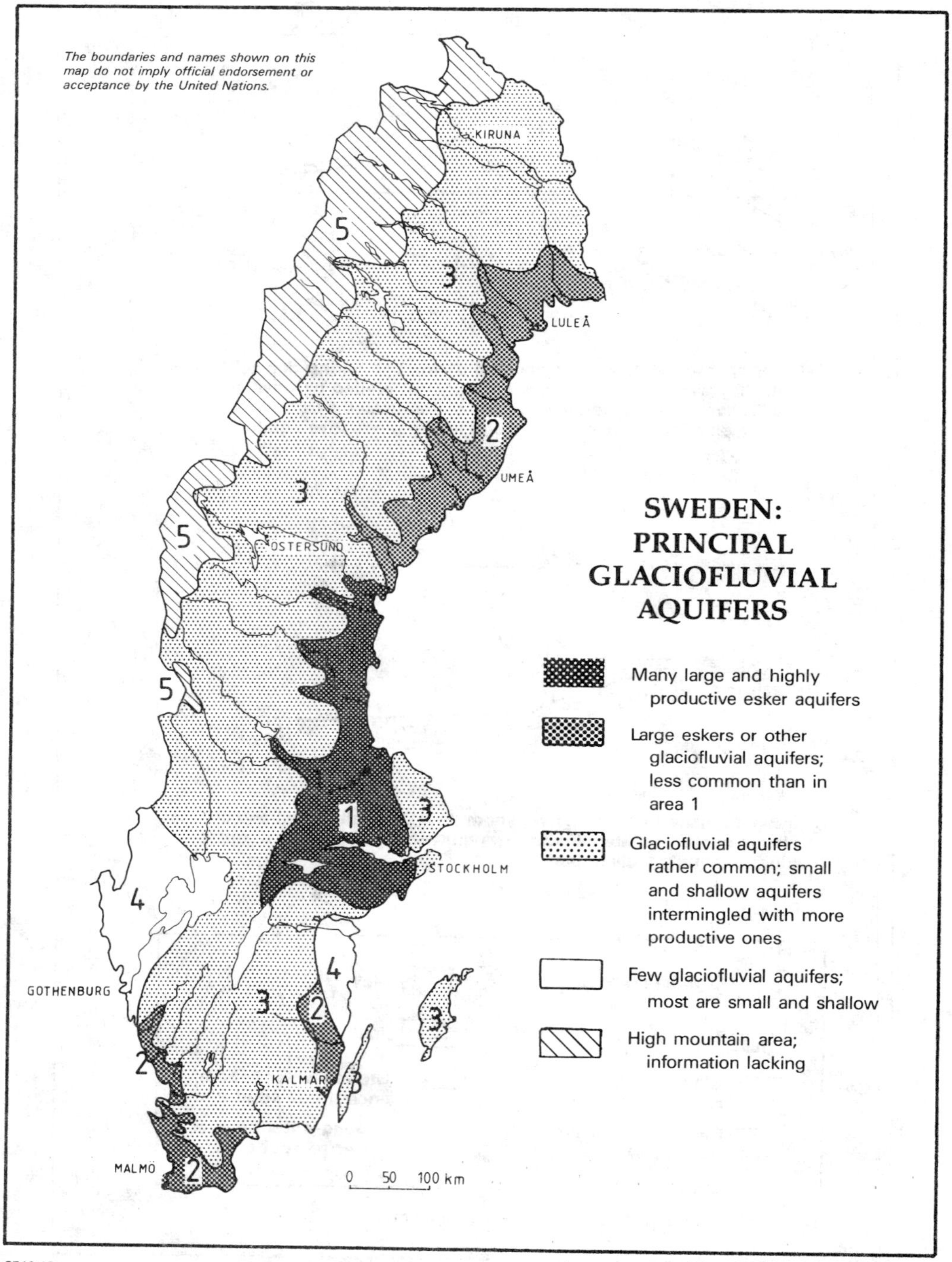

3546.43x

Source: Knutsson and Fagerlind, 1977

Figure 57

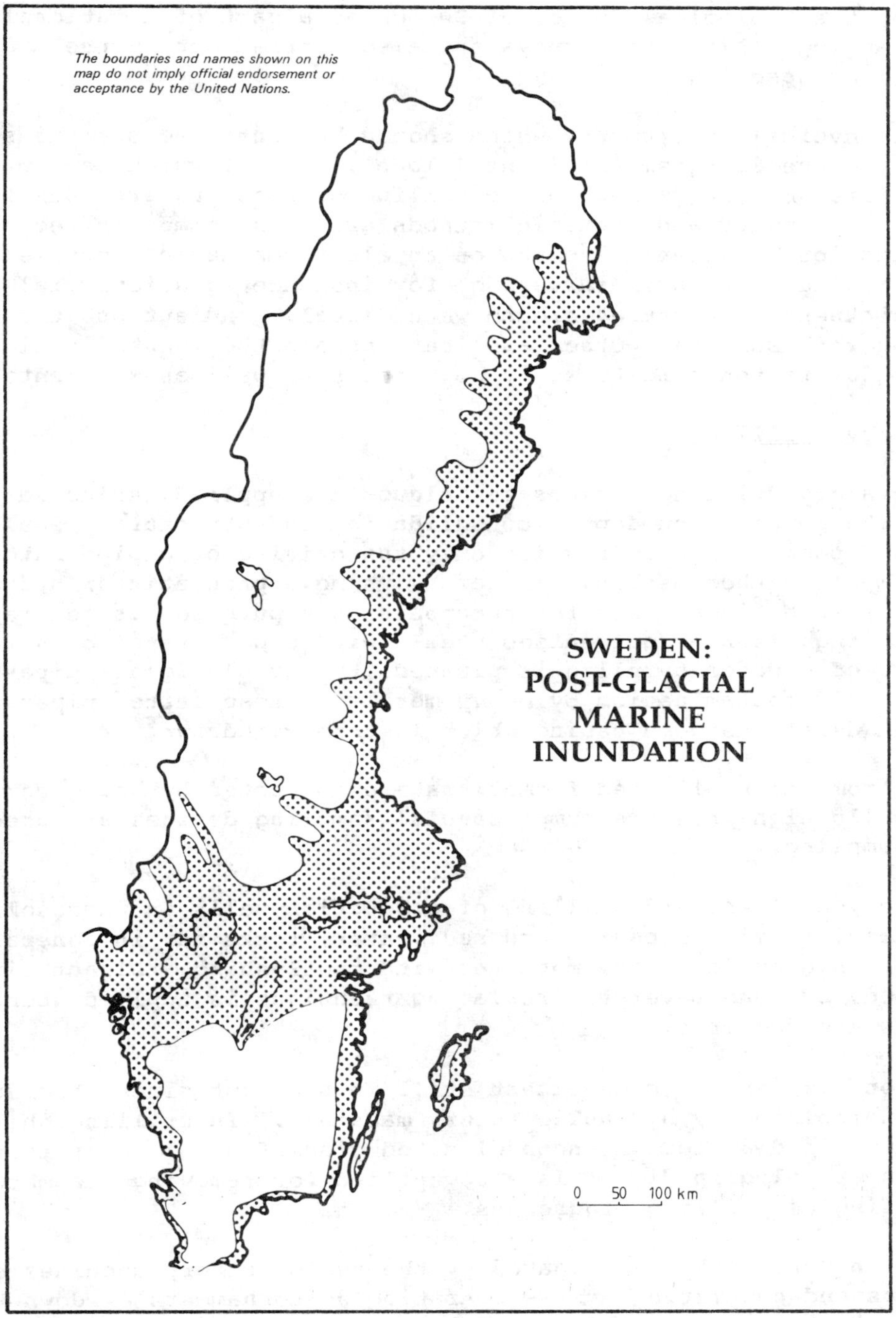

3546.44x

Source: Lindewald, 1981

(b) Geophysical surveying

In geophysical surveys, conventional ground prospection methods like seismic refraction and electrical resistivity are used. Also, data from airborne geophysical surveys may be collected. These investigations, which include magnetic, electromagnetic VLF (Very Low Frequency) and gamma-ray measurements, are carried out by the Geological Survey of Sweden as a part of a national programme for bedrock mapping. Airborne surveys are also carried out by the mining industry during mineral prospecting.

Other geophysical instruments which should be mentioned are the surface operated VLF and the Slingram (horizontal loop), both of which have proved to be sensitive to water-bearing zones in crystalline bedrock and are thus finding increasing use. Gravity and magnetic methods are less common in regional ground-water investigations. Recent research on an electromagnetic instrument (georadar) has given promising results in surveying, for instance, glaciofluvial aquifers concerning thickness, structure, ground-water level, gradient and the topography of the underlying rock surface. Other applications are the identification and location of major tectonic features in hard rocks as well as sedimentary rocks.

(c) Exploratory drilling

In exploratory drilling various techniques are applied, adjusted to aquifer conditions, like penetration depth, consolidation and structure. Shallow glaciofluvial deposits usually require only the driving of a pipe into the ground by repeated impact either mechanically or by using a pneumatic or hydraulic pump. The diameter of such pipes, used for reconnaissance purposes is generally 50-75 mm, the material being plain or galvanized steel, with a perforated or screened pipe end. When periodic water sampling is planned, PVC or plexiglass pipes may be used to reduce or avoid contamination by heavy metals. These latter pipes are installed to the desired depth inside a casing which is then withdrawn.

Samples from unconsolidated formations are collected by using compressed air or by a hydraulic high-pressure pump; special sampling devices are used for undisturbed sampling.

Exploratory drilling and drilling of production wells in unconsolidated aquifers in glaciofluvial deposits and sedimentary rocks is, in general, performed by using modern hydraulic rotary machines with mudflushing. Direct circulation is most often used, but the reverse circulating technique is applied when large-diameter wells are required.

Percussion rigs are to some extent still in use, but since the 1960s they have gradually been replaced by hydraulic rotary machines. In drilling shallow wells more than 600 mm in diameter in unconsolidated glaciofluvial deposits, a drilling technique using a polygrip digger is also applied for removing the material from inside the casing as drilling progresses.

Drilling in hard rock is dominated by the use of rotary machines equipped with hydraulic pumps and pneumatic hammers operating as tophammers or down-the-hole hammers. This technique is mostly combined with a new (ODEX) method of penetrating the overburden. With this method an eccentric reamer drills a hole slightly larger than the casing tube, which is continuously inserted in the borehole through the overburden, and finally a couple of metres into the underlying hard rock. This drilling technique is also applicable as an alternative to the previously mentioned methods of drilling test holes or constructing wells in unconsolidated formations.

Other subsurface investigations applied as a complement to drilling are geophysical borehole logging and various kinds of aquifer tests. Borehole logging may include resistivity, spontaneous potential, induced polarization, radioactive logs (gamma and neutron), temperature, conductivity, Eh and pH registrations.

(d) Borehole logging

Recent research projects related to deep-seated repositories for spent nuclear fuel in Sweden have given rise to the introduction and development of various instruments and other equipment specially designed for borehole logging. Examples of such equipment adapted for 56-mm boreholes of great depth are devices for water sampling from fixed sections of the borehole and to record borehole deviation, natural gamma radiation, fracture density, temperature and conductivity. Special instruments are used for injection and pumping tests and piezometric measurements. The use of packers makes it possible to investigate selected sealed-off sections of the borehole.

On a regional scale the hydraulic properties of the aquifers are calculated from various conventional pumping test techniques like step drawdown and constant yield tests designed for single and multilayer aquifers of unconfined and confined leaky and non-leaky conditions. Tracer techniques are in some cases used for estimating the velocity and direction of ground-water flow.

(e) Data collection

Hydrogeologic data have been collected and stored at the Geological Survey of Sweden for several decades. To emphasize the importance of such information, a Water Act came into force in 1976, requiring water drilling companies, consultants and others to send the results from well drillings, ground-water investigations to the Survey. The major reason for this regulation on a central record system, which is open to the public, is to increase the knowledge of the ground-water conditions in the country. Incoming data, now processed by computer, have proved to be an important source of reference in hydrogeologic mapping, community planning, ground-water exploitation, environmental investigation and control. The records contain such data as type of drilling machine, type and length of casing, diameter of borehole, pump used, yield, applied method and elapsed time of pumping, type of screen and water quality. The number of boreholes made in Sweden is estimated at 200,000. About 90,000 boreholes are registered at the Survey and the corresponding number of chemical analyses stored and processed is about 6,000. Other examples of ground-water related data banks at the Survey are the registration of springs and municipal water supplies. The data processing technique applied allows plotting of maps of various scales with selected information on location of boreholes, depth of the soil cover, yields and other data.

Ground-water technical reports prior to 1976 are being collected and registered as a part of the ongoing hydrogeologic mapping programme.

Other sources of information related to ground water are available at a number of other government agencies, regional and local authorities, universities, research institutes and consulting firms.

(f) Monitoring

A ground-water monitoring network is administered by the Geological Survey of Sweden. In this network the variations of ground-water level, temperature and chemical composition are followed in a nation-wide network of observation and sampling stations. The project started in the late 1960s. The number of stations is about 400 for ground-water level measurements, about 40 for temperature and about 200 for chemical sampling. The main aim of the project is to collect data for reference and environmental control. The project is partly connected with the National Environmental Monitoring Programme of Sweden.

(g) Modelling

In ground-water modelling techniques, three main types of model may be distinguished; the mathematical, the analog and the physical method. Of these three types the analog and physical models have been used to only a limited extent in Sweden, leaving the ground-water modelling technique essentially with mathematical models. Of the two-dimensional mathematical models, the analytical method is limited due to its restrictions in solving ground-water problems where variations of, e.g., aquifer thickness, permeability, transmissivity, storage coefficient and complicated boundary conditions are to be considered.

Of the numerical mathematical models, the finite difference method (FDM) is less used than the finite element method (FEM). The applicability of the finite element method which is dependent on computer services, is currently increasing with the development of computer capacities. Apart from computer programs available abroad, programs are also developed within the country, adapted to the special aquifer conditions prevailing in Sweden.

In the application of the finite element simulation method, the so-called GEOFEM, developed by Chalmers University of Technology in Gothenburg in co-operation with the University of Gothenburg, and the AQUIFEM are commonly used.

Other models include the integrated finite difference method, used as a transport model and a three-dimensional model, based on the porous media concept, but directed to consolidated aquifers with secondary permeability developed as a joint and fracture system.

In considering surface water runoff aspects in regional water and ground-water problems the so-called HBV model is available. This model, introduced and developed by the Swedish Meteorological and Hydrological Institute, exists in various operational versions. The model consists of subroutines for snow accumulation and melt, a soil moisture accounting procedure and routines for ground-water recharge, storage and discharge.

3. Development

In Sweden, more than 7 million inhabitants are supplied from municipal water works. The rest generally draw water from private wells.

The total water production of the municipal water works amounted to 942 million cubic metres in 1983. Of this quantity 53 per cent consist of surface water and 47 per cent of ground water. Of all the population about 55 per cent uses ground water and the rest surface water. Some of the industrial supplies are also included in the production figures (about 13 per cent), although most industrial plants with a large water consumption have their own supply, mainly from surface water. Irrigation in Sweden uses roughly 35 million cubic metres of ground water per year. That is less than a tenth of the need of the municipal water works.

Construction of municipal ground-water works in Sweden began towards the end of the nineteenth century - in the town of Uppsala, for instance, in 1875. Artificial recharge was soon introduced through the pioneering methods of J. G. Richert. He first succeeded in realizing his ideas in the city of Gothenburg in 1897-1899.

The first successful drilling in Archean bedrock took place in 1894. After the first 28 drillings it was noticed by a leading scientist that ground water was generally struck at depths between 30 and 35 m.

The oldest type of well for shallow aquifers is the dug well. Many such wells are still in use but few new ones are being dug. The old dug wells were initially timbered or lined with brick or stone, later giving way to installations with concrete lining. In constructing such wells, gravel with clay on top is backfilled around the concrete pipes and gravel at the bottom to control sand and silt entry. Other examples of low-cost ground-water extraction techniques include driven or jetted wells, usually with diameters of 50-75 mm.

Modern drilling techniques have been introduced and led to exploitation of somewhat deeper water resources in hard rocks, sedimentary rocks and unconsolidated aquifers. Well point installations are frequently used for lowering the water table furnishing dry excavation for foundations and other subsurface construction operations. Natural springs, of which the largest ones are associated with glaciofluvial deposits, are often encased and used as individual or municipal water supplies. Most such springs yield between 1 and 20 l/s, but some give 50-100 l/s.

Boreholes in hard rocks and sedimentary rocks for individual water supplies are usually made with an uncased diameter of 110-140 mm. The casing, penetrating the overburden, normally has an inner diameter of 128-165 mm. The casing is drilled a few metres into the bedrock and the transition zone between the soil cover and the bedrock sealed by grouting. If the bedrock is heavily fractured, the borehole may be lined with a perforated casing. The slot size is about 10 mm.

Of a total of about 20,000 boreholes drilled during a recent two-year period some 30 per cent were drilled with the purpose of extracting energy from the ground water or from the heat flow of the bedrock by means of a heat pump.

Inclined boreholes are sometimes drilled with the objective of encountering and penetrating a maximum number of fissures or fracture zones. The drilling angle is normally 30° or less.

High-yield wells constructed in the overburden for municipal water supplies occur in a variety of sizes, depending on such conditions as grain size distribution, permeability, yield, pump dimensions and gravel packing. Most high-yield wells in unconsolidated glaciofluvial formations are constructed with a casing diameter in the range of 200 to 400 mm. At the more favourable sites even larger diameters are applied.

In coarse homogeneous aquifers a natural gravel pack is often considered sufficient. Artificial gravel packing is also common and the gravel pack is usually constructed with a thickness in the range from 70 to 100 mm. Gravel packing is often considered beneficial to the well construction in order to reduce problems associated with segregation, bridging, plugging and sand pumping. This kind of well generally produces up to 100 l/s but larger yields are not uncommon.

In the construction phase of boreholes and wells, various kinds of material are used. Most casing tubes are made of mild or stainless steel. High-yield municipal wells are usually equipped with stainless steel screens owing to their durability and resistance to corrosion. PVC casings and screens are becoming common in smaller and individual water supplies. The screen types most widely used are of the continuous slot type.

The development of pumps over the last decades has resulted in a gradual introduction of submersible centrifugal pumps, to a large extent replacing piston, jet and turbine pumps. Ground-water resources in Sweden are also locally exploited for air-conditioning and industrial cooling.

In regions of insufficient ground-water resources, artificial recharge is often applied. Surface water is pumped into basins excavated in glaciofluvial deposits. In the Uppsala area, problems of insufficient surface-water resources have been solved by transferring water from one catchment area to another. Collector wells have not been regarded as appropriate to Swedish conditions and few such wells are in operation.

As already mentioned, boreholes in crystalline Precambrian bedrock normally yield 0.1-1 l/s. To stimulate the capacities of low yielding boreholes, hydraulic fracturing has proved successful. The use of explosives has been replaced mainly by high-pressure pumping of water into the open hole between packers.

4. Problems

Downward movement or subsidence of the land surface has been noticed in several places where different kinds of underground construction have lowered the water tables or piezometric surfaces.

The major reasons for such ground-water depletions include extraction for water supplies, and construction of tunnels, storm sewers and other underground pipelines.

Urbanization and township development have in various ways disturbed the natural ground-water balance. In larger cities like Stockholm the development of the subway network, including the construction of numerous tunnels for conduits, cables and communication under the city indirectly causes dewatering of the bedrock and soil cover with a subsequent cracking in old buildings and subsidence of pavements and street sections. The amount of subsidence is in some cases on the order of 0.5-1 m.

The quality of the ground water in Sweden is generally good. Contamination problems are not common, but occur locally in connection with waste disposal sites, bad waste-water handling, industrial plants and other hazards.

Problems with high nitrate concentrations in the ground water are common in some agricultural areas in southern Sweden. The problem occurs especially in shallow aquifers connected with cultivated and fertilized sandy soils. To avoid the problem, deeper wells and wells in other aquifers have been used. Restrictions on the use of fertilizers in some areas have been considered.

Much attention has been given the problems of acid ground water (pH below 6). The low pH values may partly be connected with the increasing amounts of airborne acid pollutants from industrial Europe although most of the acid ground water probably is so owing to natural geologic and climatic causes. Acid ground water is especially common in those parts of south-western and southern Sweden with low lime content in the ground. The low pH of the ground water contributes to corrosion of pipes which may cause high concentrations of certain metals. The quality of the water can be improved by adding various alkaline substances to the surroundings of the well, into the well or to the water in the supply plant.

The most common natural ground-water quality problem is high content of iron or manganese. These metals are generally removed in municipal water works through oxidation and filtration. New methods including re-infiltration of the ground water to the aquifer have been much used lately. Another new method (VYR) which during the last years has shown its usefulness is a procedure where the treatment takes place directly in the ground by adding oxygen.

Problems with salt and brackish water are also rather common in some parts of Sweden. Sea water intrusion may be the cause in some coastal areas and on small islands. The main problem, however, is connected with relict salt water. As already mentioned, large areas, mainly in southern Sweden (figure 57) were covered by the sea for a longer or shorter period during postglacial time. Salt water from that time is still found both in Quaternary deposits and in the bedrock. These salt ground waters are especially common in lowlands with small water table gradients and long turn-over time. Many seem to be entrapped, with little or no connection with the surroundings. To avoid salt water contamination, deep wells and pumping of large quantities of water should be avoided in the high risk areas.

Legislation has been enacted to conserve and protect ground-water resources.

The fundamental principle concerning the Swedish Water Act is that individual property owners may extract up to 10,000 m^3/d with no regulation or permit required. If, however, the proposed withdrawal exceeds 10,000 m^3/d the Government, through its regional Water Courts, can decide whether the enterprise will be permitted. The Environmental Protection Act and the Public Health Act establish standards for the quality of ground water.

D. Training, research and international co-operation

Education in ground-water resources and hydraulics is provided at various levels in Sweden and is an important subject in water engineering, including water supply technology, drainage, foundation techniques, irrigation and energy.

Education associated with research is primarily provided at the universities and certain technical schools. Research is also carried out by government agencies and engineering companies, often supported by funds made available by certain research councils like the Swedish Natural Science Research Council.

Ground-water resource mapping is carried out on a national basis by the Geological Survey of Sweden (SGU). The current hydrogeologic mapping programme covers all counties and the maps are presented at the scale 1:250,000. A national hydrogeologic map is in preparation, to be published at the scale 1:1,000,000. Some 10 maps at the scale 1:50,000 have been published.

Other government agencies and major universities involved in ground water include the Swedish Meteorological and Hydrological Institute (SMHI), the Swedish Geotechnical Institute (SGI), the National Environment Protection Board (SNV), Chalmers University of Technology, the Royal Institute of Technology (KTH) and the Swedish University of Agricultural Sciences.

Since the mid-1960s, Sweden has co-operated with developing countries. Within the ground-water sector this co-operation has essentially been directed towards rural areas.

Some 10 countries have received assistance by the Swedish International Development Authority (SIDA). The current bilateral co-operation programme on water and its association with health and environment is limited to five countries, namely Botswana, Ethiopia, India, Kenya and the United Republic of Tanzania. In countries like India, Angola and Guinea-Bissau a ground-water component is included in other sector programmes.

Within the current international water decade, Sweden also co-operates through SIDA with several United Nations organizations including UNICEF, UNDP, IBRD, WHO and a number of research and educational institutes. The Swedish water programme in India is run in co-operation with UNICEF.

Other co-operation channels are through the Swedish Agency for Research Co-operation with Developing Countries (SAREC) and the Swedish Commission for Technical Co-operation (BITS). These organizations are among those giving financial support to water-associated projects in developing countries, including seminars and courses adapted to the conditions prevailing in such countries. Swedish universities have supported water projects in such African countries as Tunisia, Algeria, Egypt, Zimbabwe, Malawi and the United Republic of Tanzania, and in other parts of the world, including Sri Lanka and Viet Nam.

There is intensive co-operation between Sweden and the other Nordic countries (Denmark, Finland, Norway, Iceland) within the field of hydrology and hydrogeology. This co-operation is partly within the Nordic Hydrological Programme (NHP) and partly within the Nordic Association of Hydrology. The activities within the Nordic ground-water monitoring programmes, for example, have been co-ordinated by an NHP working group.

Preparation of the International Hydrogeological Map of Europe, which for Sweden's part was carried out beween 1969 and 1984, has led to new contacts and co-operation with adjacent European countries, among others the Soviet Union and Poland.

<u>References</u>

Almén, K. <u>et al</u>. Final disposal of spent nuclear fuel - equipment for site characterization. Swedish Nuclear Fuel Supply Co/Division KBS. SKBF-KBS Technical report 83-44. Stockholm, 1983.

Eriksson, E., Y. Gustafsson <u>and</u> K. Nilsson (Editors). Ground water problem. <u>Wenner-Gren Center International Symposium Series</u>. vol. 2. London. P. A. Norstedt & Söner. Stockholm, 1968.

Fagerlind, T. <u>and</u> S. Modig. A new law on the compulsory notification of data in the drilling of new wells. <u>Striae</u>. vol. 4, pp. 5-6. Uppsala, 1977.

Henkel, H. <u>and</u> L. Eriksson. Interpretation of low altitude airborne magnetic and VLF measurements for identification of fracture zones. In <u>Subsurface space</u>. Pergamon, Oxford and New York, 1980.

International Hydrogeological Map of Europe. Map sheets C 2, Trondheim; C 3, Oslo; C 4, Berlin; D 2, Haparanda and D 3, Stockholm, with explanatory notes. UNESCO, Paris, n.d.

Knutsson, G. <u>and</u> T. Fagerland. Grundvattentillgangar i Sverige (Groundwater resources in Sweden). Geological Survey of Sweden. <u>Rapporter och meddelanden</u>, nr 9. Stockholm, 1977. In Swedish.

Lindewald, H. Saline groundwater in Sweden. In <u>Intruded and relict groundwater of marine origin</u>. Swim 81. Geological Survey of Sweden. <u>Rapporter och meddelanden</u>, nr 27. Uppsala, 1981.

Morfeldt, C-O. Flooding leakage in tunnels - Method of prediction and prevention. International Congress "Tunnel 81" Düsseldorf, 1981.

Müllern, C.-F. Airborne geophysical measurements used for hydrogeological mapping. Uppsala University. UNGI Rapport, nr 53, 1981.

Nordberg, L. <u>and</u> G. Persson. The national groundwater network of Sweden. Geological Survey of Sweden. Ser Ca No. 48. Stockholm, 1974.

Water in Sweden. <u>National Report to the United Nations Water Conference 1977</u>. Ministry of Agriculture, Stockholm, 1977.

XII. USSR IN EUROPE

Area: 5,570,000 km^2 Population: 200 million

A. General

1. Physiography

The European part of the Union of Soviet Socialist Republics, the most densely populated area of the country, also includes its major industrial districts - Moscow, Donetsk, Leningrad, the Urals, the Middle-Volga - all developed and transformed through human activity. The density of population may be as high as 200 per square kilometre. Over the centuries, the economic development required the plowing of land, drainage of bogs, cutting of forests, changes in stream régimes, and many other human actions affecting the hydrogeologic environment. Ground-water conditions were also disturbed by intensive extraction for household and industrial use, mine drainage, reservoir construction and operation of large irrigation projects.

Ground water is one of the most important natural resources in the European part of the Soviet Union. Ground water is used to supply more than 60 per cent of the cities, most rural settlements and for crop irrigation. For centuries, waters of the Carpathians, the Baltic republics and the Urals with their fine health resorts, has been used in balneology. Of particular interest in recent years is the use of ground water as an energy source by means of the heat pump. When not properly managed, however, ground water can have severe negative effects in water-logging, unwanted swamps and secondary salinization of land.

Ground water, therefore, because of the many ways it affects the national economy, is given special attention and high priority in its study and use by the Government.

The European part of the USSR is geomorphologically full of contrasts. Its major feature is the great Russian (East European) Plain. To the east it is bordered by the Ural Mountains which extend for more than 2,500 km and reach a height of 1,894 m. The south-west boundary of the Plain is the Carpathian Mountains culminating at 2,061 m. Over much of the southern portion of the Plain, lowlands alternate with hilly and plateau-like highlands (figure 58).

Average altitude of the Plain is 170 m above sea level. The most elevated northern region is the Kola Peninsula where some mountains reach altitudes of 1,000 m (Khibiny Mountains). Other uplands (Byelorussian, Smolensk-Moscow, Northern Urals) have altitudes of 150-350 m.

The uplands are separated by vast lowlands (Ladoga, Ilmen, mid-Lithuanian) around lakes and in coastal areas.

The flat swampy North Dvina Plain which slopes gently towards the White Sea lies in the north-east. To the east it is confined by the Timan Ridge. The Pechora Lowland, heavily bogged and containing scattered areas of permafrost, is situated between the Timan Ridge and the Urals.

The central part of the Plain is occupied by the large Central Russian Upland (200-300 m) severely dissected by river valleys and gullies.

MAP NO. 3568.2 UNITED NATIONS
NOVEMBER 1989

The Volga Upland which in some parts rises above 300 m and gradually passes into the Oka-Don lowland to the west, stretches out along the right bank of the Volga to the south as far as the mouth of the Kama River. The Yergeni Hills (220 m) are a southward orographic continuation of the Volga Upland. In the extreme south-east of the Russian Plain lies the Caspian Lowland a part of which is below sea level. South-west of the Russian Plain the Volyn-Podolian Upland rises above 500 m as part of the Carpathian piedmont.

The Pripe and Marshes (Polesye) is a vast, once boggy but recently dewatered, lowland between the Volyn-Podolian and Byelorussian uplands.

The topography of the European USSR is mainly the result of neotectonic activity and, to some extent, of exogenic processes, in particular continental glaciation which covered most of the northern Russian Plain.

2. Climate

From north to south, the climate ranges from extremely humid and cold to warm and dry.

Annual precipitation in the north-east (Malozemelsky tundra) is 300-400 mm, increasing southward to 600-800 mm in the Baltic republics and adjacent areas, thence gradually falling to 100-200 mm in the Caspian basin. Except for the Crimea and southern Ukraine, the European USSR has a cover of snow all winter.

Mean air temperatures in January vary from -12° in the coastal areas of the northern seas to 4° C in the Crimea. Average values of air temperature in July are 10°-12° C in the north and up to 28° C in the south-east, in the Caspian basin. The climatic zones differ according to altitude. The maximum annual precipitation in the Alpine zone of the Carpathians exceeds 1,600 mm. The character of vegetation depends on the latitude and local topography. Bush and sparse woods in tundra gley soils are widespread in the northern part of the Russian Plain. Dark-conifer taiga lies south of the tundra zone as far as the line Lake Peipus to Gorky to Perm. The central part of the Russian Plain is occupied by diverse types of forest ranging from mixed to deciduous. Steppes stretch out southward in a belt 500-700 km wide. To the south-east, steppes give way to semi-desert then salt desert along the shores of the Caspian Sea.

Mountain vegetation, although determined by altitude, is fairly uniform because most of the mountains are relatively low. Zones of deciduous, mixed and dark-conifer forests are clearly delineated in the Carpathians.

3. Hydrography

The Russian Plain is drained by the large rivers listed in table 8.

Table 8. European USSR, parameters of the larger rivers

River	Basin area (in thousand km^2)	Length (km)	Annual average flow volume (km^3)
Pechora	422	1 870	130
North Dvina	357	1 302	110
West Dvina	88	1 020	21.4
Volga	1 360	3 531	243
Dnieper (Dnepr)	504	2 201	53.5
Don	422	1 870	29.5
Ural	231	2 428	11.6

The bulk of river flow originates from snow melting during the spring, but abundant summer floods occur in the Carpathians, the Crimea and the Baltic republics. Ground water greatly contributes to the flow.

The flow rate of streams is related to latitude and altitude. Flow rates in the northern part of the Russian Plain are 10 $l/s/km^2$ and more, gradually reducing to 0.2 $l/s/km^2$ in the Caspian basin. Flow rates increase from 1 or 2 $l/s/km^2$ in highlands to 30 $l/s/km^2$ in mountainous areas such as the Urals and Carpathians. The mineral content of the water in small and medium streams increases from north to south, i.e., from 70-80 mg/l in the north to 1,000-3,000 mg/l in the south-east. Stream water is mainly of the calcium bicarbonate type with higher content of chlorides and sulphates in the south-east.

The largest lakes are Ladoga (18,400 km^2), Onega (9,610 km^2) and Ilmen (2,500 km^2). The artificial lakes of Rybinsk, Kuibyshev, Saratov, and Kakhovka created on the rivers Volga, Dnieper and Don are comparable in size.

B. Geology

1. Russian platform

Most of the region is part of the East European or Russian platform. To the east it is limited by the Urals-Timan folded mountains, to the south-west by the Carpathian folded mountains, and to the south by the Scythian plate. The Russian platform joins the West European platform between the Carpathians and the coast of the Baltic Sea.

The geologic structure of the Russian platform has been studied in detail through deep and superdeep drilling, geophysical surveys and other technologies.

The Precambrian basement includes Archean metamorphic rocks with complex folded structure involving a number of blocks separated by deep faults. The sedimentary cover includes diverse formations from Proterozoic to Quaternary in age. The platform cover deposits are 15-17 km thick.

The largest tectonic units of the Russian platform are the Russian plate and the Baltic and Ukrainian shields.

The Russian plate includes several major syneclises (Moscow, Baltic, Black Sea, Caspian) and anteclises (Byelorussian, Voronezh, Volga-Kama), deep troughs of ancient rifts (the largest being the Dnieper-Donets trough), and piedmont trough (pre-Urals).

The basement plate is formed of deeply metamorphosed rocks (granite-gneiss, migmatite, quartzite).

The Moscow syneclise has the largest area. Its sedimentary cover averages 1.5-2 km thick, ranging from several tens of metres near the Baltic shield margin up to 3 km and more in the deepest parts of the structure. The basement underlying the syneclise is dissected by deep faults confining some graben-like extended depressions or aulacogens. Usually, the cover structures are inverse relative to the aualacogens. For example, cover rocks form an anticline over the Krestetsk-Soligach aulacogen.

Proterozoic deposits include sandstone, shale and argillite. Cambrian deposits are also represented by terrigenous rocks, Ordovician deposits chiefly by carbonate rocks (limestone, dolomite). Devonian deposits outcropping in the north-west part of the anteclise are mostly terrigenous. Widespread Carboniferous deposits are mostly composed of carbonate rocks. Permian deposits include dolomite and such evaporites as gypsum and salt.

Mesozoic and Cenozoic deposits cover the Paleozoic formations. The Triassic and Jurassic systems are chiefly represented by terrigenous rocks. Cretaceous deposits also contain carbonate rocks. Neogene and Paleogene deposits are common in the southern part of the syneclise and consist mostly of sand and clay.

Quaternary deposits include sand, clay, loam and other glacial or alluvial deposits. Their maximum thickness in buried valleys is between 60 and 80 m. Other structures of the Russian plate generally have a similar lithologic section.

The Baltic syneclise in the north-west part of the region is a flat structure partially beneath the Baltic Sea. It is built of deposits ranging from Cambrian to Quaternary in age. Having a thickness of 500-3,000 m, the rock types are similar to those of the Moscow syneclise. A regional sequence of laminated clay is characteristic in Cambrian deposits.

The Black Sea syneclise occurs in the south of the Ukrainian crystalline shield. In the Black Sea area the syneclise is interrupted by a large regional fault which appears to separate the Russian and Scythian plates.

The thickness of the sedimentary cover in the deepest part of the syneclise is 2-3 km or more. The Cambrian deposits above the basement rocks are terrigenous, whereas the overlying Silurian sediments are terrigenous carbonate rocks. At shallower depth occur Jurassic terrigenous rocks and Cretaceous deposits consisting of sandstone and limestone. The Cenozoic sequence includes Sarmatian limestone and Pontian terrigenous carbonate deposits.

The Caspian syneclise is a zone of continuous subsidence which started at the end of the Proterozoic. As a result, the thickness of the sedimentary cover is 17 km. A thick mass of Permian salt beds divide depression sediments into undersalt and oversalt stages. Sediments of the undersalt stage in the central part of the depression were not reached by the Aralsor superdeep well drilled to a depth of 7 km, and were studied using geophysical techniques. Jurassic, Cretaceous, Paleogene and Neogene layers are composed of terrigenous and terrigenous carbonate formations. The Quaternary is represented by transgressional deposits of the Caspian Sea. Salt dome diapirs are common throughout the depression.

In the west the central part of the Russian plate is bordered by the Byelorussian anteclise. Crystalline basement occurs at moderate depth (500 m on average). The basement is covered by a complex of Proterozoic, Cambrian, Silurian and Devonian rocks overlain by horizontal layers of Cretaceous, Paleogene and Neogene terrigenous deposits. Quaternary deposits are rather thick (to 100 m) and attributed chiefly to continental glaciation.

In the south the Moscow syneclise is bordered by the Voronezh anteclise. Basement rocks are found at a depth of 250-1,000 m. At a shallower depth is a thick mass of Devonian terrigenous deposits in turn overlain by Jurassic, Cretaceous and Paleogene sediments.

The Volga-Kama anteclise covers a vast area in the eastern part of the Russian plate. It consists of several arches (Tokmovsky, Tatarsky) separated by troughs and aulacogens. Basement depth ranges from 1 to 5 km and more. The syneclise includes diverse rock complexes of the age period from Proterozoic to Quaternary. Permian terrigenous and evaporite deposits outcrop over large areas.

The Dnieper-Donets basin is a thick aulacogen separated from adjacent structures by deep tectonic dislocations. The basin falls in a number of depressions. Devonian terrigenous carbonate deposits are less than 3 km thick and overlain by Upper Devonian salt beds. The Carboniferous sequence includes sandstone, siltstone and limestone having a total thickness of 1-3 km. Permian deposits include many salt beds and salt domes.

The Baltic shield constitutes a large crystalline mass. It includes several tectonic zones composed of heavily dislocated and metamorphosed Archean and Proterozoic rocks (gneiss, granite-gneiss, amphibolite) overlain by thin glacial deposits. The Baltic shield is dissected by large faults trending nearly north-south.

The Ukrainian shield is another large unit of Precambrian crystalline basement having similar structure. With faults having diverse trends ranging from nearly north-south to near east-west. Crystalline rocks are overlain by Sarmatian limestone, Pontian terrigenous formations and Quaternary deposits having a total thickness of 100 m.

2. Ural-Timan Mountains

The Ural-Timan folded mountains were developed in Paleozoic times. They involve three large structures, the Timan and the Urals mega-anticlinoria and the Pechora basin.

The Timan mega-anticlinorium is a system of horst uplifts and large anticlinoria separated by graben-like depressions. The outcropping cores of the anticlinoria contain Proterozoic metamorphic and Devonian volcanic and sedimentary rocks. Quaternary and Permian carbonate deposits and sediments occur in depressions.

The Urals mega-anticlinorium is a tectonic feature divided into several second-order structures such as the Central Urals uplift, and the Tagil-Magnitogorsk trough. It includes sedimentary, volcanic and metamorphic rocks dating from Precambrian to early Carboniferous. Linear folded structures are broken by intrusions of late Paleozoic granite and peridotite. A dense net of faults of various orders (chiefly N-S trending) is observed. The Urals structures passing through the Yugorski Peninsula and Vaigach Island find their continuation in Novaya Zemlya. The base of the Pechora basin is composed of highly metamorphosed Riphean rocks. The sedimentary cover is 6 km thick. Two structural stages are distinguished. The lower stage consists chiefly of Ordovician to Permian terrigenous carbonate rocks. The upper stage is represented by mostly Jurassic, Cretaceous and Cenozoic terrigenous deposits.

3. Carpathian Mountains

The Crimean and Carpathian folded mountain area is a small extension of the Alpine system of southern Europe. The Carpathian Mountains include three large tectonic elements, the pre-Carpathian marginal trough, the Carpathian folded area and the trans-Carpathian inner trough.

Precambrian and Paleozoic phyllites, argillites and shales of (a) the pre-Carpathian trough constitute the basement of the outer Carpathian zone. In the upper part occurs Miocene molasse which is underlain by Jurassic, Upper Cretaceous and Paleogene deposits. Cretaceous and Paleogene flysch greatly contributes to the structure of (b) the Carpathian folded area. Faults and adjacent structures are chiefly arc-like striking south-east. (c) The trans-Carpathian inner trough is formed by a thick mass of Miocene molasse overlying Mesozoic rocks.

The Crimean mega-anticlinorium which occurs in the southern part of the peninsula is formed by folded and faulted Triassic, Jurassic and Cretaceous carbonate and terrigenous rocks. In the north, Mesozoic deposits are overlain by flat, chiefly Paleogene and Neogene, carbonate layers.

4. Scythian plate

In the south, the Russian platform borders on the Scythian plate having a basement composed of Upper Proterozoic and lower Paleozoic metamorphic rocks. The basement is found at a depth of 1,000 m. The plate includes several secondary structures, the Crimean and Manych troughs, and the Karpinsky swell. In the south, the plate joins the Caucasus structure. The sedimentary cover is represented by a thick mass of chiefly terrigenous, with some evaporites and carbonate rocks dating from Devonian to Quaternary. The Carboniferous terrigenous carbonate rock mass is up to 7 km thick.

C. Hydrogeology

1. Historical

The history of hydrogeologic studies in the European part of the USSR is conventionally divided into three periods of the twentieth century, (a) before 1918, (b) 1918-1945, and (c) since 1946.

The first wells were drilled in the 1830s and 1840s to provide water for the cities of Odessa, Sevastopol and Kerch.

The second half of the nineteenth century was marked by the start of geologic surveys in the European part of Russia at a scale of 1:420,000. As a result, data were obtained for several regions on the depth, mineral content and chemical composition of ground water, chiefly of shallow aquifers. Also, hydrogeologic studies were carried out during soil, geographic and land management surveys led by Dokuchaev, Nikitin and others.

While hydrogeologic information was collected through specific surveys, the study of ground water developed as a component of geologic and geographic research. Investigation of the relations between hydrogeologic, geologic and geomorphic factors and processes provided development of regional hydrogeology as a science.

The results obtained over a 50-year period of ground-water study in the Russian Plain were generalized by Nikitin in 1900. In particular he delineated artesian basins of the Russian platform.

After the Russian Civil War, hydrogeologic studies were intensified. General survey at a scale of 1:420,000 in the Lower Volga (Kamensky, Plyusnin et al.) and survey at a scale of 1:42,000 in the Ukraine started at the end of the 1920s.

Il'in (1925) first suggested a scheme of ground-water zoning in the European part of the USSR and a report on its ground water was published by Semikhatov. Works dealing with the hydrogeology of particular regions, including the Ukrainian crystalline rock mass and the Dnieper-Donets basin, were also published. Such works were based on scarce hydrogeologic data complemented by hypothetical considerations of the authors which in the end were confirmed to a great extent.

Since the early 1930s hydrogeologic investigations have a stable position in the system of the state geologic studies. Initiated were a regular ground-water exploration for municipal water supply, deep oil drilling in the Volga-Kama oil-gas province associated with studies of deep aquifers in sedimentary formations, and other activities which provided data in the region and were the basis for elaboration of basic hydrogeologic concepts. Publication of a multivolume edition of Hydrogeology of the USSR started in 1939. Several volumes were issued before the Second World War.

During that war, hydrogeologic studies were limited to solution of emergency problems.

From 1946, regions-hydrogeologic study of the European part of the USSR was considerably expanded. Ground-water exploration for water supply, theoretically demonstrated in works by Plotnikov, was carried out on a large scale.

Combined mapping on a scale of 1:500,000, using a procedure elaborated by Zaitsev, started in 1948-1949. Since 1960, a regular hydrogeologic survey of the USSR mapped on a scale of 1:200,000 has been under way by geologic organizations using procedures developed by VSEGINGEO.

Hydrogeologic and engineering geology surveys on a scale of 1:50,000 have also been conducted since the early 1970s. Much information on deep aquifers is obtained as exploration for oil and gas proceeds. Geophysical, geochemical, isotope and other modern techniques are routinely applied in hydrogeologic studies and more recently remote-sensing techniques have revealed previously unknown hydrogeologic features.

By 1964-1972 the Hydrogeology of the USSR in 45 volumes was completed and the Hydrogeological and Engineering Geology Atlas of the USSR compiled.

Some specific studies are to be mentioned such as monographs on the Russian platform hydrogeology by Ignatovich (1949), on the Volga-region by Afanasiev, on the south-west Russian platform by Babinets, and on the Urals by Budanov. Over the past years compiled were the international Hydrogeological Map of Europe on a scale of 1:1,500,000 (Editor-in-Chief: Karrenberg). A map of sub-surface flow in Central and Eastern Europe on a scale of 1:1,500,000 (Editor-in-Chief: Konoplyantsev) and the hydrogeologic maps of the Ukrainian SSR and the Baltic republics are available. Additional sources of information are given in a list of references at the end of this country paper.

Ground-water resources are monitored through a network of more than 10,000 observation wells. Since the mid-1970s, of particular urgency has been study of human impact on ground water (ground-water pollution in particular).

2. Hydrogeologic zoning

Hydrogeologic zoning distinguishes structures or their parts which affect the distribution, transfer and discharge of ground water.

Zoning is carried out using a system of collaterally subordinated taxonomic and hydrogeodynamic units. This leads to consideration of each zoning unit as a distinct reservoir (varying from global and regional to local and elementary) containing ground water moving from recharge to discharge areas.

Hydrogeologic zoning involves the following series of taxonomic units:

(a) Region, a system of ground-water basins,

(b) Province, a multilayered ground-water basin,

(c) Subprovince, a confined water basin,

(d) Area, a group of basins of regional unconfined or subconfined-water flow,

(e) Subarea, a basin of regional unconfined or subconfined-water flow,

(f) Region, a group of basins of local unconfined or subconfined-water flow,

(g) Subregion, a basin of local unconfined or subconfined-water flow,

(h) Site, a group of basins of elementary unconfined or subconfined-water flow, and

(i) Subsite, a basin of elementary flow of unconfined or subconfined water.

Taxonomic units of the first two orders are given in figure 59.

The scheme of zoning in the continental part of the USSR shows structures of platforms (with multilayered aquifers in plates and fracture-water storage in shields), structures of orogenic (epi-platform, epi-geosynclinal) belts with fracture-multilayered, fractured, and multilayered aquifers) and in intermontane basins multilayered aquifers.

Hydrogeologic structures of continental platforms fall within regions, i.e., systems of multilayered and fractured aquifers in shields depending on the age of the folded basement, age of the sedimentary cover and the character of recent tectonic activity. Each region has in common a platform development of the sedimentary rocks which determines, in combination with tectonic activity, the zonation and trend of hydrogeologic processes. The following four regions are established within the European part of the Soviet Union.

(i) East European: a system of multilayered aquifers having a basement of Proterozoic crystalline rocks, a sedimentary cover of Upper Proterozoic to Cenozoic rocks and structural variations and blocks reflecting later tectonism.

(ii) Scythian: a system of multilayered aquifers with a folded Paleozoic basement, a Mesozoic-Cenozoic sedimentary cover, thick aquicludes, inherited geostructural basement elements and basic lineaments which account for block structure in the sedimentary cover.

Hydrogeologic structures of orogenic belts are also subdivided into regions, depending on the age of the major orogenic cycle and dislocation of fold structures. This affects the type of aquifer, trend and intensity of the hydrogeologic processes. These regions are represented by systems of basins of multilayered fracture and multilayered aquifers. The following regions are delineated within orogenic belts.

(iii) Urals-Timan: a system of fracture, fracture karst, fracture-multilayered water basins associated with Baikalian, Caledonian and Hercynian orogenic belts and further neotectonic uplift (generally 200-400 m, 1,000 m at some sites).

(iv) Carpathian-Crimean: a system of multilayered-fractured, fractured, and fractured-karst aquifers associated with the Alpine orogen with a range of neotectonic strains to 7,000 m.

The boundaries between platform and orogenic regions are in large regional faults. A contact line between the rocks of sedimentary cover of the platform and an orogen is taken as a boundary, assuming no faults in the upper part of the sedimentary cover.

Figure 59

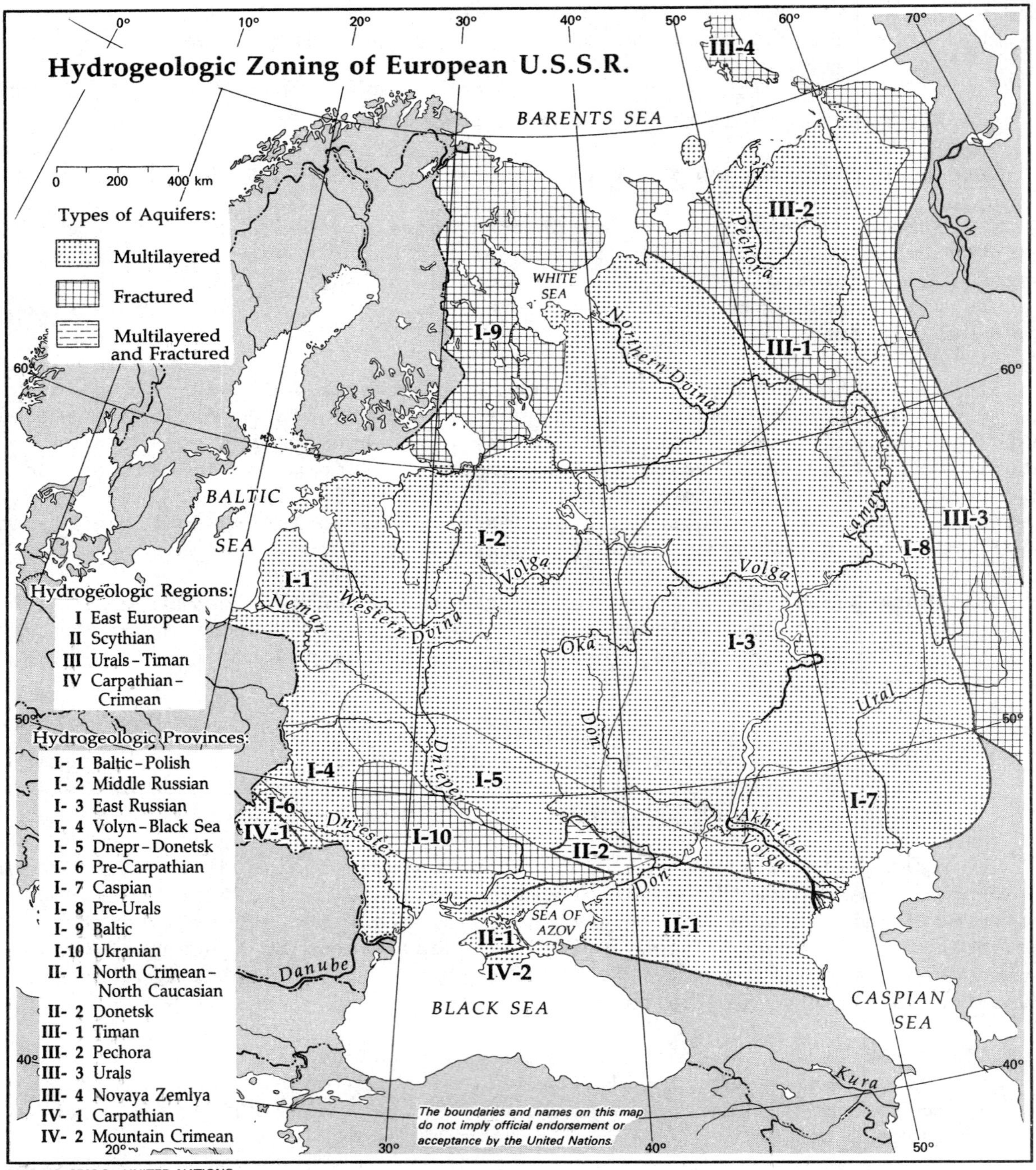

MAP NO. 3568.3 UNITED NATIONS
NOVEMBER 1989

Provinces, i.e., multilayered ground-water basins distinguished by a general trend in the flow of ground and surface water towards regional and global discharge areas, are considered to be a taxonomic unit of the second order of zoning.

In platforms, provinces are usually represented by multilayered aquifer basins with various combinations of hydrogeodynamic zones and a prevailing flow of ground water toward discharge areas. In shield areas, provinces are represented by basins of fractured and multilayered aquifers confined to bedrock and overburden.

The East European system of basins involves the following basins of multilayered aquifers: the Baltic-Polish (with discharge into the Baltic Sea), mid-Russian (with discharge into the Baltic Sea and Arctic Ocean), East Russian (with discharge into the Caspian Sea), the Dnieper-Donets (with inner discharge into streams), Yolyn-Black Sea (with discharge into the Black Sea), pre-Carpathian (with discharge into the adjacent Volyn and Black Sea basin), the pre-Urals (with discharge into the East Russian multilayered basin), the Caspian (with local discharge in salt dome areas). Also, multilayered fractured and overlying multilayered aquifers are delineated. Baltic shield basin (with discharge into the White and Barents seas and large lakes), the Ukrainian shield basin (with discharge into streams).

The Scythian region includes (1) the North Crimean and North Caucasus basin of multilayered aquifers discharging to the north, and (2) the Donets basin of chiefly fractured-multilayered aquifers discharging to the south and east.

The Urals tectonic region includes the folded mountain provinces of Timan, the Urals and Novaya Zemlya with chiefly fractured aquifers flowing into adjacent platform structures, and the Pechora basin of multilayered aquifers discharging into the Arctic Ocean.

The Carpathian-Crimean region is within the Alpine orogenic zone with (1) the Carpathian basin of multilayered (rarely fractured) aquifers discharging toward the East European and West European platforms, and (2) the Crimean Mountains basin of mostly fractured karst discharging into the Black Sea.

The North Dvina, Leningrad and Moscow basins of confined multilayered aquifers constitute the Mid-Russian basin. The East Russian basin comprises the Volga-Kuma, Volga-Khopersk and Syrtov basins of confined multilayered aquifers. The Pripyat and Dnieper basins of multilayered aquifers are parts of the Dnieper-Donets basin. The first volume of Hydrogeology of the USSR describes the above areas.

3. Principal aquifers

(a) Basis of classification

Type, permeability and water content of rocks are considered when defining aquifers, aquifer complexes and other units.

Distinctions are made of (a) the type of permeability: pore fracture and karst aquifers, (b) the permeability coefficient: permeable layers (aquifers) having hydraulic conductivity greater than 1 m per day, semi-permeable layers (semi-aquifers) with hydraulic conductivity of 10^{-4} to 1 m per day, and impermeable layers (aquicludes) having hydraulic conductivity of less than 10^{-4} m per day.

Based on the volume of water stored, the following taxonomic units are recognized: (1) permanent regional aquifers and semi-aquifers, (2) sporadic regional aquifers being permanent at particular sites throughout the area, (3) permanent linear aquifers, (4) periodic (seasonal) regional aquifers, and (5) regional aquicludes which do not contain free water.

Major taxonomic units delineated in the section of sedimentary, effusive, sedimentary-effusive rocks are aquifers, semi-aquifers and aquicludes or their complexes. Fractured aquifer zones and impermeable zones occur in the areas of crystalline highly metamorphosed rocks. Also, local aquifers in fault zones, their hydrogeodynamic role (recharge, faults, discharge faults) are distinguished on the maps.

With the use of appropriate graphic devices the hydrogeologic maps become volumetric and structural and show clearly the sequence of aquifers and aquifer complexes.

Taxonomic units are given in detail on maps as a scale of 1:200,000 and larger. A general description of the major regional taxonomic units in hydrogeological areas of the first order is presented in figure 60.

(b) Regions and aquifer complexes

(i) Russian (East European) region

The sedimentary cover overlying basement rocks of the East European region includes a complex system of commonly multilayered aquifers.

In the northern and middle parts of the Russian plain ground water is distributed mainly in glaciofluvial, glacial and alluvial sediments. In the southern, non-glaciated area, aquifers are in alluvial and eolian-diluvial (loess) sediments. In the Caspian basin the most widespread aquifers and aquifer complexes are in marine sediments.

The aquifers and aquifer complexes of Quaternary sediments are mainly phreatic, rarely artesian. The hydrogeologic parameters and characteristics (mineral content, chemical composition, water table) vary from north to south.

In the northern and central parts, glaciofluvial and alluvial, aquifer complexes are the most widespread and important. Glaciofluvial sediments are composed of inequigranular gravel sands and form several aquifers divided by morainal loam, gravel and boulder deposits. The total thickness of the aquifers amounts to dozens of metres, some reaching 150-200 m (e.g., in the Pripyat and Desna woodlands). Total dissolved solids average 500 mg/l mainly calcium bicarbonate. Well yields are usually 1-2 l/s. In the Baltic republics, the Karelian ASSR, central areas of the RSFSR and Byelorussia, the ground water from glaciofluvial sediments is widely used for water supply.

Alluvial sediments include sand, gravel, loam and clay. Their permeability decreases downstream. Ground water occurs at shallow depth, TDS amount to 500-1,000 mg/l, mainly calcium bicarbonate. Wells yield an average of 1-3 l/s.

In the valleys of the Volga, the Dnieper and other large rivers pockets of brackish water (sulphates and chlorides) are common. Less productive (less than 0.5 l/s) are glaciofluvial and glacial sediments composed of sand and clay.

Figure 60. European USSR, Stratigraphy of principal aquifers

AR = Antecambrian; C = Carboniferous; D = Devonian; J = Jurassic;
K = Cretaceous; N = Neogene; O = Ordovician; P = Paleocene; P = Permian;
PE = Crystalline Basement; PR = Cambrian-Infracambrian;
PR = Infracambrian; PZ = Paleozoic; Q = Quaternary; S = Silurian;
T = Trias

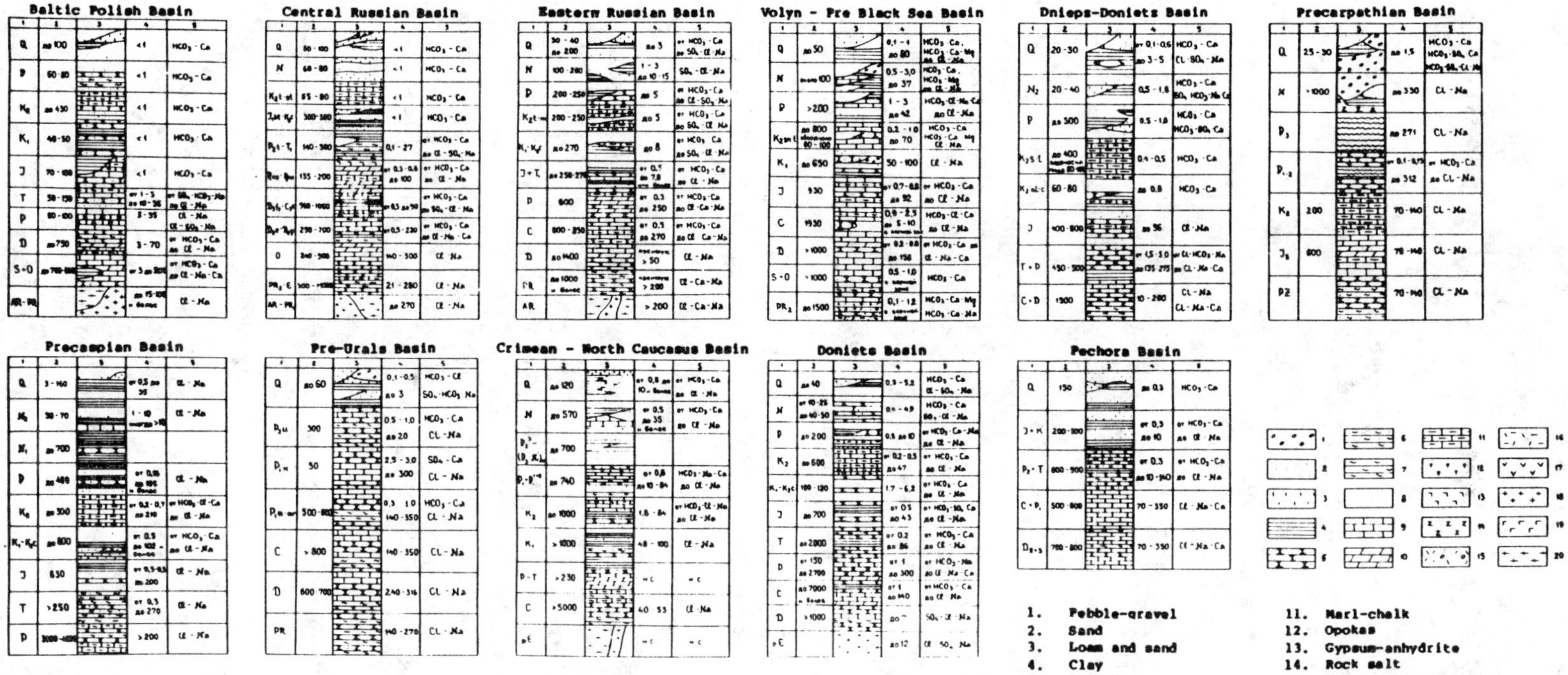

Columns:

1 = age of aquifer; 2 = thickness (metres); 3 = lithology; 4 = total dissolved solids (grams/litre); 5 = prevailing chemical contents of water

1. Pebble-gravel
2. Sand
3. Loam and sand
4. Clay
5. Sandstone
6. Siltstone
7. Argillite, clay shale
8. Flysch
9. Limestone
10. Dolomite
11. Marl-chalk
12. Opokas
13. Gypsum-anhydrite
14. Rock salt
15. Sedimentary effusive
16. Effusive acidic
17. Effusive basic, ultrabasic
18. Intrusive acidic
19. Intrusive basic
20. Metamorphics

Waters are of the calcium or calcium-magnesium bicarbonate type. Aquifers and aquifer complexes of the alluvial sediments in the non-glaciated parts of the Russian Plain are similar. However, the mineral content of the water, particularly in the valleys of small rivers, may range from 1,500 to 3,000 mg/l, which makes it unsuitable for drinking. In the southern area, loess 10 to 12 m thick (exceptionally 20-30 m) are widespread. TDS exceeds 3,000 mg/l and few wells yield more than 0.5 l/s.

Marine sediments include layers and lenses of sand, loam and clay. Sodium chloride waters there commonly have mineral contents of 3 to over 30 g/l. Perched fresh water is used for irrigation but wells yield 0.1 to 0.3 l/s at most.

a. Neogene aquifers

Neogene aquifer complexes are widespread on the southern slope of the Russian Plain. In the eastern part are sands and clays, whereas the south-western part is composed of limestones. Gypsiferous rocks occur at the boundary with the pre-Carpathian depression.

Slopes and watersheds of the higher parts of the Central Russian upland, on both banks of the Volga, contain sand aquifers within siltstones and clays. Most wells yield 0.5-1 l/s, some 4-5 l/s. Most saturated are the Sarmatian and Tortonian limestones. The total yield of some water intakes amounts to 85-100 l/s. Water is chiefly fresh and of calcium or sodium bicarbonate type. Neogene limestones in the Black Sea depression may reach 100 to 300 m in thickness, their water becoming artesian. As a result of the arid climate and poor mixing, the water is highly mineralized.

In the Caspian depression Neogene ground water has a mineral content of 3-20 g/l. Neogene fresh water is a source of domestic water supply.

b. Paleogene aquifers

Paleogene aquifer complexes are widespread in the southern part of the region, especially along the Volga, in the Caspian depression and the Ukraine. The profile shows sandy-clayey soils, opoka and marl in Paleocene and Eocene sediments. Oligocene clay in some structures constitute a regional aquiclude.

The upper sediments occur at high altitudes and many are drained. At the top and the foot of deep synclinal structures may be aquicludes, favouring the development of confined waters. Water heads in the Dnieper-Donets basin reach 300-330 m.

Shallow ground water is mostly fresh, bicarbonate type, but brackish in areas of low rainfall. With depth, the water type gradually changes into brackish and salty, with TDS of 90 g/l and the chemical composition of sodium chloride type in the Caspian depression. In the salt dome structures here, the water mineral content may be highly variable. The Paleocene sediments are water-saturated. In the Dnieper-Donets basin and in the Baltic republics, well yields average 1-2 l/s. In the Black Sea basin the yields of flowing wells in Eocene sediments are in the range of 3 to 7.5 l/s.

In the Volga area, well yields are 10-16 l/s of fresh water; in the Caspian basin, 0.5-6.5 l/s. Ground water is widely used for water supply in the Dnieper-Donets and Volga-Kama basins.

c. Cretaceous aquifers

The Upper Cretaceous complex composed of chalk and marl is widespread in the southern part of the platform. As a rule, aquifers are in the upper fractured zone and are 80-100 m thick. The underlying layers are of such low permeability as to be considered aquicludes. Water in recharge areas is mostly fresh or slightly brackish, becoming saline and brine at depth, in the Caspian depression. Well yields vary from 0.1-0.6 l/s in the Caspian depression, to 14-28 l/s in zones of tectonic dislocation of the Volga-Podolian and Black Sea basins.

The lower-upper Cretaceous aquifer complex occupies a considerable part of the Russian Plain and is composed of sands and sandstones, with lenses and interbeds of clay. The distribution of this complex is vaster than that of the chalky marl. Most saturated are the Albian and Cenomanian sediments from which wells yield 5-10 l/s. Owing to the presence of clayey interbeds, many of the aquifers are confined. The mineral content and chemical composition of Cretaceous ground water vary from fresh, calcium bicarbonate in recharge areas to brines in deep structures of the Caspian basin. Fresh ground water from these complexes is extensively used for urban water supply.

d. Jurassic aquifers

Jurassic sediments are widespread in the central and southern parts of the platform. The Lower and Middle Jurassic sediments are terrigenous and grey. The Upper Jurassic sediments are largely bituminous clays and glauconitic sands.

In the Dnieper-Donets basin, the Moscow and Baltic syneclises and the Volga areas, impermeable clays of Kimmeridgian to Upper Callovian age create a boundary between the Upper Jurassic and Middle to Lower Jurassic aquifers. The Upper Jurassic aquifer is composed of sand, siltstone with interbedded clay and some limestone; it is mostly confined. The specific yield averages 1 l/s, but is locally greater. In the central Dnieper-Donets basin and around Kiev, wells tapping coarse sand beds at depths of 200-275 m yield 15 to 35 l/s.

The mineral and chemical composition of Jurassic ground water depends on recharge and discharge conditions; in the recharge areas it is fresh but in the central deep synclinal structures, brackish or saline.

e. Triassic aquifers

Triassic aquifer complexes occur in many parts of the East European region but are developed limited to isolated structures. The sediments occur chiefly at great depth and have been studied only where they occur at relatively shallow depth, e.g., in the Moscow syneclise and in the south-east part of the Volga-Kama anteclise. The complexes are composed mostly of speckled clay with thin interbedded sandy aquifers. In the Caspian basin, where the Permian-Triassic sequence is thick, the Triassic sediments are clay and siltstone with intercalated sandstone and lenses of conglomerate. The Triassic sediments are poorly productive and over much of the territory the ground water is confined. The concentration and chemical composition is highly varied depending upon ground-water flow conditions.

In the northern part of the Russian plain, water at shallow depth is fresh, but the mineral content increases with depth. The mineral content of brines in the Caspian basin may reach 250-300 g/l.

f. Paleozoic aquifers

The Upper Permian aquifer complex is present in the northern, central and eastern parts of the platform and is composed of sand with speckled clay, sandstone and marl. Many of the soils are gypsiferous. In the south-east, Kazanian sediments contain gypsum, anhydrite and rock salt. In the Dnieper-Donets basin the Permian complex is composed of clayey sand. The ground water is mostly confined and is at depths reaching several hundreds of metres. In the Volga-Kama basin, artesian springs from Upper Permian sediments yield some 40 l/s. Shallow waters are usually fresh and calcium bicarbonate, rarely brackish or sulphate. The mineral content of the confined water of sodium chloride type may exceed 50 g/l. Depending on the lithology and hydraulic properties, well yields range from 0.1 to 60 l/s (e.g. the well in Koryazhma).

Kungurian (Lower Permian) sediments composed of evaporites and carbonates are regarded to constitute a regional aquiclude, the thickness of which is 2 to 3 km in the Caspian basin.

An Upper Carboniferous to Lower Permian aquifer complex includes limestones and dolomites; in the Dnieper-Donets depression, mostly clayey sand. Over most of the Moscow syneclise the complex contains fresh artesian water of calcium bicarbonate type widely used for urban water supply. Well yields range, on average, from one to a few hundred litres per second. In the Volga-Kama and Dnieper-Donets basins, dissolved solids increase with depth to 250-300 g/l. Such waters are sodium-potassium chloride type. The complex is separated by the Vereisky aquiclude from the Middle Carboniferous clayey sand complex. In the central parts of the platform water is fresh. In the south, east and south-east it is highly mineralized (up to brines).

The Middle and Lower Carboniferous aquifer includes limestone and dolomite, the Lower part includes clayey sand. In the central and western areas of the Moscow basin the water is fresh, calcium bicarbonate type. The mineral content increases abruptly with depth. Well yields of the karstic limestone range from 0.1 to more than 4 l/s.

The Upper Devonian to Lower Carboniferous aquifer complex is composed of carbonate sediments and is well studied in the western part of the Russian Plain where it occurs at a shallow depth. Its water is generally fresh. Well yields reach 90 l/s (Bryansk).

The Middle and Upper Devonian aquifer contains sand aquifers; TDS widely varies from 0.5 to 320 g/l. Fresh water is confined to the southern slope of the Baltic shield, Voronezh and Byelorussian anteclises. The water content of the complex is widely variable. The yields of flowing wells reach 75 l/s.

The Middle Cambrian to Middle Devonian aquifer complex includes some aquifers mainly confined to the following strata.

1. Middle Devonian terrigenous carbonate sediments (well yields, 0.2-2 l/s);

2. Lower and Middle Devonian sands and sandstone of low storage capacity;

3. Silurian limestone marl and dolomite. Water content is low but some wells yield 1-5 l/s. In Latvia and Lithuania the Silurian strata are practically impermeable;

4. Ordovician limestone, dolomite and marl. Most saturated are karstic rocks; and

5. Middle Cambrian and Lower Ordovician sands and sandstone.

Shallow waters are fresh and the calcium bicarbonate type. In the deeper parts of the Moscow basin brines have TDS reaching 290-300 g/l. The Lower Cambrian aquiclude in the western and central areas of the platform is composed of blue clays 60 metres thick. The Upper Proterozoic to Lower Cambrian aquifer complex is composed of terrigenous rocks, chiefly sandstone, of different formations and series. The water content of the complex is well-studied at shallow depths on the slopes of the Baltic shield and Byelorussian anteclise. The mineral content is 0.5 to 12 g/l. Well yields do not exceed 2 l/s. In the central areas of the platform well yields sharply decrease (water inflow with drawdowns do not exceed 0.1 l/s). The mineral content may exceed 285 g/l.

g. <u>Water in the Archean-Proterozoic basement</u>

The crystalline rocks of the Baltic and Ukrainian shields, Voronezh and Byelorussian anteclises are water-bearing in the zone of exogenetic jointing most widespread to depths of 80-100 m. The water mineral content to depths of 200-250 m usually does not exceed 0.3 g/l in the Baltic shield and increases to 1 g/l in the Ukrainian shield. Saline and brine waters with the mineral content of 200 g/l are spread throughout the Krivoi Rog basin of the Ukrainian crystalline shield and some structures of the Baltic shield. Well yields usually do not exceed 1 or 2 l/sec. The water content increases in the disrupted zones where the well yields reach 5-10 l/s.

Exceptionally, ground-water data were taken at great depth in crystalline rocks in the Kola superdeep well (over 12 km). More intensive rock jointing is established in the depth intervals of 0.8-4.6 and 4.6-8.9 km where the rocks are virtually impermeable. In the depth interval of 5,325-5,345 m, hydraulic conductivity is about 10^{-7} m/day, that is much higher than the lower zone of exogenetic jointing.

According to Borevsky, Vartanyan, <u>et al</u>. (1984) formation of a fissured zone is due to metamorphic processes, i.e., transition of a greenschist facies into epidote-amphibolite during which hydraulic fractures occur as a result of water release. TDS increases with depth to 200-300 g/l. At the lower levels of a well profile carbonic acid is dominant in gas composition.

(ii) <u>Urals-Timan region</u>

The Urals-Timan region comprises the Pechora, Kanino-Timan and the Urals hydrogeologic provinces.

a. <u>Pechora artesian basins</u>

The Pechora system of artesian basins include multilayered aquifer and aquiclude complexes. Water exchange decreases with depth, and the water is in stable equilibrium at the lowest levels.

The basins are chiefly composed of Quaternary alluvial, glaciofluvial, glacial and marine sediments which contain phreatic water. Most productive are the alluvial and glaciofluvial sediments consisting of inequigranular sand with interbeds of loam and sandy loam. Well yields vary from 0.9 to 5.0 l/s. Frozen and semipermeable marine and glacial sediments are largely developed along the coast.

Quaternary ground water is fresh calcium bicarbonate with a mineral content of 0.5 g/l. Along the sea coast the mineral content is higher as a result of eolian salt and sea water intrusion, and the water is brackish to salty.

Pre-Quaternary aquifers are developed in terrigenous and carbonate rocks. The terrigenous rocks are Cretaceous to Upper Permian in age. Upper Aptian to Oxfordian layers are represented by siltstone and clay with interbeds and lenses of sand and sandstone containing artesian water. Well yield is 0.3-0.4 l/s. More productive are the Callovian and Bathonian deposits where sand with interbeds of siltstone and clay is dominant. Here, the yield of artesian wells is 5-10 l/s.

The hydrogeology of the Permian deposits has been studied in connection with exploitation of coal deposits. Artesian aquifers are confined to fractured sandstone, conglomerate and, in some cases, siltstone. Water yields decrease with depth; for example, yields in the Vorkuta are 0.5 l/s at a depth of 100-110 m and does not exceed 0.1 l/s 100 to 200 metres deeper.

Carbonate rock strata including limestone, dolomite, marl were formed from the Ordovician through Early Permian. Along the folded margin of the Pechora syneclise such deposits are at relatively shallow depths. Ground water is discharged through springs with yields of 0.4-0.7 l/s. Fresh ground water occurs at depths of 100-200 m, and in the Permian coal beds, at 300-400 m.

The salt content at depths of 1,500-2,000 m is 220-240 g/l.

b. <u>Kanino-Timan fracture systems</u>

The Kanino-Timan system of fractured aquifers is a hydrogeologic structure constituting a transition zone between platform and geosyncline. Platform conditions for water permeability still exist in the grabens. In some aquifers the water-exchange decreases and salt content increases with depth.

Well yields of carbonate rocks range from 1 to 50 l/s. In terms of hydrogeologic properties, each graben or block is relatively independent. In the area of horsts, the Proterozoic, Paleozoic sedimentary, sedimentary-volcanic metamorphic formations are exposed, which bear water only in zones of exogenetic jointing and tectonic dislocations. Ground water occurs, depending on relief, at depths of 30-50 m and is discharged through numerous springs yielding about 1 l/s each. The ground water of the exogenetic fissured zone is calcium bicarbonate with 100 to 300 mg/l.

c. <u>Urals fracture systems</u>

The Urals system of fractured aquifers is located within the folded mountain structure of the same name, over a distance N-S of 2,500 km from tundra to semidesert. Ground water is associated with exogenetic fissures in sedimentary, effusive, intrusive and metamorphic rocks of differing composition and age from Proterozoic to Permian. Average thickness of the zone is 25-30 m in some places to 50-60 m. The karstic zone of the Lower Permian and Carboniferous carbonate rocks is thicker reaching, in the middle and northern Urals, a depth of 1,000 m. Ground water is discharged through numerous springs at rates of less than 1 l/s in non-carbonate rocks, more than 100 l/s in karstic limestone.

Well yields in the zone of exogenetic fissures zone is 1-2 l/s, in fractured zones, 5-10 l/s. Most saturated are the karstic limestones.

The average value of water inflow from carbonate rocks into the Kizel basin is 1,000-1,900 m^3 per hour, with a maximum of more than 3,000 m^3 per hour.

The mineral content and chemical composition of ground water differs uniformly from north to south. In the northern Urals the water is calcium bicarbonate and calcium-sodium type with TDS of 200 to 300 mg/l. To the south the mineral content increases to 500-1,000 mg/l, turning the water into a mixed cation bicarbonate type.

Ground water is widely used for water supply. But it also creates problems in the mining of coal and bauxite. Powerful pumping facilities are required to drain the mines.

(iii) Carpathian-Crimean region

The Carpathian and Crimean mountain region is subdivided into separate independent hydrogeologic areas. The first one involves the artesian basins located in front of and behind the Carpathians and the complex ground-water system of the Carpathian Mountains. The upper part of the pre-Carpathian basin contains mostly Quaternary alluvial ground water with TDS of 1,000 mg/l which is widely used for water supply. Neogene and Paleogene sedimentary rocks are chiefly composed of clayey and salt-bearing sediments with a low water content. Well yields average 0.1-0.2 l/s. Even at shallow depths, the water may be saline. In the trans-Carpathian basin ground water is confined to alluvial sediments and the weathered zone of crystalline rocks. Well yields in the alluvial sediments are 5-6 l/s, in the weathered zone, 1 l/s. The water is fresh calcium or sodium bicarbonate type with TDS of 1,000 mg/l.

The Neogene molasse sequence includes a number of artesian aquifers. Most saturated are pebbly sand and volcanic rocks, with well yields of 3.5 l/s. In areas of tectonic dislocation, well yields may reach 17-25 l/s. In the Carpathian Mountains, the Paleogene and Cretaceous flysch sediments are locally water bearing but with yields not exceeding 0.5 l/s. Most productive is a Paleogene sandstone yielding 5 l/s. The mineral content increases gradually with depth from 1-3 to 100 g/l. The water changes from sodium bicarbonate to sodium chloride type.

The ground-water system of the Crimean Mountains is composed of Upper Jurassic limestone aquifers, and Middle Jurassic to Triassic poorly permeable and impermeable shales.

Quaternary alluvium and proluvium has low storativity. The water is chiefly fresh, calcium bicarbonate. The Upper Jurassic limestone contains semi-confined water with intensive water exchange and its mineral content is less than 500 mg/l to a depth of 1,000 m. The water discharges through numerous springs yielding 100-150 l/s in karst areas. It is widely used for water supply.

(iv) Scythian region

The Scythian region comprises the Crimean-North Caucasus and Donetsk Ridge multilayered basins. The first is characterized by regional extension of the Maikop impermeable clays up to 1,000 m thick, dividing the hydrogeologic profile into two. Above the Maikop stage is considerable fresh ground water in the Neogene carbonate sediments (Crimean Plain), terrigenous and terrigenous carbonate sediments of the Quaternary and Neogene, and Neogene sand (Ciscaucasia). Below the Maikop stage, the water is highly mineralized and confined.

In the Donets basin, most of the fresh water is in the Upper Cretaceous chalky marl strata and is exploited by numerous water intakes. In Paleozoic sediments, the water is highly mineralized. Fresh water occurs in the peripheral areas of the basin. Water flows into mines at rates from 20 to 200 m^3 per hour.

4. Mineral waters

(a) Types

Mineral water is widespread in the European part of the USSR and has long been used for therapeutic purposes, both for drinking and balneology. There is a great variety of mineral water types, generally categorized as mineral water not defined by specific component or property but with mineral content of at least 2 g/l; and mineral water identified by specific component, e.g., iodine, bromine, iodine-bromine, hydrogen sulphide, radon, ferriferous or carbonated. Characteristics of the main mineral water types are given in table 9.

(b) Occurrence

The formation and distribution of the various mineral waters depend on interactions of geologic structure, tectonic, geochemical, hydrogeodynamic and hydrogeothermal conditions. In this context, zoning of distribution area of mineral water is based on structural geology principles bearing on hydrogeochemical and hydrogeodynamic characteristics. The basic taxonomic units of zoning are provinces, areas and regions. By a province, one understands a territory characterized by uniformity in chemical and gaseous composition, conditions and basic mechanism of mineral water formation. Provinces are subdivided into hydrogeologic areas corresponding to large tectonic structures which are characterized by uniform processes affecting the formation of mineral water resources.

Within a given area, several regions are identified with specific geologic structures, development, sedimentation conditions, i.e., hydrogeodynamic, hydrogeothermal and hydrogeochemical conditions.

The major part of the European USSR represents a province of platform artesian mineral waters characterized chiefly by the gases methane, nitrogen-methane, and in the upper parts of the profile, nitrogen. The following areas and regions within them are delineated (figure 61): (1) East European area: Baltic Sea, Mid-Russian, East Russian, the Dnieper-Donets, Black Sea, Caspian regions; (2) Pechora area; (3) Scythian area: Crimean, Azov-Kuban, East Ciscaucasus regions; (4) Pre-Carpathian area.

Table 9. European USSR, principal mineral waters

Prevalent gaseous composition	Basic ions	Water types	Mineral content (in g/l)	Places of usage and their number on the scheme in figure 61
		I. Non-specific components		
N_2	SO_4-Ca	Krain	2-3	1) Krainka, 2) Nizhnee Ivkino
	SO_4-Ca-Mg	Zvenigorod	2-3	3) Zvenigorod, 3) Dorohovo
N_2	SO_4-Ca-Mg-Na	Moscow-1	3-5	3) Moscow, 3) Monino, 11) Kashin, 3) Ostankino, 4) Chernovtsi
N_2	SO_4-Cl-Na-Ca	Feodossia	4-5	5) Feodossia
N_2	SO_4-Cl-Na-Ca-Mg	Izhevsk Novoizhevsk	4-5 14.0	6) Izhevsk, 7) Hilovo, 8) Novoizhevsk, 9) Truskavets
N_2	Cl-SO_4-Na	Solvychegodsk	5-25	10) Solvychegodsk, 11) Kashin, 12) Uglich, 2) Nizhnee Ivkino
N_2	Cl-Na	Nizhnie Sergi	5-10	13) Nizhnie Sergi
N_2	Cl-Na	Mirgorod	2-5	14) Mirgorod, 15) Minsk
N_2	Cl-Na	Yevpatoria	5-10	16) Yevpatoria
N_2	Cl-Na-Ca	Staraya Russa	15-25	17) Staraya Russa, 18) Pyarnu
N_2	Cl-Na	Krasnoussolsk	25-35	19) Krasnoussolsk
N_2	Cl-Na	Druskininkai	5-60	20) Druskininkai, 21) Palanga, 17) Staraya Russa
N_2	Cl-Na	Kashin	180-220	11) Kashin No. 3
N_2	Cl-Na	Boenski (Moscow III)	250-270	3) Moscow

Table 9 (continued)

Prevalent gaseous composition	Basic ions	Water types	Mineral content (in g/l)	Places of usage and their number on the scheme in figure 61
		II. <u>Carbonated mineral water</u>		
CO_2	HCO_3-Na	Dilizhan	2-5	Luzhanskaya No. 1
CO_2	HCO_3-Na	Borzhomi	5-8	Polyana Novaya
CO_2	HCO_3-Na	Polyana-Kvassova	8-12	22) Polyana-Kvassova, Polyana-Kupel, Utsera
		III. <u>Sulphide mineral water</u>		
N_2	SO_4-HCO_3-Ca	Hilovo-Kemeri	2-3	7) Hilovo, 23) Kemeri
N_2	SO_4-HCO_3-Ca	Nemirov		9) Nemirov
CH_4, N_2	Cl-Na	Sukhumi	5-15	24) Yeisk
N_2, CH_4	Cl-HCO_3-Na	Sernovodsk	2-5	25) Sernovodsk
N_2, CH_4	SO_4-HCO_3-Ca-Mg-Na	Sergiyevsk	2-5	25) Sergiyevsk Mineral Water, 9) Luben Veliky, 26) Sinyak, Sherbintsi
N_2	SO_4-Cl-Na-Ca-Mg	Kluchevski	3-5	Kluchevskaya, 19) Krasnoussolsk
CH_4, N_2	Cl-Na	Ust-Kachka	35-150	28) Ust-Kachka, Pokhvistnevo, Ishimbai
N_2, CO_2	Cl-Na	Nukut	35-150	27) Syzran

Table 9 (continued)

Prevalent gaseous composition	Basic ions	Water types	Mineral content (in g/l)	Places of usage and their number on the scheme in figure 61
		IV. Radon mineral water		
Na_2, O_2	HCO_3-Ca-Mg	Lipovka	to 0.5	Lipovka, 32) Zhytomir
CO_2	HCO_3-Cl-Ca-Na	Hmelnik	2-6	31) Hmelnik
N_2	Cl-Na	Ust-Kut	to 150	19) Krasnoussolsk
		V. Water with high content of organic matter		
CO_2, N_2	HCO_3-Ca-Mg	Naphtussya	0.8-1	9) Naphtussya, Skhodnitsa, Undori
		VI. Arsenious mineral water		
CO_2	HCO_3-Cl-Na-Ca	Gornaya Tisza	2-5	Gornaya Tisza
		VII. Ferriferous mineral water		
N_2	SO_4-HCO_3-Ca-Mg	Martsialny	0.4	29) Martsialnaya No. 1
N_2	HCO_3-Cl-SO_4-Ca	Polustrovsky	0.27	30) Polustrovskaya, 29) Martsialnaya No. 3
N_2	Cl-Na-Ca	Belaya Gorka	9.0	33) Belaya Gorka
N_2	Cl-Na	Vologda	97.0	34) Vologda, 3) Dorokhovo
N_2, CH_4	Cl-Na-Ca	Ust-Kachka	271.0	28) Ust-Kachka

Figure 61

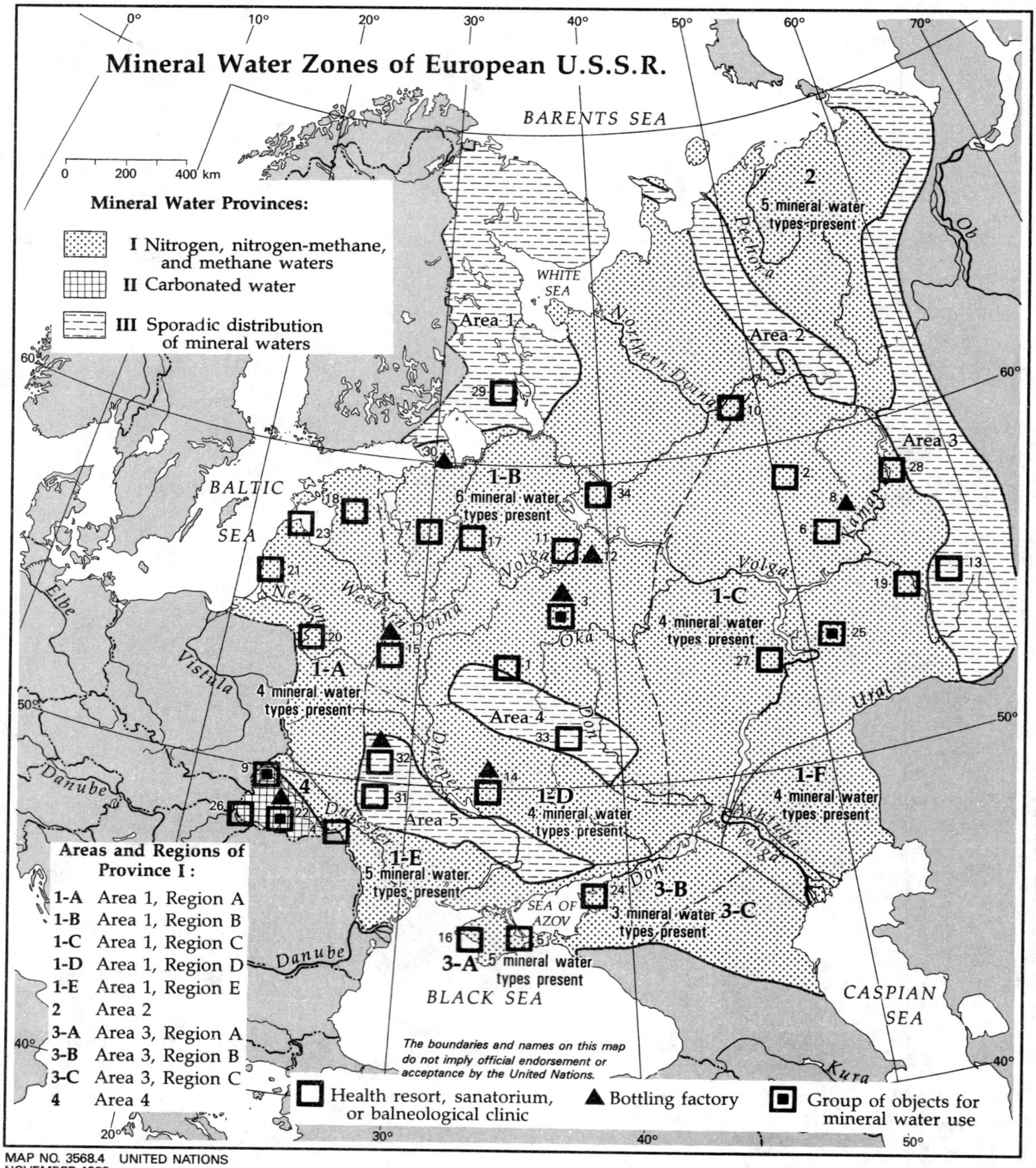

MAP NO. 3568.4 UNITED NATIONS
NOVEMBER 1989

The second province of carbonated mineral water covers the Carpathian Mountains area. Beyond the provinces, mineral water is sporadically distributed among the areas of fresh ground water. Such is the case in the crystalline shields (Baltic, Ukrainian, Voronezh) and the Urals-Timan mountains.

Mineral water of the East European area is confined to the platform sedimentary cover and piedmont troughs and has a stratified occurrence. Its mineral content is diverse. Fresh ground water in the upper layers is used for domestic water supply, the lower salty one (from brackish water to brines) has diverse therapeutical properties. The upper aquifer with a thickness of 300-500 m (except for the fresh water zone) generally has a mineral content of 10-15 g/l, varied ionic composition and is used as medicinal and table drinks, the lower aquifers hold brines with different salt content ranging from sodium chloride in the upper part to calcium-sodium chloride and sodium-calcium in the deepest zones, which are used in balneology. Methane and nitrogen are the principal gases in the ground water, but some local zones contain hydrogen sulphide. Considerable mineral water resources with non-specific properties or components and iodine-bromine brines are concentrated there.

Mineral water with non-specific components and properties and mineral content of 2-35 g/l is distributed throughout, occurring in various rocks just below the fresh water aquifer at a depth of a few tens to a few hundreds of metres. The well yields for this water usually amount to several litres per second, but the demand from a single well does not exceed 1 l/s. The total mineral water resources potential is many times the amount actually used. According to their prevalent chemical composition they are subdivided into the following geochemical types (also, table 9).

Sulphate waters (chiefly calcium and magnesium-calcium, occasionally sodium) are widespread, and have a mineral content of 4-5 g/l in the Mid-Russian basin and are typical of relatively shallow Carboniferous and Devonian sediments (to depths of 200-300 m). They are widely used as table and medicinal water particularly at the resorts of Krainka and Dorokhovo. In the Baltic Sea basin this water type is used at the resorts of Baldon and Likenai.

In the East Russian and Pechora areas sulphate waters with a mineral content of 2-8 g/l and diverse cation composition occur mainly in the Permian, occasionally in Devonian, Carboniferous and Mesozoic gypsiferous sediments to a depth of 700 m. They are not widely used (Nizhnee Ivkino). There are many sulphate waters of a complex cation composition. Among them is the table water called Moscow Mineral which has been bottled for more than 30 years. Of great importance are sodium sulphate waters of the Feodossiya type developed in the Crimea and some locations in the Mid-Russian basin.

Sulphate-chloride, chloride-sulphate sodium and calcium-sodium waters with a mineral content of 5-15 g/l of the Mirgorod, Staro-Russian, Izhevsk types are found in all basins occurring between the sulphate and chloride-sodium water zones (the Mid-Russian, East Russian, East Ciscaucasian basins) and in the marginal parts of the Dnieper-Donets, Black Sea, Caspian Sea and South Pechora basins. This water type is widely used in the central areas and especially the spa resorts of Kashin, Uglich, Monino, Soligalich and bottling factories in Uglich and Izhevsk in the Dnieper-Donets and Crimean regions.

Bicarbonate-chloride waters with a mineral content of 15 g/l have a more narrow distribution as compared to the previous ones, but their medicinal value is high. The waters are known in the Kerch Peninsula, marginal parts of the Dnieper-Donets, Caspian Sea and Tersk-Kum basins.

A more rare type for platform basins is the sodium bicarbonate water extracted by isolated wells in the Novgorod region and Pripyat basin.

Sodium-chloride waters with a mineral content of 10-35 g/l are widespread and usually occur below brines at a depth of several hundred metres. They are used for balneological purposes at resorts, sanatoria, balneo-clinics (in the Mid-Russian basin - Kashin, Yaroslavl; East Russian basin - Izhevsk Mineral Waters, Krasnoussolsk, Sol-Iletsk; Dnieper-Donets basin - Mirgorod). This mineral water has a long history of balneological application thanks to its relatively shallow occurrence. Its natural discharge is observed, for example, in the areas of the Zhiguli-Pugachev arch and Soligalich.

Sodium-chloride brines (over 35 g/l) are distributed throughout the artesian basins of the region, forming a thick zone with a depth range of a few hundred metres to the crystalline basement. Brines are considered mineral waters of non-specific components and properties where they contain iodine, bromine, hydrogen sulphide in amounts less than defined for balneotherapeutic use. The total mineral content of these brines is commonly not greater than 100 g/l. In this case, well yields vary from the tenth fractions to 3-5 l/s. Ground water is confined (up to artesian) and available in abundance. Such brines are used in sanatoria and balneotherapeutic establishments, located in the Mid-Russian basin (Kashin, Dorokhovo) and Baltic Sea area (Pyarnu, Druskininkai).

Iodine (in concentrations exceeding 5 mg/l), iodine-bromine (over 5 mg/l of iodine and over 15 mg/l of bromine) and bromine (over 25 mg/l) waters are used for drinking and balneotherapeutic purposes. For drinking purposes the water with a total mineral content up to 35 g/l is used, provided that it is diluted to 10 g/l. For balneotherapeutic purposes the brines are diluted to 30-40 g/l.

Iodine and iodine-bromine potable water has a scanty distribution in the Baltic Sea basin, some areas of the Mid-Russian, Azov-Kuban, Black Sea and Dnieper-Donets basins and is not commonly used.

Iodine-bromine brines widely used in balneotherapy for external use are developed in the deep (commonly over 500 m) artesian basins. Chemically they are of NaCl, Ca-Na or Na-Ca types. In most basins the brines occupy a zone 2,000-3,000 metres thick.

In the Mid-Russian basin this zone is composed of sediments dating from the Permian to Late Proterozoic ages. The mineral content of the brines is 260-290 g/l (Moscow, Soligalich, Lubim), iodine to 28 mg/l and bromine to 800 mg/l. They are used at the resorts of Kashin, Monino, Dorokhovo, Arkhangelskoe and Solnechnogorsk.

Of iodine-bromine type are the brines of the Azov-Kuban and East Ciscaucasian basins where the iodine content in the ground water of Akchagyl, Maikop and Upper Cretaceous sediments at a depth of 300-1,700 m reaches 80-100 mg/l and the bromine content - 440 mg/l. The boric acid content is high. Use of these brines is still limited.

Also, iodine-bromine brines occur in the Black Sea basin. They are associated with Upper Cretaceous-Paleocene aquifers with irregular water content and characterized by high iodine and low bromine content.

The iodine-bromine brines are confined to the deep aquifers the hydraulic properties of which are relatively low. Most well yields are but a few litres per second but some are considerably greater. The ground water is highly confined and the potential resources are great.

Potable bromine waters occur, in some cases, near the bromine brines (Serpukhov, Borissoglebsk).

Bromine balneological waters (i.e., highly mineralized brines) are widespread at a depth of a few hundred metres to 2,000-3,000 m in the Baltic Sea, Mid-Russian and East Russian basins where they are confined to various aquifer complexes of Proterozoic to Permian ages. Their mineral content is 50-300 g/l with a bromine content to 1,300 mg/l. They are used at the resorts of Pyarnu and Ust-Kachka.

The total available resources of bromine brines are many times any foreseeable need. They may be tapped easily wherever they occur.

Hydrogen sulphide waters (H_2S and HS - over 10 mg/l) are of great value. They are formed, according to current understanding, as the result of bacterial reduction of sulphates. Two groups of water are distinguished: shallow and deep origin. The shallow group involves ground water with a low mineral content in the zone of active water exchange (Kemeri - in the Baltic Sea basin, Nemirov - in the Carpathian basin). The deep group includes ground water confined to bituminous and oil-bearing rocks and has higher hydrogen sulphide concentrations, to hundreds of milligrams per litre.

In the East Russian basin, at depths of 50 to 300 m, the hydrogen sulphide concentration in the Carboniferous and Permian ground water in places reaches 200-1,000 mg/l, for example in the Tatar and Zhiguli-Pugachev arches, and the pre-Urals). In the Dnieper-Donets basin the sulphide concentration in the Devonian bituminous and oil-bearing sediments reach 400 mg/l. Hydrogen sulphide waters occur in the Caspian Sea, Pechora and pre-Carpathian basins.

Hydrogen sulphide waters are concentrated in great amounts in the European territory of the country. Those best known are the resorts of Sergievskie Waters, Goryachi Kluch and Kemeri.

Mineral waters with a higher organic content occur chiefly in the pre-Carpathian basin (Shodnitsa, Truskavets, Undory). The well-known Naftusya water (Truskavets) contains carbon dioxide, in addition to specific organic matter up to 30 mg/l. Mineral waters with organic matter to 16 mg/l occur in the Dnieper-Donets basin at Berezovye near Kharkov. The mineral fraction of this water usually does not exceed 1,000 mg/l.

Carbonated mineral waters are developed in the Carpathian folded area where they are confined to the Carpathians and to the trans-Carpathian artesian basin. The carbonated waters are cold, mainly bicarbonate and bicarbonate chloride varying in composition (to 12 g/l). Their natural discharge areas are related to fractures.

Carbonated waters are widely used in sanatoria and spas for balneological cure, and bottling factories; they are also locally popular. A variety of carbonated waters are distinguished; many are like the well-known Arzni, Yessentuki, Borzhomi, Narzan types but have their own names, for example Polyana-Kvasova water. Most popular are sodium bicarbonate waters of the Borzhomi type with a mineral content of 4-11 g/l (Polyana, Polyana Kvasova, Solnechnoe Zakarpatie). In some cases the carbonated waters are also ferriferous and arsenious.

Ferriferous waters having more than 20 mg/l of iron have a purely local distribution. In the first province, high iron content is typical of the fresh or low-mineralized waters (Moscow). The second province includes ferriferous carbonated waters of the Carpathians. Ferriferous waters occur also in the Baltic Sea and Voronezh crystalline rocks (Polustrovo, Linetsk) where they have been used for a long time.

Radon waters have a strictly local distribution and are commonly associated with radon intake into low-mineralized (to 1 g/l) waters with shallow circulation from intrusive or emanating rocks, e.g., eluvium of the crust of weathering. Their manifestations are observed in the crystalline rock masses of Voronezh and Georgiu-Dezh; the Baltic Sea, Porossozero, Kartashi, the Ukraina-Hmelnik and Zvenigorodka.

Well yields are 2-3 l/s. The available exploitable reserves are limited but sufficient for known needs.

Thus, the European USSR is rich in various mineral waters. Most considerable with regard to amounts and use are the drinking waters of non-specific components and properties and balneological brines, including iodine-bromine and bromine. Considerable are the hydrogen sulphide and carbonated water resources. Currently, there are more than 500 locations where mineral waters are used for therapeutic purposes or are bottled. Some 150 additional locations have been identified and further developments can be expected.

5. Thermal waters

In the European territory of the USSR the Crimean Plain and the trans-Carpathian trough are of the greatest interest for thermal water development.

The thermal waters of the Russian platform, Crimean Plain and trans-Carpathian trough are briefly described below.

(a) Russian platform

The East European Precambrian platform is characterized by low values of geothermal gradient and heat flow as compared with Paleozoic and later platforms. The Ukrainian and Baltic shields, the East European platform basement show stable low geothermal gradients (1-1.1° C per 100 m) which is accounted for by low heat flow in Precambrian structures and high thermal conductivity of igneous and metamorphic rocks of the basement. The sedimentary cover of the Russian platform is mainly composed of Paleozoic and partially Mesozoic and Cenozoic rocks with a typical average geothermal gradient of 2° C per 100 m that is attributed to the prevalent carbonate rocks with high thermal conductivity. The mean values of heat flow amount to 0.9-1.25 $\mu cal/cm^2/s$. These values vary in area and are notable for positive and negative anomalies of -0.4 to 0.6 μcal/cm^2/s due to different (including hydrogeological) factors (Mavritsky et al., 1983).

The minimum heat flows of 0.85-0.9 μcal/cm^2/s are specific to the Ukrainian and Baltic crystalline shields.

Most considerable negative anomalies are found in the Belaya Tserkov and Krivoi Rog zones of the Ukrainian shield, within the Byelorussian anteclise and the Baltic syneclise. Positive anomalies are formed in the areas of Oka-Tsnin, Zhiguli, Don-Medveditsa-bars, Uhta dislocation.

Mavritsky has subdivided the East European platform, according to hydrogeothermal conditions, into two régime types: low (Baltic and Ukrainian shields) and moderate (East European platform). The thermal water horizon, with temperature exceeding 40° C, is at considerable depth (1.5-2 km) on the Russian platform.

The East European artesian area is a confined water system consisting of large artesian basins having diverse interrelations that is the Pechora, Moscow, the Dnieper-Donets, the Pripyat and the Caspian basins.

Over the major part of the artesian area, a Devonian rock complex composed of terrigenous carbonate strata contains water with a mineral content of 100-300 g/l and temperature up to 60° C. Brine thermal waters are widespread also in the Permian aquifer complex of the Pechora, Caspian, Dnieper-Donets and pre-Urals basins.

Saline and brine thermal waters with temperature exceeding 40° C are distributed in Mesozoic sediments of the Baltic-Polish and Caspian artesian basins. Owing to poor filtration in the rocks, well yields are only 1-2 l/s. Water temperature at the well head is only about 40° C but at depths of 2,500-3,500 m it may reach 75°-85° C.

Currently, the prospects to use thermal waters of the East European artesian basins are minimal. In the future, subthermal waters with the temperature of 20°-40° C may be employed using heat pumps.

(b) Trans-Carpathian trough

The trans-Carpathian trough dating from the late Paleogene period is composed of a thick (to 2,000 m) sequence of Neogene molasse with salt-bearing layers overlying a complex, folded, Paleogene and Mesozoic basement. Volcanogenic rocks are widespread.

The trans-Carpathian trough holds a particular place among inner troughs of Alpine depressions. It is characterized by very high geothermal gradients, to 4° C per 100 m and higher.

Heat flows here are 2-2.5 times higher than other similar structures as evidenced by anomalous stress conditions of the thermal field in the region.

In the trans-Carpathian trough, Neogene, Mesozoic and Paleozoic sediments contain mineral waters confined mostly to the Chop-Mukachevo and Solotvin depressions separated by the Vygorlat Guta volcanic ridge.

The Chop-Mukachevo depression is characterized by considerable heat at depth, with water temperatures increasing from the edges to the centre of the depression with its thick sedimentary cover where clays of low heat conductivity prevail. The highest temperatures occur in the areas of Zaluzh (146° C at a depth of 2,620 m) and Mukachevo (157° C at a depth of 3,343 m). The isothermal surface of 50° C occurs at depths of 600-700 m, rising in Beregovo to depths of 520-585 m. In the area of the Uzhogord cross-uplift, where the Paleozoic basement approaches the surface, the 50° C isotherm falls to a depth of 800-900 m and at the periphery of the depression - to 1,000-1,400 m. In the Solotvin depression the heat field is lower than that of the Chop-Mukachevo depression. The 50° C isotherm occurs at depths of 1,000-1,200 m, except at the Vyshkov thermal anomaly where the temperature of 50° C is found at a depth of 855 m.

The main thermal water resources of the Chop-Mukachevo and Solotvin depressions are contained in Neogene, Mesozoic and Paleozoic sediments.

Mesozoic and Paleozoic thermal waters of the Chop-Mukachevo depressions were sampled in territories of the Uzhgorod uplift and Chop area. The aquifers here are composed of limey clay and shaly layers. The water is confined. Chemically the waters are bicarbonate and sodium chloride with a mineral content of 48-60 g/l of which more than 90 per cent is diluted carbon dioxide. Their gas content is high.

The main thermal aquifers are distributed in Neogene sediments (Helvetian-Tortonian and Sarmatian complexes).

The Helvetian-Tortonian aquifer complex is developed throughout the Chop-Mukachevo and Solotvin depressions. Aquifers include tuff, tuff gritstone, siltstone and sandstone interbedded with clay. The depth of occurrence ranges from 200 to 1,000 m in the Chop-Mukachevo depression and to 500 m in the Solotvin depression. The thickness of Helvetian-Tortonian sediments is generally in the range of 700-1,000 m, in some places 2,000 m. The water is confined. The yields of flowing wells range from 10 or 20 m^3 per day (Uzhgorod) to 900 m^3 per day (Tereblya).

Thermal waters are chiefly sodium-calcium chloride with a mineral content ranging from 2 to 65 g/l and rising in the Zaluzh area to 240 g/l which is due to upward brine seepage. The mineral content of ground water in the Solotvin depression is 130-150 g/l due to salt in the Tereblya series. Helvetian-Tortonian waters contain high amounts of boron, bromine, iodine and iron. Their gas content is mainly nitrogen-methane in the Vyshkov area and carbon dioxide in the Solotvin depression.

Thermal water temperatures in Neogene sediments of the Solotvin depression range from 40° C at a depth of 600 m to 105° C at a depth of 2,950 m and 25-80° C at spring openings.

The Sarmatian aquifer complex is distributed in the Chop-Mukachevo depression and comprises beds of sandstone, siltstone, andesite and tuff interbedded with clay and argillite. Depths of the complex range from a few dozens of metres to 500-600 m. Waters with temperatures of 40-45° C are discharged at a depth of 500 m. The thickness of the Sarmatian sediments reaches 1,000-1,500 m and that of water-bearing layers, 10-320 m.

The Sarmatian thermal waters are confined, with well yields varying from 70 m^3

per day (V. Gorezdivka settlement) to 2,000 m^3 per day (Ivanovka village). Higher yields are related to zones of tectonic fractures. Water-saturated are contact zones of tuffs with sedimentary rocks (e.g., Began, Beregovo).

The Sarmatian thermal waters are mainly of the sodium-calcium chloride with a mineral content of 20 to 60 g/l (Uzhgorod), reaching in the Zaluzh area 200 g/l due to the influence of underlying Tortonian salt beds. Commonly, waters have high contents of iodine, bromine and boron. The prevalent gases are nitrogen, methane and, near the effusive rocks of the Vygorlat-Guta ridge, carbon dioxide.

The Helvetian-Tortonian and Sarmatian sediments represent a single confined water system as shown by almost identical static levels and chemical compositions. The hydrodynamic régime is governed by numerous tectonic fractures.

In the trans-Carpathian artesian basin, prospects for thermal water development are offered by the Helvetian-Tortonian complex, but are less promising in the Solotvin depression due to high mineral content (over 100 g/l).

The best prospects (based on temperature, mineral content and reserves) for thermal water production in the Chop-Mukachevo depression are Neogene sediments in the areas of Uzhgorod, Beregovo and Irsha.

(c) Crimean Plain

The Crimean Plain is the most promising region for thermal water development. Its territory is included in the western part of the Scythian plate with its typically stressed hydrothermal régime. The heat flow in the Crimean Plain is equal to 1.35 $\mu cal/cm^2/s$; on the Tarhankut uplift, 1.75 $cal/cm^2/s$ (Smirnov, 1980). The geologic sequence includes sediments from Jurassic to Quaternary over a rock basement of different age periods, Precambrian, Paleozoic and early Mesozoic.

Basement rocks are found at depths from 300 m in the Simferopol anteclise to 850-1,800 m on the Novosselov uplift, and at depths of 4,000-5,000 m in the Sivash trough.

Prospects for thermal water use are in Lower Cretaceous and Danian-Paleocene aquifer complexes. Neocomian-Aptian waters are confined to inequigranular sandstones and conglomerates. Depths of their occurrence range from a few metres on the Simferopol uplift to 2,400 m in the northern part of the peninsula, exceeding 2,500 in the areas of Berezovo, Novo-Alekseevo and Nizhnegorsk. In the Novoselovo uplands, the piezometric level is 70-130 m above ground surface; at Novo-Alekseevo it is close to ground surface.

The yields of flowing wells reach up to 2,590 m^3 per day (Alma basin) but average 1,000 m^3 per day. Transmissivity is 20-100 m^2 per day. Temperatures at spring openings range from 25° to 70° C and at depth, from 40° to 180° C. The mineral content of water is 1-78 g/l. Mineral composition is from sodium bicarbonate and sulphate to sodium-calcium chloride. The gaseous content is nitrogen or methane.

Danian-Paleocene waters are confined in fractured limestone and marl 30-70 m thick. Piezometric levels are 20-50 m above ground surface. The water content of the Danian-Paleocene aquifer complex varies from 3,000 (Alma basin) to 1-5 m^3 per day (Krasnaya Polyana), decreasing with depth and distance from the recharge area.

Transmissivity ranges from 20 to 80 m^2 per day; temperatures are 40-120° C. Waters vary from sodium bicarbonate to sodium-calcium chloride type. The mineral content ranges from 1 g/l in northern Crimea to 20-25 g/l on the Tarhankut peninsula. Mineralized waters are gas-saturated with carbon dioxide predominant.

Available resources of thermal water and heat were estimated for Lower Cretaceous and Danian-Paleocene aquifer complexes. For the Danian-Paleocene complex the available resources of thermal water and geothermal energy amount, using fountain methods of production, to 23 thousand m^3 per day and 438 thousand Gcal per year respectively.

For the Lower Cretaceous complex the thermal water resources also produced by these methods amount to 40 thousand m^3 per day and geothermal energy of 870 thousand Gcal per year.

Using pumping methods of production, with dynamic well levels decreased at the end of the estimation period to 600 m the resources of thermal water and heat increase to 240 thousand m^3 per day and 4,700 Gcal per year relatively.

In the Danian-Paleocene aquifer complex waters with mineral content of 10 g/l are found in the Kerch hydrogeothermal zone where its resources amount to 1,000 m^3 per day using fountain methods of well production and 18,000 m^3 per day using pumping methods.

In the Lower Cretaceous strata the thermal waters with the mineral content of 10 g/l occur in the Simferopol zone where their resources amount to 3,000 m^3 per day using fountain methods and 34,000 m^3 per day using pumping methods.

The Ciscaucasus territory holds a multilayered confined aquifer system consisting of a series of interconnected artesian basins: Azov-Kuban, Tersk-Kum (Tersk-Caspian) and the hydrogeologic area of the Stavropol arch.

The best prospects for practical utilization of thermal waters are the "Chokrackaragan" and Lower Cretaceous aquifer complexes spread almost all over the Ciscaucasus territory. The first complex is best known in the Tersk-Sunzhensk basin, at the southern side of the West Kuban trough and the Daghestan Plain. Its territory is composed of marine terrigenous, partly carbonate sediments. The total thickness of sediments ranges from 20-50 m (in the northern part of the western Ciscaucasia) to more than 900 m in central areas of the west Kuban and Tersk-Caspian troughs. Waters of the complex range from fresh to brine with mineralization of 0.5-65 g/l. Chemically they are nitrogen-methane sodium bicarbonate and sulphate type; sodium bicarbonate and methane and sodium chloride type.

The water conductivity reaches 100-200 m^2 per day in eastern Ciscaucasia, decreasing in the west to 20-50 m^2 per day. The highest transmissivity is found in the Grozny and Makhach-Kala areas, i.e., 150-300 m^2 per day.

All the wells are flowing. Artesian flowing water heads reach 100 m, rarely 300 m. Yields of flowing thermal-water wells may reach 2,000-3,500 m^3 per day. The water temperature at well head ranges from 50°-70° to 100° C.

The Lower Cretaceous aquifer complex has proved irregular. The best prospects are in the Aptian, Albian, Cenomanian and partly Barremian sediments. The deepest parts of the Azov-Kuban and Tersk-Caspian basins have not been studied.

Saline and brine methane and sodium chloride waters with mineralization of 10-150 g/l are predominant. Within the Azov-Kuban basin and Stavropol arch mineralization is lower (10-90 g/l) than in Tersk-Caspian basin (10-150 g/l). Temperatures in the Neocomian water-bearing complex vary between 40° and 170° C reaching the highest value in the Praskoveisk well at a depth of 3,160-3,170 m. At most deep wells, the water temperature exceeds 100° C, decreasing to 50°-70° C in the Cavminvody ridge.

Piezometric levels are, in general, above ground surface with heads reaching 300 m. The maximum surplus pressure of 315 atmospheres was observed at the Galugaevsk well number 1 at the depth interval of 5,354-5,427 m, showing tectonic isolation of the studied area. Well yields vary widely from 100 to 1,000 m^3 per day, achieving more than 3,000 m^3 per day in the piedmont zone of the Adygei ridge and East Kuban trough. Water conductivity ranges from 10 to 50 m^2 per day increasing to 100-150 m^2 per day in the East Kuban basin.

D. Exploration and development of ground water

1. Classification of resources

In the Soviet Union, ground water which can be used for practical purposes is considered a mineral resource. However, ground water is the only mineral resource which under certain conditions, is renewable. This includes an increase in ground water by recharge, i.e., inflow of surface water and ground water of adjacent aquifers, a decrease in evaporation from the ground water with water-table decline, diverse water-management activities including artificial recharge of ground-water resources.

Specific features of ground water that distinguish it from other mineral resources have predetermined the need for defining the concepts to (a) the amount of water in the aquifer, (b) the amount of water flowing into the aquifer under natural conditions due to various water-management activities and ground-water withdrawal, and (c) the amount of water which can be extracted from the aquifers by appropriate water extraction systems.

Currently, numerous classifications of the ground-water resources have been established in the USSR and other countries. Generally such classifications differ in terminology. The classification of Bindeman is the most widely used in hydrogeologic studies of the USSR. According to this classification, ground-water reserves and resources fall within four categories: (1) natural, (2) artificial, (3) recharge or inflow during extraction, and (4) available (safe yield).

(a) Natural storage

Natural storage is the amount of free water in pores and fissures of water-bearing rocks. Natural reserves of the confined aquifers also include elastic storage. The latter means the quantity of water which can be extracted from aquifers with lowering of the ground-water level at the expense of elastic properties of water and rocks. Natural storage of ground water is measured in volumetric units.

Related ground-water resources represent the value of ground-water recharge under natural conditions through infiltration of precipitation, seepage from streams, leakage from the over- and underlying aquifers, and inflow from adjacent areas. Natural resources are equal to the sum of all recharge balance elements of the given aquifer. They are expressed in discharge units and can also be determined by the sum of discharge balance elements (e.g., evapotranspiration, spring runoff, seepage into surface-water courses and ponds, and leakage to adjacent aquifers).

(b) Artificial storage

Artificial storage is the amount of water in the aquifer generated as a result of irrigation, seepage from water reservoirs, and artificial recharge (replenishment) of ground water. Related resources include discharge of water percolating into the aquifer from canals and reservoirs, in irrigated lands, as well as with intensification of ground-water recharge.

(c) Recharge

Recharge or inflow during development is a discharge of water flowing into the aquifer under intensified replenishment of ground water induced by pumping through an increase in the seepage from adjacent aquifers, reduction of evaporation from the surface of phreatic water due to an increase in its depth from the earth's surface in case of coincidence between the recharge and discharge areas.

(d) Safe yield

The concepts of safe yield and available resources of ground water are equivalent. Safe yield is the amount of ground water which can be developed by means of extraction devices, in technical and economic terms under a given production and water quality satisfying the standards during the estimated period of water consumption. Thus this value represents a water intake capacity and is expressed in discharge units (usually thousands of cubic metres per day). Also, environmental implications of the ground-water withdrawal must be considered with the safe-yield assessment. Currently, the term safe yield is applied to the prospects for ground-water use to satisfy the demands of particular consumers. In describing general potentials for ground-water extraction in any large region, as well as in considering ground water as part of the total water resources, the term available resources is preferred.

No doubt to solve the problems concerning the use of ground water, an assessment of safe yield or available resources is of paramount importance, as only the results of such estimation can allow for the construction of well fields. At the same time, the evaluation of natural reserves and resources of ground water may have a particular significance towards settling such general hydrogeologic problems as determination of the water-exchange rate, of ground-water age, and of subsurface flow into rivers, seas and oceans.

The relations of the safe yield and other types of reserves and resources can be expressed by the following balance equation:

$$Q_s = \alpha_1 Q_n + \alpha_2 \frac{V_n}{t} + \alpha_3 Q_a + \alpha_4 \frac{V_a}{t} + Q_r, \quad (1)$$

where

Q_s = safe yield (available resources) in cubic metres per day,

Q_n and Q_a = natural and artificial resources, respectively, in cubic metres per day,

V_n and V_a = natural and artificial reserves, respectively, in cubic metres,

Q_r = recharge during production, in cubic metres per day,

t = period of time for which the safe yield is calculated, in days, and

α_1, α_2, α_3, α_4 = coefficients of the use of natural resources, natural reserves, artificial resources and artificial reserves, respectively.

This balance equation shows only a scheme of principle for the safe-yield formation over the entire period of production t. Depending on hydrogeological conditions, various sources (different types of reserves and resources) prevail in the safe-yield formation, with a contribution of particular sources being changed in time.

Utilization of ground water in the USSR is regulated by the classification of the safe yield and predicted resources developed by the State Commission on Mineral Resources under the Council of Ministers. This classification contains common principles for calculation as well as determination of the storage availability for national economy depending on the knowledge and reliability of the reserves. In accordance with the classification the safe yield falls into four cateogries A, B, C_1 and C_2 depending on the knowledge and production of ground water. Also predicted resources of category P are distinguished.

Like other mineral resources ground water may be considered as a deposit in reservoirs. A ground-water reservoir is a volumetric underground area where natural or artificial factors favour ground-water extraction in amounts sufficient for its intended uses.

Based on the geologic-hydrogeologic environments which determine regularities in the safe-yield formation, procedure for exploration and the safe-yield assessment, reservoirs of ground water in the European part of the USSR used for domestic water supply can be subdivided into the following five major types: (a) of river valleys, (b) artesian basins, (c) limited-area structures or masses of fissured and karst-fissured rocks, and zones of tectonic dislocations, (d) of desert and semi-desert sands, and (e) supramoraine and intermoraine glaciofluvial deposits.

Mineral, thermal and industrial waters are stored in specific reservoir types.

From the concept of safe yield it proceeds that the safe-yield estimation includes determination of the possible withdrawal under a given drawdown or predicted drawdown during an estimated or unlimited long-term period, provided that ground-water quality satisfies the standards for the entire period. In this case, the impact of ground-water extraction on surface flow, vegetation and other environmental components must be assessed. Thus, the safe-yield assessment pursues the following main objectives:

1. To determine the potential yield of an artificial ground-water outlet under a given drawdown,

2. To choose the best withdrawal system in technical-economic terms - wells, plain or radial galleries, number of water points, spacing, yields, maximum depth of dynamic level,

3. To prove that water quality will meet standards during the entire period of production,

4. To establish rates of water level decline in the cone of depression, and

5. To evaluate potential variations in surface flow, surface deformation and other implications of ground-water extraction.

The assessment of safe yield in the USSR and other countries is accomplished in two basic ways:

1. At local sites, to provide water supply of some consumers and irrigation of particular lands, and

2. In large hydrogeologic regions for exploration planning and proving schemes of water conservation.

The safe yield at local sites is evaluated in relation to the most rational withdrawal scheme. If the ground-water extraction is to be accompanied by artificial recharge that must be considered in the safe-yield estimation. The safe yield is calculated in relation to existing natural conditions and water-management situation. Given developed projects on changing natural environments (e.g., construction of reservoirs and canals), the influence of measures planned on the safe yield, as well as the need for reassessment of the available resources after implementation of these measures, should be determined.

Safe-yield calculations must include an estimation of the impact of a designed water intake on variations in the surface, spring and drainage flow.

The estimated period of development must be established, if natural (including elastic) storage of the aquifers is the main source of the safe-yield formation, and water is extracted in an irregular way. Usually, this period is defined by production (project) organization. The period of water-intake functions to be used in assessing the safe yield is conventionally assumed to be 25 years, provided that an estimated period is not fixed and a water intake is planned for the indefinite long term. This must exceed the depreciation period of and water extraction works sufficient for reassessment of the safe yield before the end of this period, using the data on long-term production, or to supply additional water to a particular consumer. In addition, the safe yield should be assessed for a 50-year period of development for regional evaluation.

Acceptable drawdown values are determined by hydrogeologic, technical, economic and ecologic factors. It must be such that the residual water column in a well can provide the designed yield and keep the pumping facilities in good working order. Also, the maximum possible water rise under pumpage must be considered for deep aquifers.

In all cases, the acceptable limits for drawdown should be chosen with regard to the potential changes in the natural environment, as well as the need for environmental control, for example, in the case of dewatering of shallow aquifers which results in low yields and degradation of vegetation.

The safe yield can be evaluated using hydrodynamic and hydraulic balance methods or by a method of hydrogeologic analogy, or by a combination of these methods.

The safe-yield estimate by a hydrodynamic method involves calculations of water withdrawals with the assumed initial and boundary conditions and parameters of the aquifer in a study filtration area using geofiltration equations. It can be done through mathematical modelling using computers.

The main advantage of computer methods is the ability to forecast variations in the ground-water table with development, i.e., the possibility to demonstrate the availability of the safe yield over time.

The safe-yield assessment with a hydraulic method involves determination of an estimated yield of a water intake or predicted drawdown of wells using empirical data derived directly from an experiment and considering the combined impact of various factors which define a water-intake operation.

Unlike hydrodynamic methods, the use of which is characterized by a selection of the estimated relations and determination of the basic hydrogeologic parameters following the schematic boundary conditions of an aquifer, hydraulic methods provide for an assessment both of a type of the relation and its major parameters on the experimental data. Therefore, the combined impact of various factors (e.g. heterogeneity of a layer, existence of various limits, resistance to water flow in the near-well zone, violation of a linear seepage law) is considered. Hydraulic methods should be used where hydrogeologic conditions are complicated and marked by heterogeneous seepage zone, as well as by hardly detected sources of the safe-yield formation.

The safe-yield assessment by water balance method implies determination of the ground-water discharge which can be obtained by water extraction points within the given area during a predetermined period of extraction at the expense of particular sources of the safe-yield formation presented in the right-hand-side part of the equation. In this case, an individual evaluation of each of the possible sources of the safe-yield formation is made followed by a summation of these sources.

Usually, application of the balance method is complicated by failure to determine the use factors of the ground-water resources, as well as the amount of inflow by means of the balance equation. The balance method provides for only an average value of variations in the ground-water table for a given area. The drawdown of a particular water well cannot be established.

At the same time, only the balance method permits determination of the contribution of particular sources to the safe-yield assessment of the reliability of the ground-water storage calculated using other methods. Hence, use of the balance method in combination with hydrodynamic, hydraulic and other methods of the safe-yield estimation is indicated in most cases.

2. Forecasting quality

Prediction of potential changes in the quality of ground water extracted is one of the most important problems of the safe-yield assessment. Variations in the ground-water quality at points of extraction are attributed to the inflow of ground or surface water which is naturally unsuitable, or artificially polluted. Naturally unsuitable is water having higher natural mineral content or containing particular components in amounts which exceed potable water standards (mineralized ground water, sea water, mineralized water of streams and canals).

By artifically polluted is meant that the ground and surface water has experienced qualitative change from industrial and municipal wastes or agricultural contaminants.

Flow of polluted ground water to a water extraction point depends upon the type of water intake, boundary conditions of the aquifer, specific features of natural flow of ground water, and hydraulic heterogeneity of rocks in plane and section. Natural flow of ground water is the most important factor.

Ground water may move to a well from all parts of the aquifer. If any layer contains polluted water, it will reach a water intake sooner or later, even decades after the original pollution.

In such cases, ground-water contamination at a water well will depend on flow speed, as well as on the further changes in the ground-water quality.

Natural movement of ground water (i.e., flow conditions) in a layer triggers formation of a local recharge area of a water intake separated from the remaining part of an aquifer by the so-called neutral flow line. The size of a recharge area depends on the ratio between water withdrawal and discharge of natural flow, as well as on the type of water intake and boundary conditions of the aquifer.

Given the occurrence of polluted water in an aquifer, the forecasting of the ground-water quality at a well includes:

1. Estimation of the potential for the flow of unsuitable water in the cone of influence of the well,

2. Assessment of time needed for unsuitable water to reach the well, and

3. Evaluation of variations in the ground-water quality at the well after irruption of unsuitable water.

The above assessments are carried out relative to particular hydrogeological conditions. Forecasting of changes in the water quality is necessary, if the time elapsed in the movement of unsuitable water to a water well is less than the planned period for well operation. Three major types of variations in the water quality at a well occur, depending on the hydrogeology of an aquifer:

1. Increase in the mineral content of water, in excess of acceptable standards,

2. Limited growth of the mineral content, reaching values slightly above acceptable standards and salinization of the water intake, and

3. Limited increase in the mineral content followed by reduction of mineralization.

Movement of unsuitable water to a well causes loss in operation and usually results in installation breakdown. However, in regions having scarce resources of fresh ground water, a moderate delivery of slightly brackish water by a well enables growth of the safe yield suitable for domestic water supply, provided that the quality of the water extracted meets standards.

Usually, changes in the chemical composition of ground water as a result of ground-water development correspond to the deterioration of the ground-water quality.

Zones with special restrictions on human activity which provide for protection of ground water against pollution are to be established for well-fields used in domestic water supply.

3. Environmental impact of ground-water withdrawal

Ground-water development has a particular impact on other environmental components. Therefore, estimation of such effects to compensate or eliminate negative effects may in itself justify a ground-water resources assessment. Water withdrawal and lowering of the water table may lead to the following environmental changes:

1. Changes in the surface flow, especially flow of small rivers,

2. Decline of water table accompanied by degradation of vegetation, drainage of swamps and depletion of lakes,

3. Land subsidence under secondary consolidation of drained rocks and compaction of depressed sands and clays, and

4. Activation of karst phenomena.

Stream flow is heavily dependent upon ground-water extraction. Usually, ground-water development results in a slowdown of subsurface runoff and to an induced recharge by stream-water seepage to an aquifer. In this context, surface-water resources are depleted. Sometimes, with production of aquifers having no active hydraulic relations to streams, the long-term increase in surface flow can result from disposal of the spent ground water into rivers.

Maximum changes in the surface flow occur during the development of the ground-water reservoirs in river valleys where surface water is the major factor contributing to safe yield. Value of the stream-flow reduction and its time variations depend on a large number of factors including lithology and such hydrogeologic parameters as transmissivity and specific yield of aquifers, distance between a water intake and a river, and different ways of the ground-water discharge, including evapotranspiration and direct flow into rivers. In some cases, surface flow is constant, if the amount of ground water extracted does not exceed evaporation loss. In other circumstances, ground-water withdrawal can affect flood flow alone, whereas a low-water flow remains unchanged. This situation is typical of aquifers having a large regulating storage, i.e., when the safe yield of a low-water period is attributed to the storage which is further replenished by floods. Thus, a set of specific factors must be considered in estimating the effect of ground-water extraction on surface flow. The methods of analysis may include hydrologic, hydrodynamic and analog methods.

Most frequently used in the USSR are hydrodynamic methods. Analysis is easy under simple hydrogeologic environments as compared to the methods of mathematical modelling necessary for complex situations.

Studies show that the flow of small streams alone is heavily affected by ground-water withdrawal. In some cases ground-water extraction results locally in the drying up of small rivers during low-water periods.

Ground-water development may change the landscape owing to the lowering of water levels which affects the moisture content of soils and rocks in the unsaturated zone.

Induced land subsidence is to be expected as a result of large-scale ground-water withdrawal carried out over long periods. Subsidence dynamics depends on the intensity of water extraction, decline of the water table, compressibility of rocks, thickness of a compressed mass, geologic structure and morphology of an area. Secondary consolidation of loose rocks is likely to happen with pronounced declines of the water table as well as compaction of geologic formations.

Negative environmental effects of land subsidence are of limited scope in the European part of the USSR. This process is observed chiefly in coal basins although land subsidence there is usually triggered by abrupt rock collapse. The safe-yield assessment in areas of potential land subsidence implies prediction of the above process using information on the effective load variations, rock compressibility and thickness of a compacted layer.

Ground-water development in karst regions accounts for sinks in the land surface. Surface karst forms (sinks, ponors) are well-known, but over the past 20-25 years the formation of new sinks has sharply accelerated, particularly in urbanized and intensively developed karst areas, as a result of human activity and especially the extraction of fissure-karst water. A deformation of land surface occurs in carbonate rocks underlying a mass of sands and clays. Usually, carbonate rocks are paleokarst, marked by cavities which were later filled with loose uncemented deposits.

Activation of karst processes occurs in two ways: (a) regeneration of paleokarst cavities by the suffosion of features accumulated there, and (b) development of new cavities.

Karst-suffosion processes are stimulated by intensive withdrawal of ground water followed by hydrogeologic change. Water extraction leads to disturbance of existing natural hydrogeodynamic conditions, variations in the direction and rate of ground-water flow and formation of new sources of ground-water recharge. The abrupt effect of karst water withdrawal can be illustrated by the example of a water intake in the river valley of the Maly Reft, the Urals, where production wells were drilled close to the channel (20-30 m). As a result of large flow gradients, sinks were formed in the river channel and fine earth was brought to the production wells.

4. Reconnaissance and exploration

Ground-water reconnaissance and exploration are a combination of studies aimed at determination of ground-water reservoirs, assessment of the safe yield and obtaining the data to be applied to the design and construction of water wells.

Currently, the major stages for reconnaissance and exploration followed in the USSR: (a) reconnaissance, (b) preliminary exploration, (c) detailed exploration, and (d) pilot production. Reconnaissance and exploration of the ground-water reservoirs include the following main types of field studies: specialized geologic-hydrogeologic survey; drilling; filtration tests; geophysical, hydrologic, laboratory, water-balance, topographic and geodetic investigation; stationary observations of the natural and disturbed régime of ground water, examination of the functioning of wells, water and soil sampling, specific research (e.g., isotope, tracer, hydrogeochemical, hydrogeothermal) and the proving of artificial replenishment of ground water. The range of studies at each step is determined by the amount of data available, water demand and specific hydrogeologic environments.

The main objective in reconnaissance is the delineation of aquifers and areas promising for further exploration, and forecasting of the safe yield within these areas to make a decision concerning the need for, and priority of, further study of prospective areas. Here, a future exploration area must be determined. Prospecting involves processing of the regional hydrogeologic information available, specialized geologic and hydrogeologic surveys at scales of 1:200,000 to 1:50,000, using drilling, filtration tests, geophysic, hydrology and other studies.

Preliminary exploration is the most important step in water investigations. Its main objectives are the study of specific hydrogeologic conditions of the ground-water reservoir, estimation of basic components of the safe yield, and preliminary assessment of the total amount of the safe yield.

Preliminary exploration must answer the question of the quality of water which can be extracted. A combination of drilling, pumping tests, geophysics and other studies during preliminary exploration must provide the data to enable description and preliminary quantification of the safe yield, the determination of aquifer parameters, verification of water-quality acceptability in terms of its probable use, a preliminary estimation of hydrogeologic, technical and economic factors of development, and the preparation of specifications for wells and pumps.

Detailed ground-water exploration is carried out in areas considered suitable for future development as evidenced by the results of preliminary exploration. Detailed exploration tests the designs for construction and operation of wells, and studies the ground-water reservoir to evaluate the safe yield all of which are necessary in choosing the best scheme of water intake. Detailed exploration must determine the most practical means to extract the required amount of ground water.

Detailed exploration is completed on assessment of the safe yield for approval, when necessary, by the State or territorial Commissions of the USSR on mineral resources in accordance with regulations. This approval is the basis for allocations to the construction and operation of water intakes.

Pilot production of ground-water reservoirs is carried out during construction and operation of water intakes to confirm the availability of water, to re-assess the ground-water resources using production data, and to obtain research data for further application in exploration of similar ground-water reservoirs elsewhere. Studies of the latter step are constantly conducted during development of the ground-water reservoir and include regular observations of wells, ground-water dynamic levels in observation wells, ground-water quality and temperature, and the working order of wells.

Depending on the character of hydrogeologic environments, water demand, hydrogeologic knowledge of a study area or reservoir, some investigative steps can be omitted or combined with other research steps. Reconnaissance is not necessary in well-studied regions and a water-intake site or an aquifer can be selected using the data available. Simple hydrogeological conditions allow for combination of preliminary and detailed exploration.

5. Development of ground-water resources

(a) Planning

The choice of a ground-water development scheme is proved by technical and economic calculations depending on particular factors:

1. Determination of production areas and priority for their development,

2. Number, location and specification of wells,

3. The need for artificial recharge, and

4. Decision made on extraction, transportation and treatment of water.

Usually, such a technical and economic assessment, as well as the choice of type of water intake, is determined by a common procedure which includes traditional methods for estimating the efficacy of economic measures. Each type of a water intake (R_i) can be evaluated using the following expression:

$$R_i = EI_i + C_i + L_i \text{ (in rubles)} \qquad (2)$$

where:

E_i = standard investment gain currently assumed to be 0.15,

I = investments for construction and completion of water supply system, (in rubles),

C_i = annual production expenditures, (in rubles),

L_i = losses suffered by other branches of national economy from water-well operation, (in rubles), and

i = type index.

An optimum well is one having a minimum value of R_i. Calculation of investments (I) includes summation of expenditures for:

1. Construction and completion of water wells or other capping structures, as well as a network of observation wells,

2. Construction of reticulated network with pumping stations and intermediate tanks,

3. Construction of roads, power and communication lines for water-intake maintenance,

4. Construction and completion of infiltration basins or absorbing wells, settling ponds, pumping stations or gravity mains (for artificial recharge),

5. Construction and completion of stations for water treatment, removal of iron and disinfection,

6. Charges for expropriation of agricultural lands and forests to be used as water-intake areas, sanitary zones and roads, expenditures for recultivation of lands which are utilized over the short term, replacement of structures which are not acceptable within sanitary zones, and

7. Completion and fencing in water-intake sites, construction of administrative and service buildings and structures.

Annual production expenditures (C_i) include:

1. Wages of workers and administration involved in a water-supply system with regard to local fee coefficients and taxes for social insurance,

2. Depreciation for renovation and overhaul of major industrial structures,

3. Expenses for maintenance of buildings, structures and equipment,

4. Cost of electricity, heat and chemical reagents, and

5. Overhead.

Annual production expenditures of water schemes which provide for an artificial recharge of ground water include expenses for maintenance of artificial-recharge structures (e.g., conduits, gravel-sand packs, infiltration basins).

Usually, distribution networks are not considered in calculation of investments and production expenditures, as the expenses for their construction and use do not depend on a source of water or a water-intake scheme. In many cases the construction and operation of water-supply systems lead to reduction of productivity of natural resources used in agriculture, forestry and fishery. Reduced productivity can be attributed either to the environmental changes (lowering of water tables, drying up of springs and small streams, or to regulations currently in force for the zones of sanitary control such as the prohibition of chemical fertilizers and pesticides, restricted forest cutting). As a result, industries which use these natural resources suffer annual losses in an amount equal to the sum of estimated profit gained from unavailable products. Judging from the principle of a national economic approach to an assessment of efficiency of any measure, in comparing types of water schemes having diverse environmental impact, losses suffered from operation of wells by other branches of national economy are added to water expenses.

In some cases, environmental changes such as swamp drainage, or reduction of mine water inflows can be beneficial. In such cases, additional gains which compensate expenditures for this type of a water-intake scheme must be considered.

The major economic parameters of the water development scheme finally proposed (investments, annual production expenditures cost per cubic metre of water) are presented in standard form (neglecting the losses suffered by other industries).

(b) Pumps and related equipment

In the USSR water is extracted from wells using various kinds of pumps - air-lift, submersible, rotary, horizontal-flow rotary, ejector and Cornish.

Most frequently used are air-lift pumps. They enable water extraction from small-diameter wells having deep dynamic levels and high content of sand and clay. The simple design of air-lift pumps allows quick installation and absence of moving parts in a well makes air-lift pumps reliable and durable. Well yield is easily controlled with air-lift pumpage which facilitates the setting up of various schedules for water withdrawal.

Some limitations of air-lift pumps are their low efficiency and restrictions on use of these pumps in wells drilled to particular depths which depends on the coefficient of immersion of a pump nozzle below the dynamic water level.

Air-lift pumps are used for test pumping in wells whose depth is 1.5-1.8 times that of the dynamic water level during pumpage.

Pumpage in wells having deep dynamic levels (to 200 m) and high yield is accomplished with submersible pumps which are highly efficient, light and easy to install. This is of particular importance for field studies. The above pumps can be mounted in wells with small doglegs. They cannot, however, extract water containing solid particles. A wide range of such pumps are manufactured in the USSR.

Horizontal-flow rotary pumps are used for water pumpage under the lowering of the dynamic water level to a depth not greater than 6 or 7 m from the ground surface. These pumps have high efficiency, high capacity and can extract water containing solid particles. Horizontal-flow rotary pumps of K and NDv types are widely used.

K-type pumps are horizontal-flow, one-step rotary, console-type with a one-way-inlet impeller. Currently, 13 versions of these pumps are manufactured to provide yields of 4.5 to 360 m^3/h under pressures of 8.8 to 95 atm. Pumps are electrically driven using elastic coupling or V-belt pulleys.

In addition to K-type rotary pumps, various modifications of ND-14 one-step pumps are also applied to water extraction. Pumps of ND-type are driven by elastic coupling.

Usually, water is extracted using NDv pumps having a capacity of 90-360 m^3/h, full pressure of 22-30 m and motor power of 14-75 kW.

Ejector pumps come in diverse modifications: NV-108 (for well diameter of 108 mm), UNV-127/168 (for well diameters of 127, 146, 168 mm) and NV-89 (for well diameter of 89, sometimes 108 mm). These pumps enable water withdrawal without the use of casing if a hydraulic packer is installed, as well as the measurement of the dynamic level while pumping. The need for application of a drill pump is a shortcoming of ejectors.

Cornish displacement pumps are used for water extraction in wells with a dynamic level depth of 100 m. Efficiency of these pumps is high, although their capacity is low (6-20 m^3/h).

(c) Well screens

The following screens are used in the USSR for test pumpings and water-well operation:

1. Perforated and edge screens having a single filtering surface and made of diverse pipes,

2. Filters having filtering material: mesh or cloth fabric attached to pipe frames,

3. Specific structures either having no pipe frames or having pipe frames supplemented by such other elements as wire spirals, bars, rings, nozzles, and

4. Gravel filters (prepared at the ground surface and then sunk to the well bottom, or prepared right at the well bottom) using gravel packs which are in contact with rocks and replace mesh or other filtration surfaces.

(d) Data collection

Registries for natural resources are an information base of the State Nature Management and Environmental Control in the USSR. That the State Water Registry be supplied with the data of the State Water Record is specified in the Water Act of the USSR and its constituent republics.

The State Water Registry falls in three subsystems with a constant information exchange: surface water, ground water and water use subsystems. Computer-assisted information systems are being adopted for the water registry. The section for ground water in the State Water Registry assists in hydrogeologic studies aimed at a regular assessment of the dynamics of ground-water bodies. Ground-water bodies include basins, aquifers and reservoirs. State of the ground-water bodies is estimated by the changes in parameters which can describe natural resources and the safe yield of a ground-water body, as well as water tables, ground-water temperature and the physico-chemical characteristics of ground water.

Information to be included in the State Water Cadastre is as follows:

1. Log books of the state of potential and prospective available resources of ground water in hydrogeologic regions (artesian basins, phreatic basins),

2. Logs of ground-water reservoirs, which contain general description of the reservoir, characteristics of geologic profiles, results of hydrogeodynamic and hydrogeochemical testing, data on the major hydrogeologic parameters of aquifers and separating semi-permeable layers, information on the explored ground-water storage, methods of assessment and formation sources, and characteristics of water intakes,

3. Logs of exploratory, observation and production wells and springs, and

4. Observation data of the ground-water régime (water table, yield, temperature and physico-chemical characteristics of the ground water).

Using the above data, ground-water divisions of the State Water Registry are expected:

1. To analyse, systematize and store the data on ground-water bodies, ground-water resources, régime, quality and utilization, to assess periodically the state of ground-water bodies,

2. To describe the safe yield and predicted resources of ground water, as well as ground-water quality by hydrogeologic regions, economic areas, Soviet Republics and the country as a whole,

3. To prepare for publication the information of the State Water Registry (Section of Ground Water),

4. To maintain computerized data banks,

5. To supply organizations, enterprises and interested offices with the data of the State Water Registry, and

6. To exchange data with other subsystems of the registry (surface water, water use).

The computerized information system of the State Water Registry on ground water includes three interrelated functional-technological subsystems:

1. Data acquisition and verification,

2. Data-base storage, maintenance and processing to be applied for the solution of hydrogeologic problems using specialized computer programs, and

3. Distribution of the data to users, with descriptions of the ground-water resources situation published in form of maps and information bulletins.

The System of the State Water Registry involves three interrelated levels: local, Federated Republic and All-Union.

E. Training and research

1. Hydrogeological Survey

The Hydrogeological Survey of the USSR forms a part of the State-owned sector of national economy. Fields of activity include prospecting and exploration for potable, mineral medicinal, industrial and thermal ground water, studying of its régime; water preservation and control; and cadastral surveys. The work is carried out in co-operation with the Ministries of Geology of the Russian Socialist Federated Soviet Republic (RSFSR), Ukrainian SSR, Kazakhstan and Uzbekistan, and through the geological departments of Union Republics. Mining, hydrotechnical, land-reclamation, industrial and civil engineering studies having hydrogeologic implications are carried out by the Ministries of Ferrous Metallurgy, Non-ferrous

Metals, Coal and Oil Industries, Energy and Electric Power, Land-Reclamation, Water Economy, and others. The Ministry of Geology of the USSR plans and controls hydrogeologic studies to be carried out by geologic units and departments through the Division of Hydrogeology.

Many geologic projects of administrative or territorial character include complex hydrogeologic and engineering geology, geologic and specialized surveys. Some surveys and water resources inventories are carried out by specialized hydrogeologic groups in these surveys. Among administrative personnel, there is a post of chief hydrogeologist or senior hydrogeologist of a geologic section who is a curator of hydrogeologic surveys.

The results of hydrogeologic studies are presented in reports containing basic data, working techniques, maps, conclusions and recommendations. The reports are discussed in technical scientific meetings of surveys and projects, and communicated to regional, Republic and All-Union geologic organizations to be used as needed.

Reports on ground-water exploration and resources assessment are discussed also by a State or regional Commission on mineral resources. Experts from the different departments are enlisted to review and discuss the hydrogeologic reports.

Geologic projects implement programmes of the Ministry or Division of Geology and such work ordered by other Ministries and departments as hydrogeologic and engineering geology surveys on a scale of 1:50,000 for purposes of land reclamation or industrial and civil engineering; assessment of hydrogeologic conditions for construction of large dams and other public works or mining under difficult natural environments. Hydrogeologic investigations are carried out by research institutes and thematic sub-divisions of geological organizations. The leading institute in the Ministry of Geology of the USSR is the All-Union Research Institute for Hydrogeology and Engineering Geology (VSEGINGEO), whose main aims are:

1. To elaborate the theoretical basis and methodology for hydrogeologic investigations;

2. To develop technical standards (state and branch standards, guiding principles, recommendations, instructions) for hydrogeologic studies;

3. To develop new hydrogeologic devices and equipment; and

4. To co-ordinate the research and programmes of other research organizations and geologic groups and to monitor their activities.

In the USSR Academy of Sciences, theoretical hydrogeologic investigations are carried out by the Geological Institute (GIN), Moscow, the Institute of Water Problems (IWP), Moscow, the Institute of the Earth's Crust (IEC), Irkutsk, the Geological Institute of the Ukrainian Academy of Sciences, Kiev, and the Institute of Geophysics and Geochemistry of Byelorussian Academy of Sciences, Minsk. The research and design institutes of other departments, such as the Institute of Engineering Prospecting, Moscow, Hydroproject of the Ministry of Energy and Electrical Engineering, Moscow, provide methodical aids for hydrogeologic and engineering geology investigations within the scope of these Ministries.

Having been given additional responsibilities to solve problems of environmental conservation, the USSR Geological Survey is being broadened and improved. Its budget has steadily increased over the past 20 years. The USSR has provided scientific help to the developing countries of Africa, Asia and Latin America in prospecting and exploration of potable water reservoirs, mineral deposits and in hydrogeologic mapping. This has been done through bilateral agreements (Cuba, Syria) or through projects of the United Nations system (Somalia, Guinea-Bissau). A number of African hydrogeologists have been trained in the Soviet Union.

2. Technical education

The USSR State system of training geologists and hydrogeologists includes universities and institutes, specialized secondary educational establishments, technical secondary schools, branch institutes for upgrading professional skill, and organizations to popularize achievements in geology and hydrogeology (e.g., scientific and technical mining societies).

Training courses at higher educational establishments include lectures, practical studies and on-the-job training. The latter is carried out after the third or fourth year of training. As a rule, these students work in geologic survey teams as technicians and laboratory assistants. Such on-the-job training lasts two or three months.

In the last years of training, students participate in the research work of sub-faculties together with professors and lecturers. Their responsibilities range from the solving of theoretical problems to practical application.

Higher educational establishments hold annual conferences for professors, lecturers and students which help raise the professional skills of the students. In addition to obligatory training programmes, the institutes give elective courses on current problems in hydrogeology. Such courses are given by scientists of research institutes or other higher educational establishments.

In many higher educational establishments both day and evening sessions are held, with the latter usually attended by workers in geological organizations. The level of training accorded by both sessions is similar. The need for evening training reflects the great demand from different branches of the national economy (engineering, land reclamation, water management, geology, power engineering, municipal economy) for skilled hydrogeologists. The training period for hydrogeologists is five years.

Specialized secondary educational establishments and technical secondary schools (at day and evening departments) train hydrogeologic technicians for geologic surveys and research organizations. Their training period is four years.

In higher and secondary specialized educational establishments the posts of lecturing professors are competitive with openings announced in newspapers, every five years. This competitive system encourages higher levels of professional skill.

Graduate students have as scientific supervisor an outstanding or leading hydrogeologist. They study problems defined in the research plans of an institute, take examinations, conduct experiments, participate in conferences and prepare annual reports on their activities for seminars and colloquiums in laboratories and technical scientific councils. At the end of the post-graduate course, a student must present a thesis bearing on current trends in hydrogeology and which must be approved by the Learned Council of the Institute. The defence of a thesis entails open discussion of the theory, methodology and practical results obtained by the candidate.

Provisions for raising professional skills are arranged at the head research Institutes and promote regular distribution of scientific and technical progress to practising hydrogeologists, professors of higher institutions and colleges. Lectures, made by leading scientists of the Institute, are followed by seminars and field trips. At the end of the training period students submit a paper on their subject of study and do the practical work of a specialist.

Popularization seeks to attract youth and specialists in other fields to geology and hydrogeology, to improve the knowledge of specialists and give a better understanding of geology to ordinary people.

Geological clubs are organized by students and professors of universities and research scientists in schools and Pioneer Palaces. These clubs provide general geologic information supplemented by field trips to such features as caves and springs. Those who attend such clubs are better prepared for college-level training in geology.

The USSR Exhibition of Economic Achievements and the exhibitions of the constituent Republics contribute much to the popularization of geology.

The USSR Academy of Sciences, the main higher education institutions and research Institutes have geology museums which exhibit minerals formed as a result of ground-water activity, first books in hydrogeology, portraits and biographies of scientists, instruments, hydrogeologic maps and atlases, and schemes for ground-water exploration and development.

A wide range of geology libraries exist in the country, among the largest are the Geological Library of the All-Union Geological Institute (VSEGEI), Leningrad, the Geological Library of the Earth Sciences Branch of the USSR Academy of Sciences, Moscow, the Hydrogeological Library of VSEGINEO. Good collections of geologic and hydrogeologic literature are kept in the V. I. Lenin State Library, Moscow, the M. E. Saltykov-Schedrin Public Library, Leningrad, the Public State Technical Scientific Library, Moscow, and in the several libraries of universities and higher educational institutions.

Specialized scientific, technical and popular scientific literature in geology and hydrogeology is issued by Nedra. The Nauka Publishing House issues books on hydrogeology. Research organizations publish their proceedings and present methodical recommendations for use in hydrogeologic studies. Dozens of books in different hydrogeologic fields are published in the country each year.

New hydrogeologic information is published in the major geological journals: Sovetskaya geologia (Soviet Geology), Razvedka i okhrana nedr (Exploration and protection of the interior of the Earth), the geological series of Izvestiya AN SSSR (Transactions of the USSR Academy of Sciences), Doklady AN SSSR (Proceedings of the USSR Academy of Sciences), Izvestiya VYSov (Transactions of Higher Educational Institutions) and Geologia i razvedka (Geology and Exploration).

All-Union and regional technical and scientific conferences, meetings, innovation courses and seminars contribute much to enhancing the professional skills of hydrogeologists. The above measures are arranged by the USSR Academy of Sciences, the Ministry of Geology of the USSR, the Ministry of Reclamation and Water Management of the USSR, the USSR Exhibition of Economic Achievements, and other ministries and organizations. Five to ten important meetings, seminars, innovation courses in hydrogeology take place in the country each year.

Continuing scientific hydrogeologic seminars are arranged by the Technical and Scientific Mining Society of some research Institutes.

Public lectures on hydrogeology are read by members of the Moscow Society of Nature Researchers (MOIP). Here, monographs and the better papers and materials of scientific discussions are issued.

The system of hydrogeologic training in the country is free. It is continually being improved and organized to promote co-operation among scientists and practitioners, and to improve technical and scientific knowledge of specialists.

References

Note: This is a list of references selected to provide supplementary information and does not necessarily document author citations in the text.

Bindeman, N. N. and L. S. Yazvin. Safe-yield assessment. Moscow, Nedra, 1970.

Bochever, F. M. Theory and practical methods of safe-yield calculations. Moscow, Nedra, 1968.

Bondarenko, S. S., L. V. Borevsky and A. A. Dzyuba. Specific flow of deep ground water. In Fundamentals of Hydrogeology, Hydrogeodynamics. Novosibirsk, Nauka, sibirskoe otdelenie, 1983, pp. 173-176.

Borevksy, B. V., B. G. Samsonov and L. S. Yazvin. Methods for determination of aquifer parameters using pumping data. Moscow, Nedra, 1979.

Borevksy, L. V., G. S. Vartanyan and G. V. Kulikov. Hydrogeological overview, Kola super-deep well. Moscow, Nedra, 1984, pp. 240-253.

Goldberg, V. M. Hydrogeological forecasting of ground-water quality at water intakes. Moscow, Nedra, 1976.

Ground-water flow in Central and Eastern Europe, Moscow, 1982.

Ground water in the European part of the USSR, Map. Ed.: A. N. Semikhatov, Moscow, Gosgeoltekhizdat, 1955.

Hydrogeological and engineering geology Atlas of the USSR, Moscow, Nedra, 1984.

Hydrogeological map of the Ukrainian Soviet Socialist Republic (1:500,000). Ed.: Storchak, 1983.

Hydrogeological map of the USSR (1:2,500,000). Ed.: N. A. Marinov, 1962.

Hydrogeological maps of the Baltic Republics. Ed.-in-chief: A. A. Grigyalis, Ed.: V. I. Iodkazis, 1980.

Hydrogeology of the USSR, combined volume, vyp. I. Moscow, Nedra, 1976.

Hydrogeology of the USSR, combined volume, Ed.: L. S. Yazvin, vyp. 3. Moscow, Nedra, 1977.

Hydrogeology of the USSR, vols. 1, 3-6, 8, 13-15, 27, 28, 30, 31, 35, 43 and 45. Moscow, Nedra, 1969-1977.

Ignatovich, N. K. Hydrogeology of the Russian platform. Moscow, Gosgeoltekhizdat, 1949.

Ivanov, V. V. Mineral water of the USSR. Moscow, 1974.

____________. Basic criteria for assessment of chemical composition of mineral water. Moscow, 1982.

Mavritsky, B. F. Thermal water of folded and platform areas in the USSR. Moscow, Nauka, 1971.

Mavritsky, B. F., A. A. Shpak, G. K. Antonenko and others. Explanatory note to Thermal Water Atlas of the USSR. Moscow, VSEGINGEO, 1983.

Mineral water in the USSR, Map. (1:2,500,000). Ed.: L. A. Yarotsky, Moscow, GUGK, 1975.

Nikitin, S. N. Phreatic and artesian water of the Russian plain, St. Petersburg, 1900.

Predicted and available resources of major types of mineral water in the USSR, Map, (1:5,000,000) and explanatory note. Ed.: G. S. Vartanyan, Leningrad, 1984.

Safe-yield classification to potable and technical water reservoirs, Instructions for applications. State Commission on mineral resources of the Council of Ministers of the USSR, Moscow, 1985.

Smirnov, Ya. B. Thermal field of the USSR, Moscow, Akad.-Nauk SSSR, and GUGK, Novosibirsk, 1980.

Vartanyan, G. S. and G. A. Yarotsky. Prospecting, exploration and assessment of safe-yield of mineral water fields. Moscow, Nedra, 1968.

Zaitsev, I. K. and N. I. Tolstikhin. Regular distribution and formation of mineral (industrial and medicinal) ground water in the USSR. Moscow, Nedra, 1972.

XIII. YUGOSLAVIA

Area: 255,800 km^2 Population: 23.4 million

A. General

1. Physiography

Most of the Socialist Federal Republic of Yugoslavia occupies the northern Balkan Peninsula; the north-western fourth of the country north of the Sava and Danube rivers is part of Central Europe.

The main physiographic units are the Pannonian Plain to the north, elongated mountainous areas which cover most of the country, coastal areas and the islands of the Adriatic. The Pannonian Plain at an elevation of 80 to 115 m holds the most fertile land of the country. In the mountainous areas of central and southern Yugoslavia, about 53 per cent of the territory has elevations from 500 to 1,000 m, while 18 per cent rise above 1,500 m. To the north-west, Yugoslavia enters into the domain of the Alps with sharp topographical contrasts. The highest point is Mount Triglav (2,863 m).

The Dinaric Alps (Dinarides) a south-eastern extension of the Alps, reach elevations of 1,000 to 2,000 m in the Kapela and Valebit mountains. This vast mountainous system of the Dinaric Alps develops over 700 km from the north-west in Slovenia and to the south-east in Albania, including carbonate-rock areas in Slovenia, Croatia, Bosnia, Herzegovina, Western Serbia, Montenegro and Macedonia, where the Dinarides reach the Chara-Pindus mountains (2,000 m).

To the east, the territory reaches the foothills of the Carpathian-Balkan arch which is primarily a mass of carbonate rocks. Its crestline reaches elevations of 800-1,200 m to the north and 1,500-2,000 m to the south. Mounts Suva and Stara rise above 2,000 m. The mountainous chains above - Dinarides, Carpathian-Balkan, Chara-Pindus - surround the old eroded rocks of the Serbian-Macedonian and Rhodope massifs.

The Adriatic area includes the Slovenian coastal area, Istrian Peninsula, the Kvarner archipelago and northern Dalmatian islands, Ravni Kotari, Zagora, lower Herzegovina with the lower plain of the Neretva River, the Middle Dalmatian archipelago, the Montenegro coast with Lake Skadar and the large Titograd plain. Most of the Adriatic coastal area is rocky and mountainous, mostly folded with prevailing calcitic, dolomitic and marly rocks. It is lacking in good agricultural land: only about 10 per cent of the area can be farmed.

2. Climate

Climatic conditions are primarily temperate (figure 62). There are, however, sharp variations from one part of the country to another as a result of the influence of the sea, of the continental mass and of the topography. Three types of temperate climatic conditions are therefore to be found: (1) Mediterranean, (2) Continental (Central Europe) type, (3) Mountain type, and with transitional areas between these three.

Figure 62

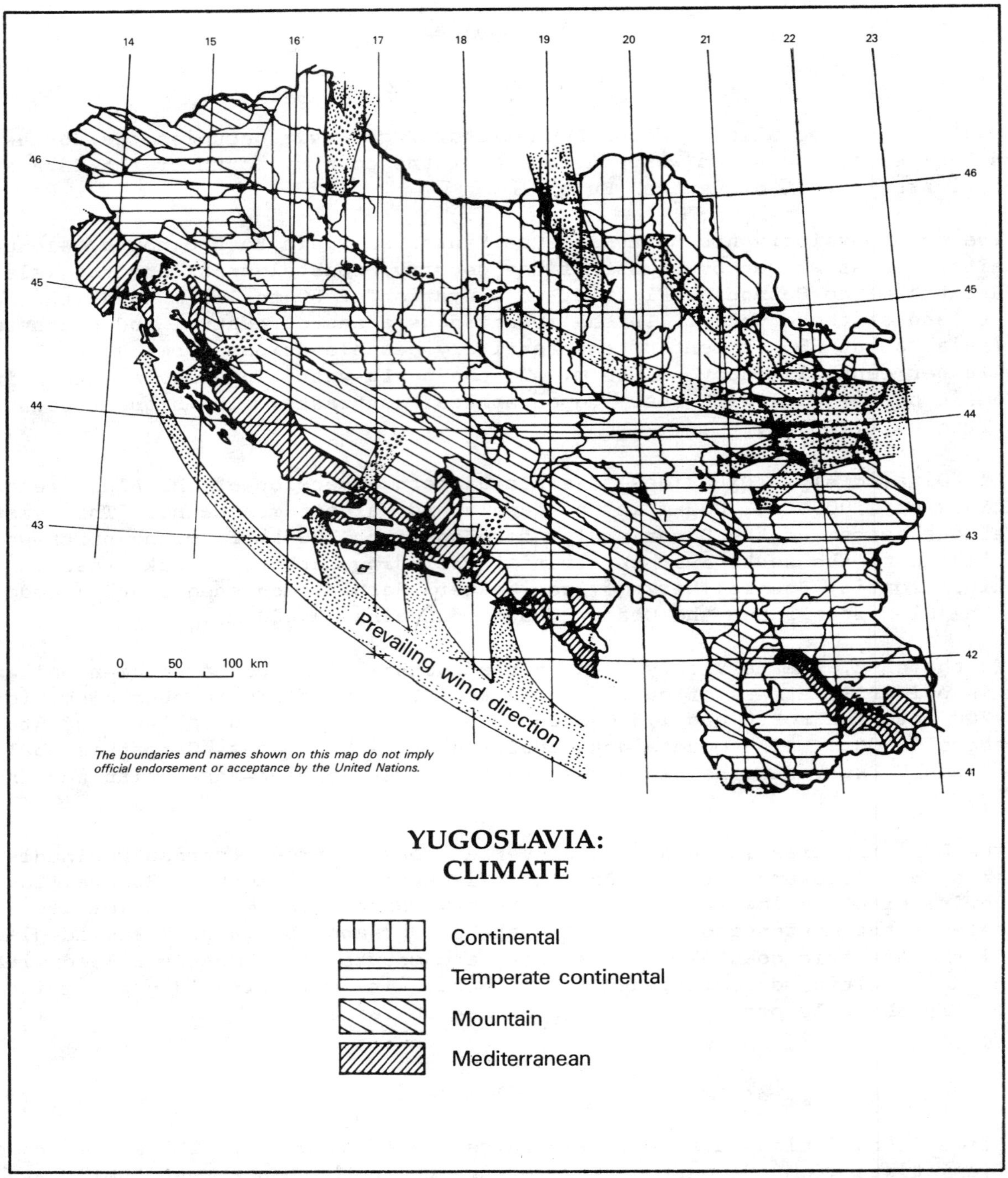

3546.50x

By and large, and on the basis of temperature and rainfall records, the climate varies from humid temperate to Mediterranean characterized by winter rains, with in between a mild humid temperate zone. In addition, large areas in both the temperate and the Mediterranean zones experience mountain climate.

To the south, on the Adriatic shores including the Istrian Peninsula and all Adriatic islands, typical Mediterranean climate prevails, and also in such coastal valleys and plains as those of Ravni Kotari, Zagora, Neretva, Lake Scutari with mild winters, moderate sunshine and irregular rainfall. On the other hand, the continental climate of the north typically developed in the Pannonian Plain is characterized by sharp contrasts between cold winter and extremely hot summer, and a largely uniform distribution of rainfall.

A high mountainous area beyond the Carnic Alps, the Julian Alps, and the Karawanken is characterized by a mountain climate with cold winters and abundant snowfall. However, this area is subjected to maritime influences to the south and continental influence to the east. Mountain climate also prevails in the higher areas of the Dinarides and those of Chara-Pindus and Carpathian-Balkans from Mount Velebit to the Chara, Stara and Suva mountains.

Temperature and rainfall are normally affected by latitude and altitude. They increase from north to south. Temperatures decrease with higher altitude (8° C per 1,000 m).

Maximum rainfall is encountered in the southern Dinarides (6,000 mm per year, the maximum for Europe). There, winters are mild with averages between 5.5° C and 9.5° C in January, and summers are hot (average in July is between 22.8° C and 26.2° C). Inland annual temperature variations are much higher. Summers are very hot (July average is 26° C with maximums exceeding 35° C) while winter temperatures may be lower than -20° C.

The average yearly rainfall of the country is approximately 1,000 mm. Major precipitation occurs in the Adriatic basin mostly to the south-west where yearly averages are 1,500-2,000 mm and above.

In all mountains, snow is found at 1,000-1,500 m and above within the November-May period. Spring snowmelts are important for aquifer recharge, especially in karstic areas.

In coastal areas and in the Pannonian Plain, the evaporation ratio is high during the summer period: reaching 1 to 1.5 m per year. In the Dinarides, most of the rainfall escapes evaporation and the evapotranspiration processes, with infiltration ratio reaching up to 60-70 per cent in arid limestone areas.

Four types of winds are common in Yugoslavia:

1. Continental cold and dry winds which blow from the north and the north-east. They bring fair weather, blowing all year but mainly from October to April in the Pannonian Plain and central part of the country.

2. Continental cold violent currents of the Bora, blowing from the karstic Dinarides down to the Adriatic coast.

3. Oceanic winds, carrying various degrees of moisture, blowing from the west, and

4. Humid and hot Mediterranean winds blowing from the south. They bring rain to the Adriatic shores (especially in autumn). They also bring snow in mountainous areas in early winter. Their effects diminish considerably to the north.

3. Surface water

Surface waters flow in the east to the Black Sea, in the west to the Adriatic and south to the Aegean Sea (figure 63). The country includes 33 major watersheds with a total runoff in the order of 240 billion m^3 (240 km^3) per year. Most of these waters flow towards the Black Sea (180 billion). The Adriatic gets some 55 billion and the Aegean about 5 billion cubic metres.

The Danube carries all the water flowing to the Black Sea. Major tributaries of the Danube are the Sava, Drava, Tisza and Morava rivers. At low ebb, the flow is about 3,500 m^3/s; up to 10,000 m^3/s in flood periods.

The Black Sea watershed covers two thirds of the country and it also carries two thirds of the entire hydrologic potential. Two large dams have been built on the Danube, several on the Drina and Sava rivers.

The Adriatic watersheds include several rivers, the main ones being the Neretva, Cetina, Souca, Krka, White Drim, Black Drim and the Zeta-Moraca. They hold about 30 per cent of the hydrologic potential but concentrated in limited areas as most of the basin is devoid of riverbeds. However, the presence of karst is an unfavourable factor in the construction of dams.

The Aegean basin (10 per cent of the country area) is drained by the Vardar and the Strumica rivers. The hydrologic potential there does not exceed 6 per cent of the entire potential of the country. Rainfall is relatively low and runoff quite irregular. Two flood periods can be observed, one in early spring, another in autumn, with a period of decreasing flow in between during the summer which may be quite extended in the Adriatic and Aegean basins.

Although Yugoslavia as a whole can be considered rich in water, availability varies greatly over the territory. In the west (Dinarides), the specific runoff coefficient can be very high, 20-25 $l/s/km^2$. On the other hand, the north, the north-east and most of the regions to the east and the south-east are less humid (10 $l/s/km^2$) while certain parts of the Pannonian Plain in the Morava and Vardar river valleys have even less water (3 to 5 $l/s/km^2$).

Important water resources potential is found in such lakes as Scutari (380 km^2), Ohrid (349 km^2), Prespa (290 km^2) and Doyran (43 km^2) which all have a tectonic origin. Smaller lakes found in the Dinaric and Alpine mountains, include the Blad, Bohin, Cercnica and Borac lakes, which are all of glacial origin.

Figure 63

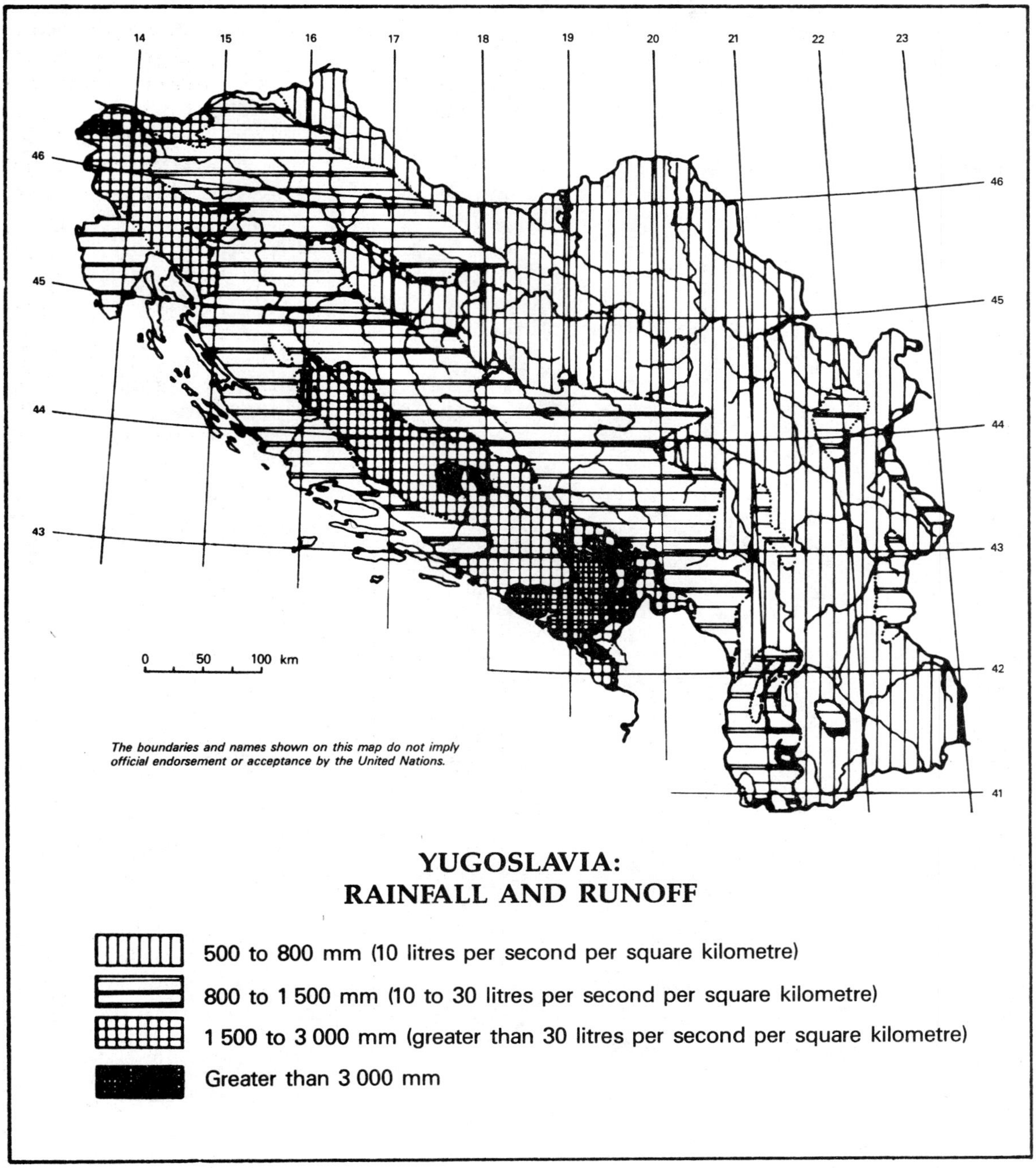

3546.51x

B. Geology

The territory of Yugoslavia can be subdivided into five major structural units, as follows (figure 64):

Figure 64

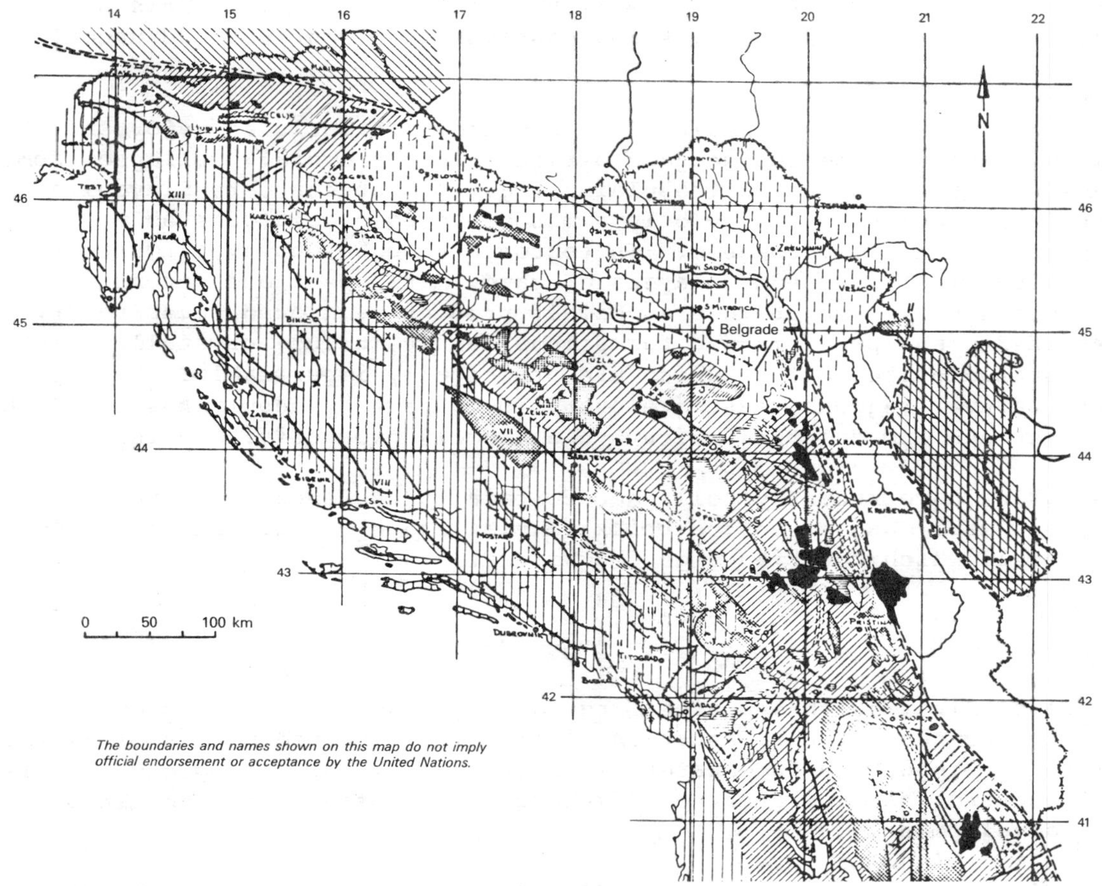

YUGOSLAVIA: GEOLOGY AND PRINCIPAL AQUIFERS

- Alpine System. Carbonate aquifers with various degrees of karstification and of moderate extent.
- Pannonian Basin. Neogene molasse and Quaternary deposits. Alluvial aquifers and extensive multilayered aquifers.
- Carpathian-Balkan Arch. Karstic aquifers.
- Serbian-Macedonian crystalline massif. Fracture aquifers, alluvial aquifers.
- Inner Dinarides and Vardar area. Karstic aquifers of varied extents, fracture aquifers, multilayered aquifers.
- Outer Dinarides. Vast aquifer system.

3546.52x

1. Pannonian Plain

The Pannonian Plain has Paleozoic and Mesozoic substrata overlain by Cenozoic and Quaternary deposits, including large alluvium fills related to the major rivers i.e., Drava, Sava, Tisza, Tamis and Danube (figure 65).

Sedimentary strata include aquifers (sand, gravel, sandstone, limestone and conglomerate alternating with impermeable beds of clay or marl. This widespread multilayered system is 1,500 m thick and contains both confined and unconfined water.

The Pannonian Plain is a basin of subsidence. It includes major and deep fault systems active since the Neogene period. In the overall collapse, some horsts remained, towering above the plain, e.g., Zagrebacka Gora Slavoni and Fruske Gora mountains. Such horsts play only a minor hydrogeologic role as the rocks which compose them are impermeable. In tectonic troughs, however, Quaternary gravel alluvium with interbedded layers of sand, marl, peat and unconsolidated Neogene deposits play a major role as highly productive ground-water reservoirs. North-east of Zagreb, pre-Triassic formations mainly include crystalline schist, gneiss and granodiorite, all of which are practically impermeable. However, in the Krapina and Daruvar regions these are overlain in part by deeply karstified Upper Triassic limestone and dolomite which are extremely productive aquifers.

Zones which have recently collapsed contain thick Cenozoic series including sand, clay and marl; and extensive thick fluvial and glaciofluvial alluvium (mainly sand and gravel) which constitute good aquifers.

2. Dinarides

The folded Dinaride chains cross the country from the Adriatic Sea to the south-west to the edge of the Pannonian Plain. They include masses of limestone and dolomite. Although surface water is not abundant, the geologic structure, the alpine climate and the type of river discharge create favourable conditions for hydropower development.

The geologic evolution of the Dinarides has had considerable impact on the karstification process as the upward movement of limestone masses (up to 1,000 metres) was followed by intense karstification, due to changes in the base level (figure 66). In fact, in the Dinarides, karst developed in all carbonate rock masses from Miocene to Triassic.

In the region there is a clear lack of surface runoff and of natural lakes in spite of the abundance of rainfall. However, some significant permanent rivers flow to the sea such as the Zrmanja, Krka, Cetina and Neretva; and underground streams such as the Lika, Gacka and Trebisnjica. Many coastal and submarine springs occur, yielding brackish water during low-water summer periods.

In the Outer Dinarides to the south-west, which include a thick carbonate rock complex more than 2,000 m thick, the occurrence of Eocene flysch failed to restrict the karstic character of the limestone-dolomite masses (figure 67). As a result of abundant rainfall, important reserves accumulate, feeding large springs in karstic depressions, or poljés, in the foothills and along the coast. This type of aquifer is specifically related to holokarst.

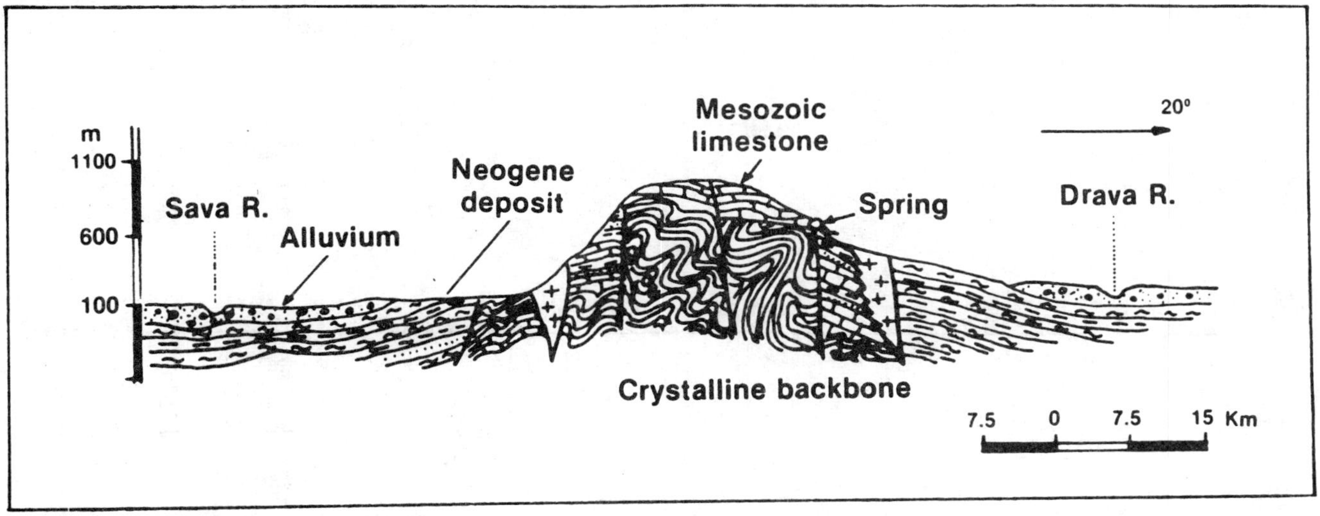

Figure 65. Yugoslavia, crystalline horst between Sava and Druva rivers

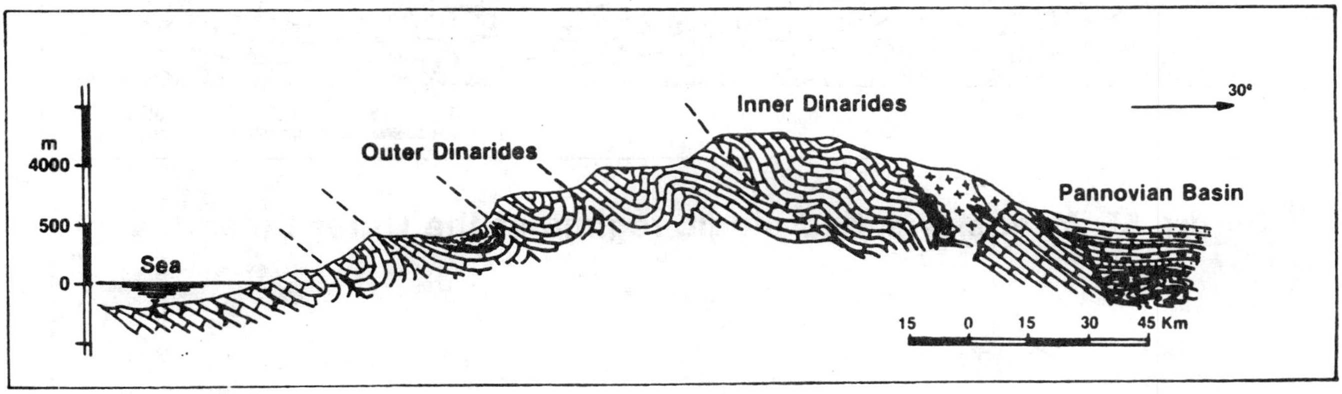

Figure 66. Yugoslavia, structure of the Dinarides

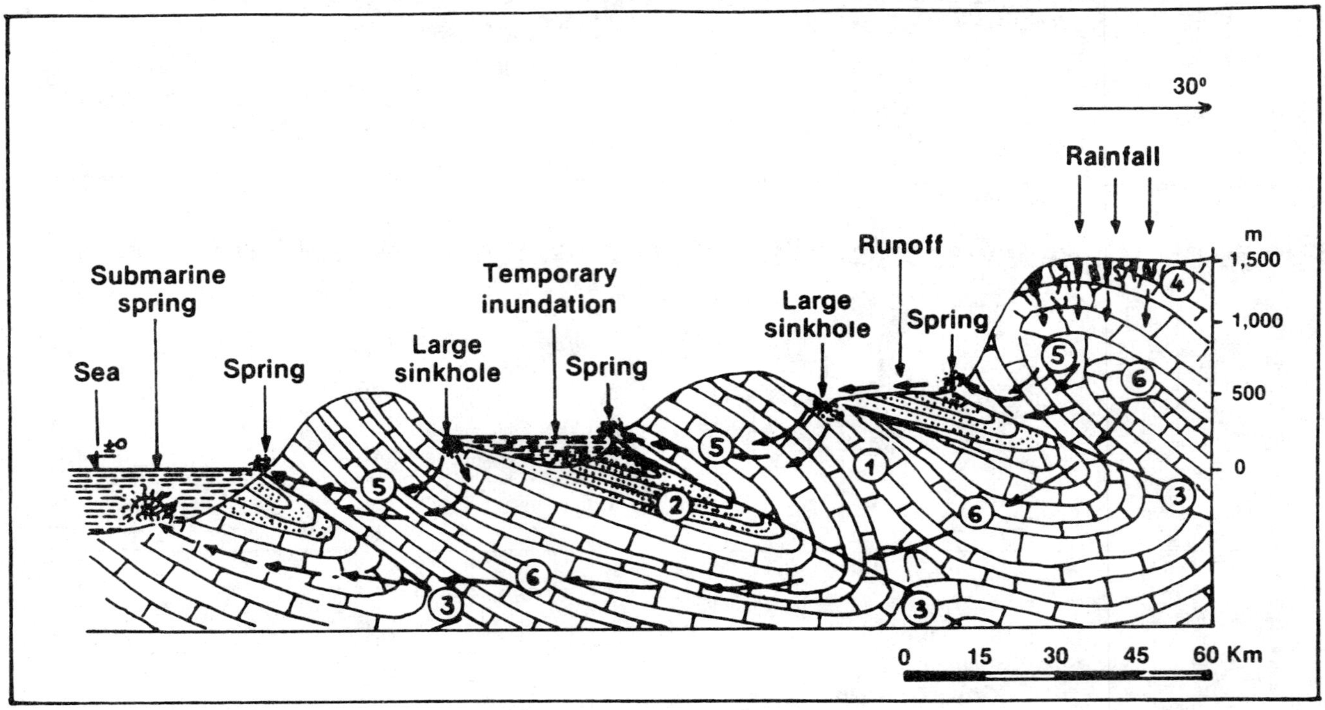

Figure 67. Yugoslavia, hydrologic régime of the Outer Dinarides

3. Vardar area

Hydrogeologic conditions are varied in the Vardar area. The geologic structure is mainly oriented NNW-SSE, that is, approximately the Dinaric orientation. The zone is bordered by deep faults.

In faulted areas and tectonic troughs, Triassic and Jurassic limestones are overlain by Cretaceous flysch. Thermo-mineral waters are found here. In southern Serbia the Vrange springs yield water at 80° C temperature. Also notable are the thermal springs of Bujanovac Vrnjacka Banja, and Smederevska Palanka.

To the south in Macedonia, the Veles series, which include mainly crystalline schist, amphibolite and quartzite, are virtually impermeable, the exception being some thin limestone strata. Effusive Cenozoic rocks such as basalt and andesite outcrop widely in some areas. Owing to their poor permeability they constitute mediocre aquifers. Good conditions for ground-water storage exist, however, in the Neogene Leskovac basin, and also in Neogene and Quaternary deposits in the Morava River valley in the form of multilayered aquifers of gravel and sand with intercalations of clay.

In the extensive plains south-east of Skopje, along the Vardar River, Neogene deposits include, mainly, compact marl which holds little ground water.

4. Serbian-Macedonian massif

The old crystalline nucleus of the Serbian-Macedonian massif which sinks to the north under the Pannonian Plain mainly includes lower and middle Paleozoic crystalline schists. The massif was folded and dislocated by intrusions of dacite and andesite.

Crystalline schists are impermeable except where faulted, deeply weathered, and in screes. The most productive layers are interbedded Paleozoic marble in the schists, as they have been dislocated by tectonic movement, and then karstified, giving way to springs. In Cenozoic formations, limited intercalations of sand and gravel constitute aquifers of local importance utilized for community water supply, especially in south-east Macedonia.

5. Carpathian-Balkan arch

The Carpathian-Balkan arch includes mainly Jurassic, Cretaceous, Cenozoic and Quaternary sediments, with some crystalline rocks (schist, mica schist, amphibolite) and volcanic rocks (granite, andesite). In Triassic and Cretaceous sediments, carbonate and clastic rocks prevail.

The arch, which to the south reaches the huge masses of Mesozoic limestone, rises in general to 1,200-1,700 m above sea level, and 2,000 m in some places. Clastic rocks occur in its middle part and, farther north, is Cretaceous flysch. Neogene deposits are found in plains and valleys of Southern Morava, Timok and Mlava.

Mesozoic limestone and dolomite are intensely karstified, thus allowing for the infiltration of rainfall. Clay and marl intercalations have been so affected by tectonic forces as to have lost their impermeable character.

The occurrence of thick marly sediments, the folded structure of carbonate rock masses and the carving of the land surface resulted in hydrogeologic units with various degrees of individualization, drainage through karstic channels being the predominant feature. This drainage may develop linearly leading to single springs. It may also be more diffuse, giving way to groups of springs. In the watershed of the Nisava River, drainage occurs in valleys where karstic springs are found, with important variations in their yield. It may also feed river alluvium, the screes serving as a transit zone for ground-water flow. The valleys which hold unconsolidated gravel constitute an axis of drainage and are, therefore, well provided with ground water.

Large karstic springs have developed as a result of a long hydrologic and geomorphologic process from the paleokarst which developed as early as the Jurassic, to the tectonic movements of the Neogene and Quaternary periods which have resulted in changes in the base level and gradient of ground-water flow.

C. Hydrogeology

1. Historical

After the Second World War, most water resources studies were oriented towards the construction of dams to provide energy and water. Government services specializing in water resources were established for water engineering studies in the federated republics and parastatal organizations. Water legislation and regulations were drafted. Water was declared a national resource to be managed and protected. The new constitution and the federal law on water approved in 1974 by the Federal Assembly provided the legal basis for water management at the level of both the federated entities and the federal state. Quality standards for rivers and effluents of treatment plants, both urban and industrial, were established.

Aiming for an efficient policy against water pollution, a major threat in several urban and industrial parts of the country, a principle was set based on water management within watersheds. This resulted in the delineation of watersheds within which water users and potential polluters are expected to co-operate with administrators at all levels, national, regional and local.

The 1974 law has resulted in changing the traditional powers of government services. At the national level, ground-water resources are dealt with by the Federal Geological Institute. At the level of federated republic, the responsibilities are with the various water authorities, services and directorates although, in certain republics, at the level of the ministry or committee that sets investment policy for water resources development and protection. In addition, they have the responsibilities of organizing, maintaining and operating ground-water observation networks, collecting data and assessing ground-water resources potential and availability from exploration to feasibility study. The Federal Geological Institute prepares water balances, evaluates exploitable ground-water resources through the hydrogeologic departments of six geologic institutes: in Ljubljana (Slovenia), Zagreb (Croatia), Sarajevo (Geoinzenjering Institute of Bosnia-Herzegovina), Belgrade (Geozavod Institute of Serbia), Skopje (Macedonia) and Montenegro (Titograd). The Ground Water Directorate of the Federal Geological Institute provides overall guidance, control and monitoring.

2. Investigation and exploration

The first ground-water investigations in 1945 were, in fact, engineering geology studies for the construction of dams. Ground-water investigations proper were initiated in areas lacking adequate surface water, on the basis of existing geologic and lithologic maps. In 1950, the Government launched a country-wide geologic mapping programme. Geologic maps at the scale of 1:100,000, 1:50,000 and 1:25,000 were prepared and a hydrogeologic map at the scale of 1:500,000. Detailed hydrogeologic studies were launched by the Federal Geological Institute in the early 1960s and completed by the State Geological Institute in 1978. They included a comprehensive inventory of water points and a hydrogeologic map at the scale of 1:200,000 covering most of the country. In 1981, a new hydrogeologic map at the scale of 1:500,000 was published.

All these maps were issued with various types of legend, according to local conditions and needs. After 1977, a commission of hydrogeologists under the Federal Geological Institute worked towards establishment of a new legend based on the experience of previous work and the guidelines of UNESCO and the International Association of Hydrogeologists, with the goal of both large- and small-scale maps that would be easy to understand. A unified programme for a new hydrogeologic map at the scale of 1:100,000 (based upon a draft survey at the scale of 1:25,000) was accepted by the federal Government in 1983, to be developed over a period of 15 to 20 years. The map was intended to show the main ground-water reservoirs and their characteristics.

Prospection and investigation methods used started with geologic surveys, hydrogeologic reconnaissance surveys and interpretation of air photographs. Later, geophysical surveys were carried out and ground-water models, both analog and digital, were devised and operated. Regional studies in carbonate rock (karstic) areas rely upon detailed structural analysis of the massifs followed by statistical, automated analysis and interpretation of observed fracturing.

At a small scale, quantitative geomorphologic methods by means of analysis of hydrographic patterns, drainage network distribution and clinography, allow the determination of infiltration and underground flow coefficients, which permit rapid assessment of water resources potential in given areas. More recently, remote-sensing methods have been applied to identify tectonic features and detect karstic systems controlling ground-water flow in fractured limestone aquifers.

Among geophysical methods, electrical resistivity probes and profiles are the most widely used. In some cases, they have successfully detected areas of low apparent resistivity in limestone-dolomite masses which correspond to karstified areas with ground-water flow. They are mainly used to determine the most promising drilling sites in sand and clay formations, in gravel fills and the weathered zones of crystalline rocks. In coastal areas they are used to locate the interface between fresh water and salt water and to find submarine or coastal springs. Induced polarization methods are utilized to determine the direction of the ground-water flow. Seismic refraction methods complement the geoelectric surveys.

In karstic regions of Yugoslavia, electrical and seismic methods have been successful in the:

1. Accurate location of fractured and karstified areas in the field, even under an overburden or unconsolidated sediments;

2. Detection of zones with high hydraulic conductivity;

3. Detection of zones clogged by clay; and

4. Determination of the thickness and lithology of the overburden.

Fractured zones, as identified on air photographs, can be confirmed in situ, and their dimension and level of karstification or clogging can be determined. Well-logging methods are also widely used, separately or in combination. They include resistivity, gamma, gamma-gamma, neutron-neutron, temperature, caliper and measurement of flow-rate.

Most alluvial and Neogene basin areas in Yugoslavia have been subjected to extensive reconnaissance drilling and are covered by a network of observation piezometers. Reconnaissance and test drilling is also necessary prior to the development of major exploitation wells and well fields. Whatever the aquifer may be, the choice of site of an exploitation well is determined taking into account the following constraints:

1. The expected yield should satisfy water needs;

2. Water quality should be good, or at the very least, acceptable; and

3. The depth to aquifer and piezometric level, and dynamic level during pumping, should not be excessive. Core drilling is carried out if hydrologic characteristics of the rocks must be tested on samples in the laboratory.

In unconsolidated sediments, large-diameter boreholes (50-65 cm) may deliver considerable yields, especially in alluvium of valleys of the Sava, Morava and Danube rivers.

In karst, water drilling operations are in many cases unsuccessful. For example on the Dalmatian coast, 11 reconnaissance boreholes and two wells which had been sited on the basis of geophysical surveys encountered no karstic channel although karstic springs on the shore nearby yield more than 100 l/s during dry periods.

Better results have been obtained by means of drainage galleries. There is a trend in Yugoslavia for systematic determination, by means of pumping tests in situ, of aquifer characteristics for permeability, transmissivity, storage coefficient and leakage coefficient. In karst, tracers are extensively used. Digital modelling is currently in use, especially for the study of ground-water resources in alluvial plains, and also for the forecasting of fluctuations of the piezometric surface in the Pannonian Plain, under the influence of pumping and artificial recharge. Specific models have been devised to study the flow in fractured rocks or karst and movement of the interface between fresh water and salt water in coastal areas.

Models have also been developed for the study of geothermal resources, in particular regarding thermal transfers and chemical exchange.

3. Principal aquifers

The aquifer types prevailing in Yugoslavia are:

1. Unconsolidated lake and alluvial deposits in the plains of the Sava, Drava, Tisza, Morava and Danube rivers;

2. Multilayered deposits in the Pannonian Plain, including artesian systems under impervious upper layers;

3. Karstic massifs of the Dinarides (Inner and Outer) and in the Carpathian-Balkan arch; and

4. Fracture zones, especially in metamorphic and volcanic rocks. Such local aquifers occur in many parts of the Serbian-Macedonian massif.

On the Adriatic shores, many karstic aquifers are subject to sea-water intrusion.

About 50,000 reconnaissance and exploitation boreholes have been inventoried. Table 10 outlines the characteristics of the main aquifer systems, here presented in more detail.

(a) Alluvial

Main alluvium fills are found in the Pannonian Plain, crossed by the rivers mentioned above. Average thickness is in the range of 30 metres over an area of about 35,000 km^2, or nearly 14 per cent of the country. In the Sava valley, the alluvium is thicker, 70 m in the Ljubljana plain, 40 m near Slavonski Brod, and 40 to 70 m in the Semberija and Macva plains (figure 68). In the Drava and Moura valleys the average thickness is 20 to 40 m, locally to 70 m.

Along the valley of the Sava, hydrogeological conditions vary widely: permeability is high (more than 2 X 10^{-3} m/s) between Pedsused and Sisak, lower (5.5 X 10^{-4} m/s) between Sisak and Slavonski Brod and between Slavonski Brod and Zupanja; it is higher to the south (6 X 10^{-3} m/s). Storativity is in the range of 10 per cent. By and large, ground-water resources in Sava alluvium are considerable and recharge conditions quite favourable.

Alluvial and lake deposits in the Danubian plain to the east (Tisza, Begej, Tamis and Karas rivers) are characterized by a widespread cover of loess that is practically impermeable and which protects the underlying gravel and sand (30-60 m) from surface pollution (figure 69). Gravel and sand layers are underlain by Neogene clays.

In the Backa region alluvial and lake sediments have a regulating storage capacity estimated at several billion cubic metres. Similar conditions are encountered in Srem along the Danube valley between Novi Sad and Belgrade. There the alluvium aquifer ranges in thickness from 60 m (Novi Sad) to 30 m (Belgrade). Typical hydrogeological characteristics are hydraulic conductivity 10^{-4} m/s; transmissivity 10^{-2} m^2/s; storage coefficient between 10 and 20 per cent. Similar values are found for alluvium aquifers near Pancevo, in the Godomine marshes and in the Kovim depression. Downstream from Belgrade and Pancevo, in valleys of the Danube tributaries Karas, Nero, Morava, Mlava, Pek and Timok, the aquifers are rarely thicker than 10 m. Some important alluvial aquifers are also found in the Sarajevo plain and the Spreca plain near Tuzla.

Table 10. Yugoslavia, parameters of principal aquifers

Aquifer type	Location	Aquifer depth (m), geology	Depth of well (m)	Specific capacity (l/s/m)	Permeability (m/s)	Transmissivity (m^2/s)	Storage coefficient (per cent)
Alluvial	Pannonian Plain	Average: 10-60 Quaternary, Pliocene	20-60	5-20 and more	10^{-5} to 10^{-4}	10^{-3} to 10^{-2}	0.1-0.3
Multilayer	Pannonian and other Neogene basins	80-300, Neogene	100-350	1-3	10^{-5} to 10^{-4}	10^{-5} to 10^{-3}	0.001-0.0001
Fractured	Dinarides (Inner), Vardar area, Serbian-Macedonian massif	Mesozoic, Paleozoic, Precambrian	50-70	0.1-0.3	10^{-6} to 10^{-5}	10^{-6} to 10^{-5}	0.01
Karstic	Dinarides, Carpathian-Balkan arch	Variable, Paleogene, Mesozoic	50-300	1-10	10^{-5} to 10^{-2}	10^{-5} to 10^{-1}	0.01-0.1

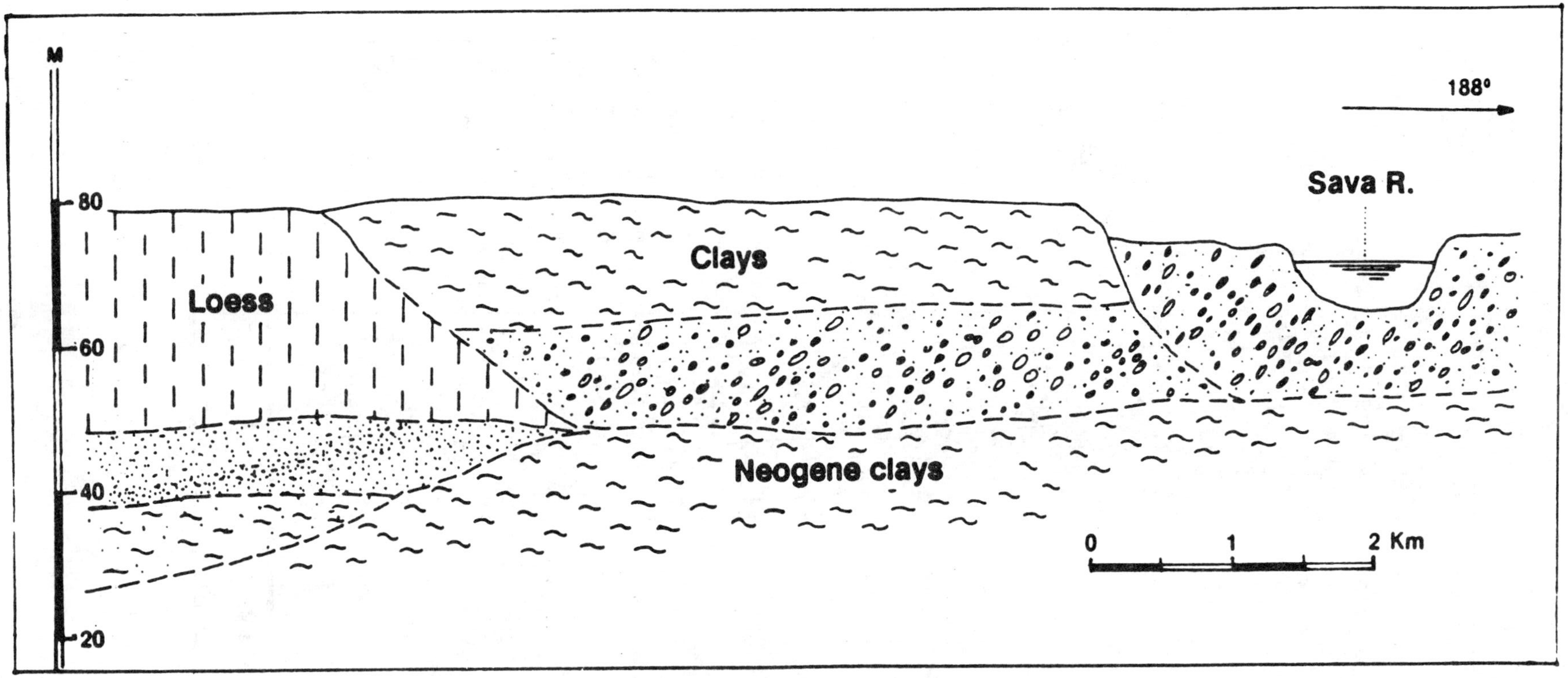

Figure 68. Yugoslavia, alluvial aquifer of Sava River near Belgrade

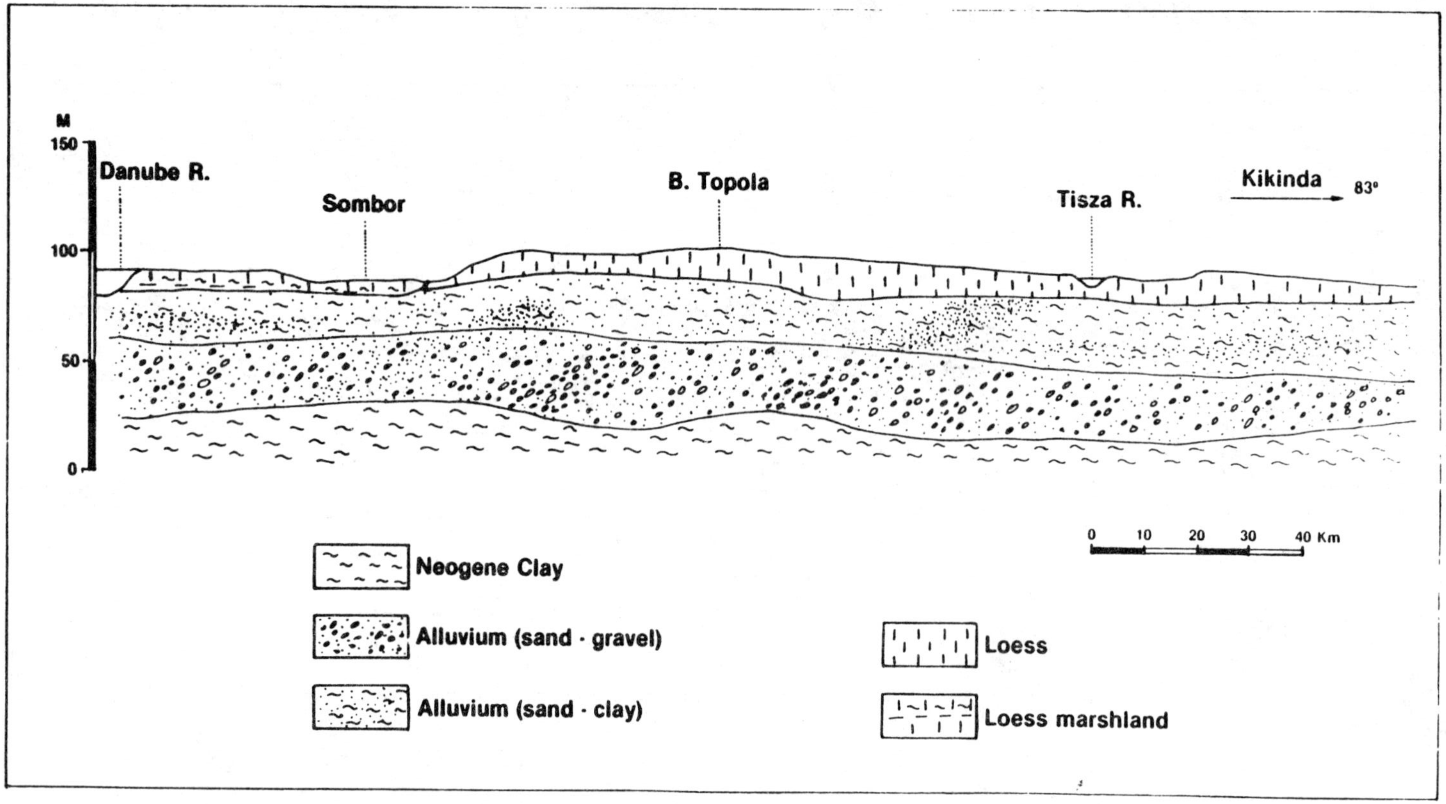

Figure 69. Yugoslavia, alluvial aquifer in Bačka region

In alluvial valleys of the Adriatic and Aegean watersheds, such aquifers are of minor importance as compared with fractured and karstic aquifers, with some exceptions for the Sòca, Neretva, Vardar and Strumica rivers.

Around Lake Scutari, a thick, high-production gravel aquifer has the following characteristics: hydraulic conductivity, 10^{-3} m/s; transmissivity, 10^{-1} m^2/s; storage coefficient 25 to 30 per cent.

(b) Multilayered aquifer systems

Multilayered aquifer systems occur in some 52,000 km^2 or about a fifth of the country. Most are located in the Pannonian Plain and in the Morava tectonic trough. They include sands of various size and, in a few cases, sandy gravel of Quaternary and Pliocene ages.

In Slavonia, artesian aquifers are found in Neogene lake and river deposits at depths of 35-60 m and 80 to 140 m. In the Pannonian Plain, multilayered aquifers are found at depths of 3,000 m. They include thick (10 m) layers of sand with intercalations of marl and clay. In some areas more than five productive layers have been identified with hydraulic conductivity between 5×10^{-5} m/s and 1×10^{-4} m/s.

In the Backa area multilayered aquifers are found at depths of 100-150 m and 170-200 m. At greater depths ground water is highly mineralized.

In the Banat region aquifer depths are at 80-120 m and 180-200 m; deeper in the north. In the Srem region aquifers occur in lenses of limited dimension, at depths of 80-300 m.

In many Neogene basins, multilayered aquifers are found with hydraulic conductivity less than 5×10^{-5} m/s, except in the Strumica and Pelagonia basins (southern Macedonia) where values are higher; aquifer depths there are in the range of 250 m.

In Leskovac (Morava) the best aquifers from the point of view of grain size (gravel) are found at depths of 40-135 m. Hydraulic conductivity is 5×10^{-4} m/s to 1×10^{-3} m/s. Depths are 80-100 m near Svetozarevo and aquifers are more sandy (hydraulic conductivity 1×10^{-5} m/s).

(c) Fractured aquifers

Fractured aquifers occur in a third of the country (85,000 km^2), namely the Inner Dinarides, the Vardar area and the Serbian-Macedonian massif.

The formations older than the Alpine orogeny include mainly schists with various degrees of metamorphism. Upper Permian formations include conglomerate, sandstone, carbonate breccia, more or less silicified and fractured. Most Paleozoic rocks are impermeable, e.g., crystalline metamorphics, phyllites and schists (Carboniferous-Devonian). At best, they give way to small springs. However, in the Vardar area, important springs occur in closely fractured schist (Cemerno, 8 l/s; Kopaonik, 14 l/s). Springs may develop at the contact between crystalline rocks and carbonate rocks with yields in the range of 1 to 5 l/s, locally 10 to 15 l/s.

In the Serbian-Macedonian volcanic massif, hydraulic conductivity ranges from 3×10^{-5} m/s to 2.5×10^{-6} m/s which is 10 times greater than those of the weathered mantle. It is also the case for peridotite, serpentine sandstone and Eocene conglomerate in areas of major faults.

(d) Karstic systems

Karst is found in several geotectonic units: the southern reaches of the Pannonian basin, the Carpathian-Balkan arch, the Dinarides (mainly the Outer Dinarides) and also in Precambrian and Paleozoic marble intercalated in the crystalline schists of the Vardar zone and in the Serbian-Macedonian massif. Main karstic aquifers are Mesozoic and Cenozoic limestones and to a lesser extent dolomites. In the Dinarides, thickness exceeds 1,000 m as a result of intense folding and karstification. Fracturing also contributes to the development of ground-water flow. Underground drainage in limestone exerts itself mainly in deep valleys where karstic springs occur with large fluctuations in yield.

In the central Dinarides are large karstic springs. In Western Serbia, two important karstic systems are present. In the Lelic system north of the Neogene basin of Valjevo-Mionica, the main springs systems are those of Gradac (0.4 to 10 m^3/s), Paklje (0.1 to 1 m^3/s), and Petnica (0.1 to 1 m^3/s). In the springs area, transmissivity is in the range of 3×10^{-2} m^2/s. In the karstic system of Tara and Zvijezda mountains, the average yield of the Peruca spring is 2.5 m^3/s.

In northern Montenegro and Bosnia, Triassic limestone and dolomite constitute a moderately folded synclinorium which gives way to important springs (0.1 to 1 m^3/s).

The great limestone massifs of the Outer Dinarides contain ground water which discharges through permanent or seasonal springs in karstic depressions (poljé) or into the sea. These vast regions are known the world over as the reference karstic system as it includes all possible karstic features: vast poljes at various levels: from sea level up to 1,100 m. Two areas of karstic poljes in western Bosnia and central Dalmatia (Kipresko, Davanjske, Livanjske, Sinjske, Dicme and Dugopolje), the other in western Herzegovina (Nevesinjsko, Gatacko, Dabarsko, Fatnicke, Plano, Popovo and Konavljansko), hold huge quantities of surface and ground water which are exploited mainly for hydropower. Several such schemes have been developed since 1960, the most significant being those of Split, Trebisnjica-Dubrovnik, Senj and Mratinje. J. Cvijic has described the morphological type of holokarst in the Dinarides where karstic phenomena have developed to the largest extent, as opposed to merokarst where karstic features are only partly developed.

As a result of dissolution phenomena occurring along joints and fractures, large channels develop and convey important yields to large springs feeding such rivers in the central Dinarides as the Una, Pliva, Bosna and Bistrica, with yields of a cubic metre per second and more. In some cases, yields may be 20 to 30 m^3/s, even 100 m^3/s during periods of high rainfall. Transmissivity can be 0.01-0.1 m^2/s up to 1 m^2/s. On the outer reaches of the Pannonian Plain and in the tectonic trough of Morava fissured karstic aquifers are also found in Miocene limestone. These aquifers are artesian to various degrees, and springs and tubewells may yield 5 to 10 l/s. Triassic limestone paleokarst aquifers on the edges of the Pannonian Plain may generate higher yields. In Nepricava near Belgrade deep wells (300 m) yield 30 to 50 l/s. Hydraulic conductivity varies from 2×10^{-6} to 5×10^{-5} m/s; transmissivity from 2.5×10^{-4} to 5.4×10^{-3} m^2/s and storativity from 3 to 5 per cent.

On the edge of the eastern Serbian Neogene basins and close to the Morava tectonic trough, Cretaceous and Jurassic limestones are flanked by relatively impermeable Neogene and Quaternary deposits resulting in springs (1 to 3 m^3/s) fed by large underground reservoirs.

Most karstic aquifers along the Adriatic (including islands) are subject to sea-water intrusion affecting submarine and coastal springs and a steadily encroaching interface between salt and fresh water as pumping continues.

Along the coast important karstic springs also occur at the contact with the Eocene flysch, for example, the springs of Jadro (near Split) and Ombla (near Dubrovnik).

4. Water quality

In alluvium, ground water is in general of the calcium bicarbonate type, with resistivity 1,200 to 1,900 ohm/cm. In many places, Danubian alluvium water contains chlorides and sulphates. In alluvium with abundant limestone pebbles (Drava River valley), total hardness is high (25 to 33 dH). Ground water also contains significant amounts of iron and manganese.

Ground water obtained from alluvium is in general suitable for human consumption from the point of view of both chemical and bacteriological content. However, alluvium is prone to pollution by industrial waste as is the case in the Drava and Sava valleys, downstream from major industrial plants and in urban areas.

In most limestone massifs of the Dinarides and of the Carpathian-Balkan arch, ground water contains mainly calcium and magnesium bicarbonate. In dolomitic rocks, magnesium prevails over calcium in the water. Total dissolved solids are in moderate amounts, 250 to 400 mg/l. These relatively high areas receive abundant rainfall which improves water quality. In most of the southern Dinarides towards the Adriatic, ground water evolves downstream from a bicarbonate type to the calcium sulphate type and as mentioned above, to sodium chloride near the coast. Concentrations in sodium sulphate and chloride increase during dry periods while they are lower during flood periods (figure 70).

In complex multilayered aquifers, water quality varies widely from one location to another, and with depth. The following types are encountered in the Pannonian Plain: calcium bicarbonate, calcium and magnesium bicarbonate, calcium sulphate, calcium and sodium sulphate, sodium and magnesium chloride. In the upper layers, that is, at depths less than 200 m in Backa, less than 300 m in Banat, waters are of the calcium bicarbonate type with 300 to 500 mg/l of TDS; but they meet drinking water standards. At greater depth, water quality is poor with high mineral content and high temperature which make the water unsuitable for normal human consumption.

In the Vardar area, thermal or thermo-mineral waters occur in several places. In the Serbian-Macedonian massif, crystalline schists contain small amounts of ground water with low TDS (100 to 200 mg/l) but excessive calcium and iron. In volcanic rocks waters are sweet, bicarbonate and TDS from 150 to 450 mg/l.

In all hydrogeologic investigations hydrochemical analyses are made. In most cases, this information is transferred into computer to be processed in the form of statistical tables, diagrams and other records.

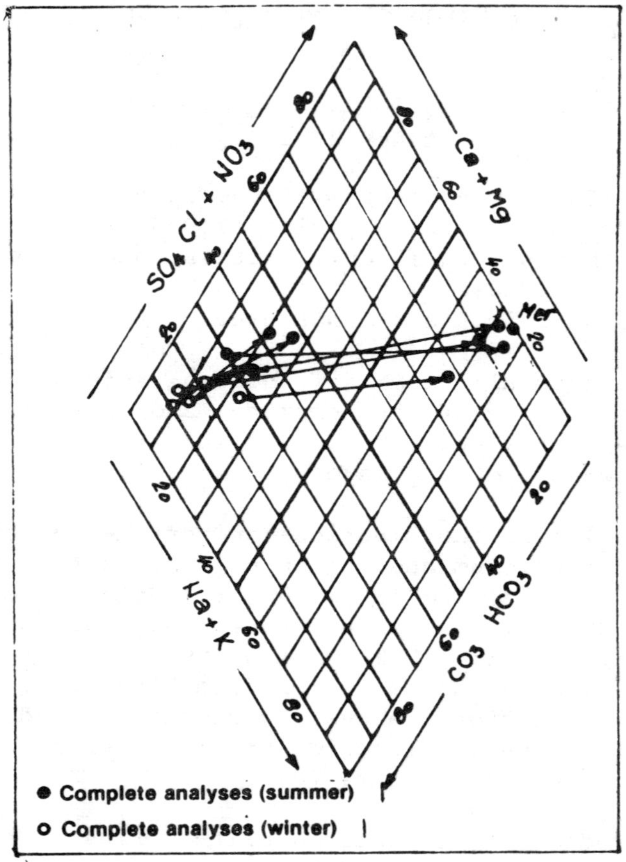

Figure 70. Yugoslavia, variations in ground-water quality, karst along coast near Split

5. Resources and assessment

Table 11 presents an overall assessment of the water balance of the country.

Table 11. Yugoslavia, water balance

Watershed	Rainfall (km^3 per year)	Runoff (km^3 per year)	Evapo-transpiration (km^3 per year)	Ground-water flow: annual reserves (km^3 per year)	Ground-water flow: yields (m^3/s)
Black Sea	163	68(+115)	90	4.5	143
Adriatic Sea	76	56	5	15	475
Aegean Sea	17	6	11	-	16
Total	(*)256	130	106	-	634

(*) Equivalent to 1 million m^3/km^2 (average rainfall 1 metre) in an average hydrologic year.

The data in table 11 should be taken as but an order of magnitude. Surface water is observed through a network of 750 gauging stations, mostly automatic, and its resources are known with reasonable accuracy. Ground-water resources also have been assessed but not evenly in all parts of the country.

Valuable information has been collected concerning the Pannonian Plain on the basis of abundant hydrologic and geologic data. Equally valuable data have also been obtained on the Sava River valley where alluvium water resources are estimated to be several hundred million cubic metres per year, and in the region between the Danube and Tisza, fluvio-lacustrine deposits contain resources of more than 1 billion cubic metres (1 km^3) per year.

In general, some 20 billion m^3 of ground water are available per year of which 75 per cent are contained in karstic aquifer systems (table 11). Considerable efforts are being made to bridge the gap in ground-water resources knowledge. The federal law of 1977, which provides for the standardization of data collection and assessment of mineral and ground-water resources, has greatly helped in the assessment of ground-water reserves in most of Yugoslavia carried out in 1981-1983.

6. Development

(a) Government services

In Yugoslavia, municipal water authorities, in addition to their responsibilities for the supply of water to households and industries are involved in ground-water management policies. Also, several scientific organizations at the national and regional levels are involved in ground-water studies.

Municipal water authorities are parastatal organizations with technical and financial attributes. As technical organizations, they prepare projects and assist in the construction and management of water installations. They are in charge of co-ordinating studies and research, and of development programmes on water resources at the level of a municipality or of hydrographic watersheds. They may also directly carry out projects for their own purposes, or as subcontractors of other Governments and parastatal organizations.

As financiers, municipal water authorities collect subventions for water resources development and improvement of water resources through such works as dams, pipelines, sewerage systems and treatment plants. This is the case for such major cities as Belgrade, Zagreb, Sarajevo, Ljubljana and Skopje.

In Yugoslavia, ground-water development by means of tubewells was initiated after the end of the Second World War, with the creation of drilling companies of various sizes using various types of equipment, generally called cable-tool and rotary rigs, most of them from either the USSR or United States. There is also in Yugoslavia an industry for the construction of light rigs.

The capacity of major rigs may reach 400 metres for boreholes exploitation. For geothermal drillings, depths of 2,000 metres have been reached. The total drilled per year for hydrogeologic and ground-water development purposes is in the range of 100,000 to 150,000 metres.

About 20 parastatal drilling companies are currently involved in drilling for ground water. The main ones are Geoloski Zavod (Ljubljana), Geotecnica and Electrosond (Zagreb), Geoinzenjering (Sarajevo), Hidrosond (Novi Sad) and Geosonda, Jugofund and Geozanod (Belgrade).

(b) Ground-water use

It is government policy to have, by the year 2000, the entire population supplied by piped water systems. The estimated water needs to be served by such means are shown in table 12.

Table 12. Yugoslavia, estimated water supply needs by year 2000

Watersheds	Community water supply	Irrigation	Industry	Protection of the environment
	(in cubic kilometres per year)			
Black Sea	2	8	8	9.6
Adriatic Sea	1.35	12	3.3	6.5
Aegean Sea	0.28	1.75	1.26	0.67
Total	3.63	21.75	12.56	17.77
Total general	56			

Water consumption in urban areas is estimated to reach 540 litres per day per person; in rural areas, 250 litres per person per day (against 350 litres and 125 litres in 1985).

About 60 per cent of the population is supplied by means of wells and drainage galleries (i.e., ground water), 30 per cent by karstic springs and 10 per cent by rivers or artificial lakes. In the Serbian Republic, 90 per cent of domestic waters originate from ground-water sources.

Ground water is extensively developed in the alluvium of the Pannonian Plain, i.e., the valleys of the Sava, Drava, Danube rivers and the karstic massifs of the Dinarides and the Slovenian Alps.

Ground water of alluvial plains in zones of Recent subsidence (Podravina, Slavonia) are extensively developed for urban and industrial supply of many cities such as Varazdin, Cakovec, Koprivnica, Virovitica, Podravska Slatina and Osijek. Reticulated systems are supplied by large-diameter wells and boreholes, in particular radial wells which yield large quantities of water. For example, in the alluvial plain near Belgrade and Novi Sad such radial wells supply 10 m^3/s and 1 m^3/s for these two cities, respectively. In the Danube alluvium, specific capacity is 3-10 l/s/m. In Semberia it may exceed 50 l/s/m. In certain areas of the Sava valley, ground-water extraction may reach 500 l/s even 1 m^3/s as is the case in Zagreb and Belgrade. Some 700 sites of ground-water extraction are found in the Neogene and alluvium aquifers of the Sava watershed. In some industrial areas, a serious risk of ground-water contamination by industrial effluents already exists, with streams contributing to pollution by infiltrating ground-water reservoirs (e.g., the city of Kraljevo). In karstic regions ground water is tapped mainly at the outlets of springs; on the Adriatic Split it is supplied by Jadro spring (2 m^3/s), at Dubrovnik by Ombla spring (1 m^3/s), at Rijeka by a karstic spring and also by drainage galleries (400 l/s in periods of low water).

In many karstic areas, however, water supply is a major problem. The surface is bare, without water points nor streams and ground water is situated at great and difficult to determine depths. Complex and detailed hydrogeologic studies are necessary to reach such water sources. In the Dalmatian coastal area, including islands, ground-water resources are developed by means of wells and galleries with yields ranging from a few litres per second up to 100 l/s in low water periods. Table 13 indicates the technical characteristics and available yields which are extracted from galleries for tourist centres of the Dalmatian coast.

Table 13. Yugoslavia, ground-water development, Dalmatian coast

Location	Well depth (m)	Length of gallery (m)	Exploitable yield in low-water period (l/s)	Cl contents (mg/l)
Dubrava	28	110	35	30
Kovca	70	70	25	30
Gustinna	82	270	110	250
Zuljana	18	30	10	350-400
Marina	18	10	5	300

In remote mountainous karst, water is obtained primarily by collecting rain in cisterns and only a few populated centres have piped systems, tapping water from artificial lakes created in the Dinaric Alps mainly for hydropower.

In coastal areas, owing to the expansion of tourism and related increases in needs for domestic and agricultural water, considerable programmes for ground-water development were initiated. These entailed considerable risk in urban areas as karstic aquifers are extremely vulnerable to pollution.

In Yugoslavia some 5.6 million hectares are cultivated at a potential of 10 million ha of arable land. Some 180,000 hectares are irrigated, for the most part in Macedonia, of which 60 per cent are irrigated by canals and 40 per cent by sprinklers. Depending on the location, crops and climate, the quantity of water needed ranges from about 2,500 m^3 per hectare per year in the Pannonian Plain, to 20,000 m^3 in some parts of the south.

The major irrigation systems use surface water from streams or artificial lakes such as that of the Danube-Tisza dam in the Pannonian Plain, the Lbar-Lepenac dam in Kosovo, and the Pelagonija dam in Macedonia. Ground-water irrigation is limited to small-scale schemes in areas where surface water is not available. In some rural areas in Kosovo and Macedonia, wells are used for community supply, irrigation and livestock. Ground-water development in drought-prone agricultural areas, which cover some 700,000 hectares of the country, is expected to expand in the years to come.

(c) Problems and solutions

In Yugoslavia, ground-water development has created serious environmental problems and little progress has been made towards integrated development and management of surface and ground waters.

Water resources are relatively abundant, but the demand has increased rapidly in recent years. In the karstic mountain areas, not enough is known on the availability of water resources. Both in alluvium and karstic areas, ground water is extremely vulnerable to pollution, naturally through sea-water intrusion in coastal areas and man-made through infiltration of domestic waste water, industrial effluents, irrigation water containing nitrates and other chemicals.

An overdraft of multilayered aquifers which resulted in substantial lowering of the piezometric surface has been observed in the Pannonian Plain and small Neogene basins. To reverse such negative effects, better control and management of ground-water development is needed and the feasibility of artificial recharge considered.

Solutions will entail an integrated approach to not only the scientific and technical aspects but also the socio-economic, legal and managerial implications of land-use planning.

(d) Artificial recharge

Detailed studies for artificial recharge have been initiated in recent years. Injection of river water into deep wells has been tried in karstic formations near Belgrade for the purpose of doubling the available yield to 150 l/s in dry periods. This, however, resulted in a 25-metre rise of the piezometric level, which might have detrimental effects on the aquifer due to the additional head thus created. It was, therefore, decided to carry out the injections in several stages separated by stoppages of three to five days for each metre rise of the water level.

(e) Integrated management of surface and ground water

There is a definite need for more study on integrated use and management of water resources. Models for economic optimization have been planned for hydrographic basins. Water-balance studies are to be strengthened, including a model for water-resources optimization. A better knowledge of water resources will allow for more efficiency in the application of water legislation. Several technical and legal measures have already been taken which are opening the way for an integrated management of water resources as follows:

1. A federal law has been issued on the standardization of methods for ground-water data collection and preservation, and also for the preservation of water balance;

2. Observation networks for surface and ground water have been established;

3. In the most advanced parts of the country computerized data banks have been established for geologic, hydrologic and hydrogeologic data storage, processing and retrieval;

4. A hydrogeologic map at the scale of 1:100,000 is planned; and

5. Deep aquifers are to be explored.

Present studies include several large-scale projects, among which are the following:

1. Project for management of the Sava River basin,

2. Project for management of the Morava River basin,

3. Project for completion of reclamation and irrigation schemes in the Pannonian Plain (Danube-Tisza canal system),

4. Project for management of the Vardar watershed and other projects on Macedonian dams, and

5. Project of development and linkage of storage dams in the central Dinarides to regulate karstic flow.

D. Conclusions

Although many hydrologic problems are under control, as in the cases related to the regimen of rivers and overall water balance, some characteristics of ground-water flow in the several aquifers need further study, especially as regards deep aquifers. Additional investigation and assessment are needed to determine the practical measures necessary for rational development of the aquifers, and for monitoring their discharge.

The ongoing inventory and assessment of water resources is intended to determine in terms of quantity, quality and costs the water resources which can be made available in various parts of the country. Future water policies should ensure that development takes into account water conservation measures and the need to preserve the environment.

Extensive development argues against projects conceived in isolation as they would inevitably interact. This has already been seen in schemes separated by relatively long distances, for aquifers separated by impermeable layers and for ground-water and surface-water schemes.

The decision to resort to ground water or surface water or both is made on the basis of study of the cost-to-benefit ratio, or of determination of the lesser cost. Large-scale ground-water development from alluvium is more economical than dams. For example, in highly productive alluvium along the Sava and Danube rivers, investment costs are in the range of $1,000 to $1,500 per litre-second (including pump and power unit). These unit costs tend to increase for wells not extracting more than 20 to 30 l/s. However, as drilling and machinery costs have sharply increased in recent years, some doubts have arisen as to the economic feasibility of developing ground water, especially in karstic areas.

Experts, consulting firms and drilling companies from Yugoslavia have been extensively involved in ground-water projects, including United Nations projects in developing countries of the Caribbean, Latin America, Africa, the Middle East and Asia. In 1983, a United Nations International Colloquium was convened in Zagreb on the subject of Technical Co-operation among Developing Countries in Ground Water Resources Development.

References

A. Publications and reports

Mijatovic, Borijove F. Exploitation rationalle des eaux karstiques. International Association of Hydrogeologists (IAH), Paris, 1975, p. 123-135.

Mijatovic, Borijove F. and others. Hydrogeology of Dinaric karst. GEOZAVOD (Institute for Geological and Physical Research), Beograd, 1984, 250 pp.

United Nations Development Programme. Regulation and management of the Sava River. Report of project, New York and Zagreb, 1975.

Yugoslavia, Federal Geological Institute. Hydrogeological study of Bosnia Herzegovina. Open-file report, in Serbo-Croatian, 400 pp.

__________. Hydrogeological study of Croatia. Open-file report, 1978. In Serbo-Croatian, 350 pp.

__________. Hydrogeological study of Croatia. Open-file report, 1980. In Serbo-Croatian, 400 pp.

__________. Hydrogeological study of Dalmatia. Open-file report, 1980. In Serbo-Croatian, 350 pp.

__________. Hydrogeological study of Macedonia. Open-file report, 1980. In Macedonian, 400 pp.

__________. Hydrogeological study of Montenegro. Open-file report, 1980. In Serbo-Croatian, 370 pp.

__________. Hydrogeological study of Slovenia. Open-file report, 1970. In Slovenian, 250 pp.

__________. Water resources in Yugoslavia Federative Socialist Republic: Utilization and protection. Water Economy Institute. Jaroslav Cerné, Beograd, 1980. In Serbo-Croatian, 64 pp.

B. Hydrogeologic maps

Hydrogeological map of Serbia, scale 1:200,000. GEOZAVOD (Institute for Geological and Physical Research), Beograd, 1983.

Hydrogeological map of Yugoslavia, scale 1:500,000. Federal Geological Institute of Yugoslavia, Beograd, 1987.

International Hydrogeological map of Europe, scale 1:1,500,000. Map sheets C-5, Berne D-5, Budapest and D-6, Athens. UNESCO and International Association of Hydrogeologists (IAH), Paris, 1974, and 1982. With explanatory booklets for Yugoslavia.

Annex

GROUND-WATER PROJECTS IN EASTERN AND NORTHERN EUROPE IMPLEMENTED UNDER THE AUSPICES OF THE UNITED NATIONS DEVELOPMENT PROGRAMME

Country and project number	Project	Executing agency	Implemen-tation period	Ground-water compo-nent*
GREECE				
GRE-59-001	Ground-water investigation study of water resources and their	FAO	1959-1963	a
GRE-65-017	exploitation for the irrigation of Eastern Crete	FAO	1965-1972	a
GRE-72-002	Studies and experiments for capturing fresh ground water for the Almyros spring of Iraklion	FAO	1972-1975	a
GRE-77-023	Water resources development of the Molai area in Lakoni	FAO	1978-1981	a
GRE-77-007	Technical support for the Institute of Geology and Mining	United Nations	1978-1982	a
GRE-85-001	Lignite deposits and karst water investigations	United Nations	1985-1986	a
POLAND				
POL-68-507	Planning the comprehensive development of the Vistula River system	United Nations	1968-1979	a
ROMANIA				
ROM-65-001	Establishment and operation of pilot irrigation station in Danube plain.	FAO	1965-1972	b
ROM-75-035	Complex development of natural resources of Danube delta	FAO	1978-1980	b
YUGOSLAVIA				
YUG-78-005	Comprehensive development of Mostarsko Blato karst area	UNDP	1977-1982	b

Country and project number	Project	Executing agency	Implemen-tation period	Ground-water compo-nent*
REGIONAL PROJECTS				
REM-71-203	Integrated development of the Vardar-Axios River Basin	United Nations	1971-1978	c
RER-79-002	International Seminar on karst hydrogeology	United Nations	1979	c

* a - Preponderant or exclusive

b - Substantial (30 to 50 per cent)

c - Minor.

WATER RESOURCES: LIST OF UNITED NATIONS PUBLICATIONS

Report of the United Nations Interregional Seminar on Flood Damage Prevention Measures and Management, Tbilisi, Union of Soviet Socialist Republics, 1969. 117 pp. English only. Publication symbol: ST/TAO/SER.C/144.

Integrated River Basin Development. Report of a panel of experts. Revised edition. 80 pp. English (French, Spanish out of print). Sales No. E.70.II.A.4.

Solar Distillation as a Means of Meeting Small-scale Water Demands. 86 pp. English (French out of stock). Sales No. E.70.II.B.1.

Triennial Report on Water Resources Development, 1968-1970. 202 pp. English. Sales No. E.71.II.A.15.

Abstraction and Use of Water: A Comparison of Legal Régimes. 254 pp. French, Spanish (English out of print). Sales No. 72.II.A.10.

Proceedings of the Interregional Seminar on Water Resources Administration, New Delhi, 1973. 204 pp. English. Publication symbol: DP/UN/INT-70-371.

National Systems of Water Administration. 183 pp. English. Sales No. E.74.II.A.10.

Ground Water Storage and Artificial Recharge. Natural Resources/Water Series No. 2. 270 pp. English, French, Spanish. Sales No. 74.II.A.11.

Management of International Water Resources: Institutional and Legal Aspects. Report of a Panel of Experts. Natural Resources/Water Series No. 1. 271 pp. English, French, Spanish. Sales No. 75.II.A.2.

The Demand for Water: Procedures and Methodologies for Projecting Water Demand in the Context of Regional and National Planning. Natural Resources/Water Series No. 3. 240 pp. English, Spanish (French out of print). Sales No. 76.II.A.1.

Ground Water in the Western Hemisphere. Natural Resources/Water Series No. 4. 337 pp. English, French, Spanish. Sales No. 76.II.A.5.

Guidelines for Flood Loss Prevention and Management in Developing Countries. Natural Resources/Water Series No. 5. 183 pp. English. Sales No. E.76.II.A.7.

Report of the United Nations Water Conference, Mar del Plata, 14-25 March 1977. 181 pp. English, French (Spanish out of print). Sales No. 77.II.A.12.

A Review of the United Nations Ground Water Exploration and Development Programme in the Developing Countries, 1962-1977. Natural Resources/Water Series No. 7. 84 pp. English, French, Spanish. Sales No. 79.II.A.4.

Efficiency and Distributional Equity in the Use and Treatment of Water: Guidelines for Pricing and Regulations. Natural Resources/Water Series No. 8. 175 pp. English (French, Spanish out of print). Sales No. E.80.II.A.11.

Water Resources Planning: Experiences in a National and Regional Context. Report of a United Nations Workshop Convened in Co-operation with the Government of Italy. 293 pp. English. Publication symbol: TCD/SEM.80/1.

Rural Water Supply. Report of a United Nations Interregional Seminar Convened in Co-operation with the Government of Sweden. Uppsala, Sweden, 6-17 October 1980. 44 pp. English. Publication symbol: TCD/SEM.81/1.

Ground Water in the Eastern Mediterranean and Western Asia. Natural Resources/Water Series No. 9. 230 pp. Arabic, English, French. Sales No. 82.II.A.8.

Experiences in the Development and Management of International River and Lake Basins. Natural Resources/Water Series No. 10. Proceedings of the United Nations Interregional Meeting of International River Organizations, Dakar, Senegal, 5-14 May 1981. 424 pp. English, French, Spanish. Sales No. 82.II.A.17.

Flood Damage Prevention and Control in China. Report of a Study and Workshop in the People's Republic of China, 16-31 October 1980. Natural Resources/Water Series No. 11. 121 pp. English. Sales No. 82.II.A.13.

Ground Water in the Pacific Region. Natural Resources/Water Series No. 12. 289 pp. English. Sales No. E.83.II.A.12.

Technical Co-operation Among Developing Countries in Ground Water Resources Development: Report of a United Nations International Colloquium Convened in Co-operation with the Government of Yugoslavia in Zagreb, 23-28 May 1983. 123 pp. English. Publication symbol: TCD/SEM.83/1.

Treaties Concerning the Utilization of International Water Courses for Other Purposes than Navigation (Africa). Natural Resources/Water Series No. 13. 98 pp. English/French. Sales No. E.F.84.II.A.7.

The Use of Non-Conventional Water Resources in Developing Countries. Natural Resources/Water Series No. 14. 278 pp. English. Sales No. E.84.II.A.14.

Ground Water in Continental Asia. Natural Resources/Water Series No. 15. 391 pp. English. Sales No. E.86.II.A.2.

Water Resources Legislation and Administration in Selected Caribbean Countries. Natural Resources/Water Series No. 16. 171 pp. English, French, Spanish. Sales No. 86.II.H.2.

Follow-up to the Mar del Plata Action Plan: Report of the Meeting. Interregional Symposium on Improved Efficiency in the Management of Water Resources, 5-9 January 1987. 47 pp. English, French, Spanish. Publication symbol: TCD/SEM.87/1.

Institutional Issues in the Management of International River Basins: Financial and Contractual Considerations. Natural Resources/Water Series No. 17. 111 pp. English, French, Spanish. Sales No. 87.II.A.16.

Application of Computer Technology for Water Resources Development and Management in Developing Countries. 95 pp. English. Publication symbol: TCD/WATER/1. 1987.

Ground Water in North and West Africa. Natural Resources/Water Series No. 18. 415 pp. English and French. Sales No. 87.II.A.8.

Ground Water in East, Central and Southern Africa. Natural Resources/Water Series No. 19. 341 pp. English and French. Sales No. 88.II.A.5.

Water Resources Planning to Meet Long-Term Demand: Guidelines for Developing Countries. Natural Resources/Water Series No. 21. 117 pp. English, French, Spanish. Sales No. 88.II.A.17.

Non-Conventional Water Resources Use in Developing Countries: Proceedings of the Interregional Seminar, Willemstad, Curacao, Netherlands Antilles, 22-28 April 1985. Natural Resources/Water Series No. 22. 515 pp. English. Sales No. E.87.II.A.20.

Ground Water Economics. Report of a United Nations International Symposium and Workshop Convened in Co-operation with the Government of Spain. Barcelona, Spain, 19-23 October 1987. 65 pp. English. Publication symbol: TCD/SEM.88/2.

Assessment of Multiple Objective Water Resources Projects: Approaches for Developing Countries. 125 pp. English. United Nations/United Nations Environment Programme. Publication symbol: TCD/SEM.88/3.

Legal and Institutional Factors Affecting the Implementation of the International Drinking Water Supply and Sanitation Decade. Natural Resources/Water Series No. 23. 121 pp. English. Sales No. E.88.II.A.21.